T0296322

TRANSMISSION ELECTRON MICROSCOPY OF MINERALS AND ROCKS

CAMBRIDGE TOPICS IN MINERAL PHYSICS AND CHEMISTRY

Editors
Dr. Andrew Putnis
Dr. Robert C. Liebermann

Transmission electron microscopy of minerals and rocks

ALEX C. McLAREN

Australian National University

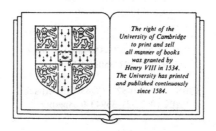

The right of the
University of Cambridge
to print and sell
all manner of books
was granted by
Henry VIII in 1534.
The University has printed
and published continuously
since 1584.

CAMBRIDGE UNIVERSITY PRESS

CAMBRIDGE

NEW YORK   PORT CHESTER   MELBOURNE   SYDNEY

CAMBRIDGE UNIVERSITY PRESS
Cambridge, New York, Melbourne, Madrid, Cape Town, Singapore, São Paulo

Cambridge University Press
The Edinburgh Building, Cambridge CB2 2RU, UK

Published in the United States of America by Cambridge University Press, New York

www.cambridge.org
Information on this title: www.cambridge.org/9780521350983

First published 1991
This digitally printed first paperback version 2005

A catalogue record for this publication is available from the British Library

Library of Congress Cataloguing in Publication data
McLaren, Alex C.
Transmission electron microscopy of minerals and rocks /
Alex C. McLaren.
p.   cm. – (Cambridge topics in mineral physics and chemistry)
Includes bibliographical references.
ISBN 0-521-35098-0
1. Mineralogy, Determinative.   2. Transmission electron microscopes.
I. Title.   II. Series.
QE369.M5M36   1991
549′. 12 – dc20   90-20039
                                                            CIP

ISBN-13 978-0-521-35098-3 hardback
ISBN-10 0-521-35098-0 hardback

ISBN-13 978-0-521-35943-6 paperback
ISBN-10 0-521-35943-0 paperback

Contents

Preface

Of the many techniques that have been applied to the study of crystal defects, probably no single technique has contributed more to our understanding of their nature, properties, and influence on the physical and chemical properties of crystalline materials than transmission electron microscopy (TEM). Although the importance of crystal defects and the use of TEM for their direct observation were recognized by physical metallurgists in the early 1950s, it was at least a decade later that earth scientists responded to many of the new ideas of the defect solid state and to the power of TEM. However, TEM is now used extensively for the direct observation of defect microstructures in minerals and rocks, and there appears to be an increasing number of earth scientists who want to use the technique or to become more familiar with the interpretation of TEM observations. This book is written for such people. However, it makes no attempt to be a practical manual of TEM or a definitive text, but rather an *introduction to the basic principles* of the technique and of the interpretation of electron micrographs and electron diffraction patterns. As such, I hope the book will also be useful to students of materials science.

The place of TEM in modern geological studies is considered in more detail in the Introduction, where I have expanded the description of the book's intent and content (usually to be found in the preface) to include a discussion of the history and role of TEM in mineralogical studies and the importance of mineralogy as one of the fundamental earth sciences.

It was Jim Boland who suggested I should write this book but, as the writing has kept me away from practical microscopy for so much of the past two years, I will not thank Jim for his suggestion – only for his contributions to our joint research and for his friendship. Other research students and colleagues with whom it has been a great pleasure to work and who have made substantial contributions to the book are too numerous to thank individually here. However, I should particularly like to name John Fitz Gerald, to whom I introduced TEM but who now teaches me;

ix

Bruce Hobbs, John Christie, and the late Dave Griggs, who were so enthusiastic and supportive of our early TEM investigations of deformed minerals; John Hutchison for giving me my first direct experience of lattice resolution imaging; Robin Turner, who taught me most of the optics I know – his initial response to most of my questions was usually "But that is obvious to the meanest intelligence"; Dick Yund for exposing me to the complexities of exsolution in the feldspars; and Mervyn Paterson and Bruce Hyde, who made it possible to me to come to ANU. I also thank the following people whose outstanding TEM investigations of minerals I have particularly drawn on for the applications chapters: Dave Barber, Pam Champness, Jean Claude Doukhan, Martyn Drury, Brian Evans, Madeleine Gandais, Dave Kohlstedt, Gordon Lorimer, Wolfgang Müller, Dave Veblen, Rudi Wenk, and Christian Willaime.

I must admit that the examples I have chosen to illustrate the application of TEM to mineralogy and geology reflect my own particular interests and that many other excellent examples have not even been mentioned. I apologize to those scientists whose work I have apparently ignored but, at the same time, emphasize that the examples chosen cover most of the types of crystal defect to be found in a wide range of rock-forming minerals. Furthermore, I believe that the examples that are given more than adequately fulfill the very important purpose of reinforcing the basic message of the earlier chapters that the *images* formed in the transmission electron microscope must be correctly interpreted in terms of an appropriate theory of electron diffraction before they can provide any useful information about the nature of the *object*.

It is with great pleasure that I thank Clementine Kraysheck and Paul Brugman for their skills in preparing the line diagrams and photographs, and Denise Devir who, with some help from Maria Davern, typed the text with phenomenal accuracy, corrected my spelling, added missing words, and never complained about my frequent changes of mind. My wife Netta deserves thanks for help with proofreading and especially for putting up with months of persistent complaints about how all this writing was keeping me away from real work.

Alex C. McLaren

Canberra
28 November 1990

TRANSMISSION ELECTRON MICROSCOPY
OF MINERALS AND ROCKS

Introduction

The physical and chemical properties of any part of the Earth depend critically on the particular assemblage of minerals that is present locally. Most of the minerals that are accessible to direct observation were formed at or near the Earth's surface, although some may have been brought up to the surface from deep levels in the crust or even from the upper mantle by processes such as volcanism and mountain building. Therefore, the minerals found at the Earth's surface may have formed over a range of temperatures from 0° to 1500°C and at pressures from 1 to more than 20,000 atmospheres. Every mineral or mineral assemblage contains within itself a record of its origin and of the physical conditions it has experienced since its formation. This record is, of course, far from perfect because later events in a rock's history often modify or obliterate the previously developed characteristics. Nevertheless, it is the physical and chemical characteristics of the minerals in the rocks exposed at or near the Earth's surface that constitute the most tangible link with the history of the Earth. Similarly, the minerals found in meteorites may contain information about the conditions that prevailed on the extraterrestrial bodies from which the meteorites originated. Thus, the study of minerals is one of the fundamental earth sciences and can be considered as the *logical* beginning of the science of geology.

Minerals are also important in another, perhaps more practical way. Since prehistoric times, people have learned to use minerals with ever-increasing skill and, in modern times, minerals and products derived from them are indispensable to our technological culture.

Modern mineralogy begain in the seventeenth century when the Dane, Niels Stensen, showed that the angles between the faces of quartz crystals were constant for all samples of this mineral, no matter what their shape and size. This realization of the significance of *crystal form* led to the science of crystallography. During the eighteenth century, new minerals were recognized and described, and attempts were made to produce a rational classification. Mineralogy and chemistry were closely linked since minerals were the chemists' raw materials, and this association led to the recognition and isolation of many new chemical elements. Mineralogy developed rapidly in the early nineteenth century, stimulated by the atomic theory and the realization that minerals were crystals with definite chemical compositions.

Although it had long been recognized that rocks were aggregates of minerals, there was considerable controversy about the nature and origin of many fine-grained rocks, like basalt. These differences of opinion were often due to the fact that the pioneer geologists could rarely identify the very small mineral grains in these rocks with the techniques available at the time. However, in 1851, this difficulty was largely overcome by Henry Clifton Sorby, who developed a technique of preparing rock sections about 25 μm thick which were more or less transparent and could be examined with ease in the optical microscope. This technique, together with the later development of the polarizing microscope, provided the geologist and mineralogist with a powerful tool for identifying the minerals in a rock and determining their optical properties. The next major advance was the discovery by Friedrich, Knipping, and von Laue in 1912 of the diffraction of x-rays by crystals and the subsequent development by W. L. Bragg of the law of crystal diffraction, from which the actual positions of the atoms in a crystal can be determined.

For many years, polarized light microscopy and x-ray crystallography have been the stock-in-trade techniques for the study of minerals. However, they have now been joined by most of the techniques currently used for investigating the structure and properties of a wide range of materials. In fact, mineralogy has become an integral part of modern materials science and, along with physical metallurgy and the study of ceramics, is based firmly on the concepts and techniques that have been developed in solid state physics and chemistry. Consequently, the distinctions that once seemed to exist between these branches of research have virtually disappeared, to their mutual advantage.

In the past decade, there has been a growing awareness that the study of mineral behavior has important implications in related disciplines in the earth sciences, even to the level where continental-scale tectonic phenomena are being considered in terms of processes taking place within individual mineral grains. Although some important properties of crystalline materials can be understood in terms of ideal crystal structures (as determined by x-ray diffraction), many of the physical and chemical properties important in geology and in technological applications are profoundly influenced by the presence, sometimes in very low concentrations, of various types of structural defects and impurity atoms. In polycrystalline materials, like rocks, the nature of the boundaries between the mineral grains may also be of paramount importance for specific properties, such as deformation, fluid migration, and chemical equilibrium.

The transmission electron microscope is ideal for studying the defect microstructures of crystals, and probably no other single instrument has contributed more to our understanding of the nature of crystal defects and of microstructurally related properties of a wide range of crystalline materials. In addition to the usual imaging and electron diffraction facilities, most modern microscopes for materials science research are equipped with an x-ray energy dispersive spectrometer for chemical analysis of very small regions of specimen. Thus, modern transmission electron microscopy (TEM) can be considered as a logical development from optical microscopy, x-ray crystallography, and electron microprobe analysis, techniques with which most mineralogists and geologists are familiar.

Probably the first significant application of TEM to an important rock-forming mineral was made by Fleet and Ribbe (1963) in Cambridge. Their observations of an alkali feldspar (moonstone) revealed a submicroscopic microstructure of alternating lamellae of orthoclase and (polysynthetically twinned) albite, which provided a detailed explanation of the complex diffraction pattern of the specimen and of its white schiller. Over the next few years, a number of other feldspars were studied successfully by the Cambridge group and also by Nissen and by McLaren and their colleagues.

In 1965 McLaren and Phakey published the first of a series of papers on several varieties of quartz. In addition to determining the structure of Brazil and Dauphiné twin boundaries, they observed dislocations in milky vein quartz and tentatively assigned Burgers vectors to them, as well as considering their relevance to the plastic deformation of quartz. These observations led to a study in collaboration with D. T. Griggs and his colleagues of the dislocations and other defects in experimentally deformed crystals of natural and synthetic quartz, from which came some clear indications about the deformation mechanisms that were operating. However, in spite of these and other successful applications of TEM in mineralogy, little work was done at the time on these nonmetallic materials compared with the very great use that was made of TEM in metallurgy. There is little doubt that one of the reasons for this situation was the difficulty of preparing specimens thin enough (of the order of 100 nm) to be transparent to the electron beam. Although such specimens were easily prepared from bulk metal samples by electropolishing techniques, no equally universal and satisfactory technique was available for non-metals. Of necessity, most of the published observations of minerals were obtained from the thin edges of tiny, crushed fracture fragments. Thus, it was not possible to examine particular regions (selected under the optical microscope) of single crystals of predetermined orientation or to select individual grains and grain-boundaries from a standard 25-μm thin section of a rock. However, in the early 1970s these serious limitations were largely overcome by ion-bombardment thinning, which was developed independently by Barber, McLaren, and others on the basis of an instrument originally designed by Paulus and Reverchon in 1961. The spectacular increase in the application of TEM in mineralogy that followed this development may be compared with the similar increase in the use of optical microscopy following Sorby's success in making petrological thin sections. The TEM work on lunar materials provided a further stimulus. By 1976, when H.-R. Wenk (1976) edited the book *Electron Microscopy in Mineralogy*, TEM had changed the aspect of mineralogy.

Although many excellent textbooks are available on the electron microscopy of crystals, most have been written by, and for, physicists or materials scientists trained in physics and thus make little or no concessions to the potential users of TEM with limited formal physics training. This book is designed specifically for geologists and mineralogists, who may have studied physics only at the first year university level; it is primarily an introduction to the basic principles of electron microscopy and is in no way a practical manual. The first seven chapters deal with the essential physics of the transmission electron microscope and with the basic theories required for the interpretation of images and electron diffraction patterns. Clearly this requires mathematics, but nothing more advanced than elementary differential and integral calculus is used in the text. A knowledge of elementary crystallography is assumed, and some familiarity with optics and electromagnetic theory (at a first year university level) would be an advantage but is not essential.

Chapter 1 is concerned with the fundamental principles of image formation by a lens. These principles were first formulated by Ernst Abbe in 1873 and are basic to the chapters that follow. According to the Abbe theory, the image of an illuminated object is the result of a twofold diffraction process. First, the Fraunhofer diffraction pattern of the object is formed in the back focal plane of the lens. Second, the light waves travel beyond this plane and arrive at the image plane where they overlap and interfere to form a magnified image of the object. The nature of the image depends on the relative amplitudes and phases of the waves that pass through the aperture in the back focal plane. If, in an optical microscope, a large number of diffracted waves pass through the aperture in the back focal plane of the objective lens, it can be assumed that the image is a reasonably faithful representation of the object. However, mainly because of the aberrations of magnetic lenses, the number of beams used to form an image in an electron microscope must usually be restricted (often to one beam only) by an aperture located in the back focal plane of the objective lens. The various *imaging modes* used in TEM depend on how many and which beams are used to form the image. These imaging modes, together with a description of the basic construction and features of a modern transmission electron microscope, are discussed in Chapter 2.

The interpretation of TEM images of crystalline materials is much less straightforward than the interpretation of optical microscope images. To interpret TEM images, we must know the diffracting conditions and the beams used to form the image as well as have a detailed understanding of the electron diffraction processes. The diffraction of electrons by perfect crystals is discussed in Chapters 3 and 4. The kinematical theory (Chapter 3) is applicable to x-rays and neutrons and to electrons, and most mineralogists and geologists will be familiar with many aspects of this theory from experience in x-ray crystallography. Because of the initial simplifying assumptions, the kinematical theory has limited applicability; in fact, it breaks down completely under conditions that are often desirable in electron microscopy. However, it is extremely useful under some conditions and provides a good introduction to the more satisfactory dynamical theory developed in Chapter 4. Conceptually, this chapter is the most difficult part of the book, and mathematically the most tedious. However, the results are essential for a proper interpretation of TEM images and an effort has been made to simplify the mathematics as much as possible and to explain with the aid of many diagrams and graphs the physical significance of the equations derived during each major stage of the development.

In the imaging modes used most commonly for imaging crystal defects (such as dislocations and stacking faults), the image is formed using only the transmitted beam or a single diffracted beam. The way in which defects are revealed in these images is discussed qualitatively in the first part of Chapter 5. This is followed by an explanation of how the mathematical description of the distortion around a defect in a crystal is incorporated into the equations of the dynamical theory in order to calculate the detailed nature of the image of the defect. Finally, the *results* of such calculations for the main types of crystal defect observed in minerals are discussed in detail.

To produce a high-resolution *lattice image* of a crystal, at least two beams must be allowed to pass through the aperture in the back focal plane of the objective lens. The basic optical principles involved in this imaging mode are discussed in Chapter 1. However, a number of factors not discussed there influence high-resolution TEM images; these are the main concern of Chapter 6.

Quantitative chemical analysis using the characteristic x-rays emitted by the specimen in the electron microscope is considered in Chapter 7. Since the basic principles are the same as for the electron microprobe (which will be familiar to most geologists and mineralogists), emphasis is placed on those aspects of the technique that are related to the use of thin foil specimens. The intensities of the characteristic x-ray peaks depend on the crystallographic orientation of the specimen with respect to the electron beam; this effect is the basis of the relatively new technique ALCHEMI (atom location by channeling enhanced microanalysis), which under certain conditions can be used to determine the location of minor element atoms in a crystal structure. This technique is potentially very useful for mineralogical studies and is therefore discussed in some detail.

The final chapters are concerned with specific applications of TEM in mineralogy and deal successively with (i) planar defects involving no change of chemistry, (ii) chemically distinct intergrowths, (iii) radiation-induced defects, and (iv) dislocations and deformation-induced microstructures. The aim is to discuss in some detail a few characteristic examples in each group rather than attempting to give an exhaustive review of what is now a very extensive literature of mineralogical applications of TEM.

1

Principles of image formation
by a lens

1.1 Introduction

The purpose of any kind of microscope is to form a magnified image of an object whose fine structure cannot be clearly discerned by the unaided eye. In the conventional optical microscope this magnification is achieved by means of two or more glass lenses. The basic characteristic of a lens that enables it to form a magnified image is its ability to bring to a focus a broad beam of light falling on it. Since many of the optical principles of image formation are also applicable to an electron microscope (in which electron beams are focused by magnetic lenses), it is appropriate that we examine these principles in some detail.

1.2 Elementary concepts of image formation by a thin lens

We begin by considering a thin double-convex lens with spherical surfaces, such as that shown in Figure 1.1. If a set of light rays parallel to the principal axis is incident upon this lens, Figure 1.1(a), then refraction at the lens surfaces will cause the rays to converge to a point F, called the *focal point*. Because the rays from a distant object are essentially parallel, the focal point F is the image of an object on the principal axis at infinity. The distance from F to the center of the lens is the *focal length f*. The focal length is a function of the refractive index of the lens material and the radii of curvature of the two lens surfaces; it is the same on both sides of the lens, even if the curvatures are different.

If the parallel rays incident upon the lens make an angle with the principal axis, Figure 1.1(b), then they will be brought to a focus at a point F'. The plane in which all points such as F and F' lie is called the *focal plane*. Since the focal plane indicated by F and F' in Figure 1.1(b) is on the opposite side of the lens from the object, it is called the *back focal plane*.

7

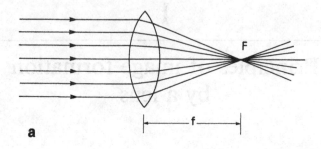

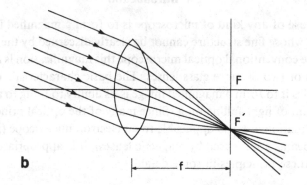

Figure 1.1. Parallel rays are brought to a focus by a thin converging lens. (a) Rays are parallel to the principal axis; (b) rays arrive at an angle to the principal axis.

If we assume the *ray model* of light, we can find the image formed by a lens for a given object by using the ray-tracing technique illustrated in Figure 1.2. The object distance d_0, the image distance d_i and the focal length f for any thin lens are related by the well-known Gaussian equation

$$\frac{1}{d_0} + \frac{1}{d_i} = \frac{1}{f} \tag{1.1}$$

The lateral magnification is defined as the ratio of the image height to the object height; hence,

$$m = \frac{h_i}{h_0} = \frac{d_i}{d_0} \tag{1.2}$$

In Figure 1.2, $d_0 > f$ and the image is *real* and *inverted*. However, if the object is placed within the focal point (i.e., $d_0 < f$), then a *virtual and erect* image will be produced, as can be seen in Figure 1.3.

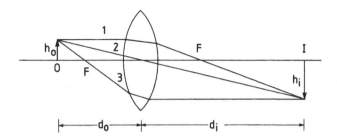

Figure 1.2. Ray diagram showing the real image *I* of an object *O*, formed by a converging lens.

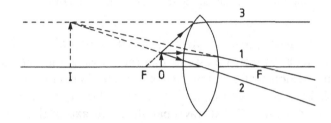

Figure 1.3. Ray diagram showing the virtual image *I* of an object *O*, formed by a converging lens.

The ray model of light is of limited usefulness. If we are to understand the fundamental processes involved in the formation of an image by a lens, we must consider the wave nature of light. The simplest form of the wave theory of light is based on a geometrical construction known as *Huygens principle,* which is usually stated as follows:

All points on a wavefront can be considered as point sources of secondary spherical wavelets; after a time *t*, the new position of the wavefront is given by the surface that is tangent to these secondary wavelets.

Even though Huygens principle makes no assumptions about the nature of the wave, the principle is extremely useful, especially when stated in a precise mathematical form (see Section 1.3).

Now suppose the object in Figure 1.2 consists of a row of small holes spaced a distance *d* apart in an opaque screen that is illuminated by light from a small (1 mW) helium–neon laser. This light is monochromatic, and the normals to the wavefronts (rays) are essentially parallel. Furthermore, such a light source illuminates the object coherently; that is, at any instant

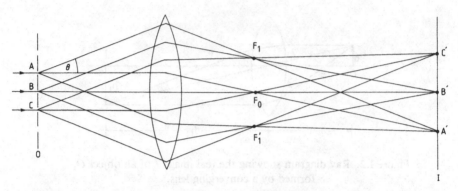

Figure 1.4. Ray diagram showing the relationship between the image *I* and the Frauhofer diffraction pattern of an object *O*.

the waves incident upon any two points on the object are strongly correlated in phase. In fact, we assume that the waves leaving the holes in any given direction at any instant are *in phase*. For simplicity, we consider only three holes *A*, *B*, and *C*, as in Figure 1.4. First consider the three waves from *A*, *B*, and *C* that travel parallel to the axis of the lens; they are brought to a focus at F_0 on the axis in the back focal plane. Because they have traveled the same optical path length they are in phase at F_0 and produce a bright spot, which is, in fact, the image of the light source. Now consider the three waves from *A*, *B*, and *C* that travel in a direction that makes an angle θ with the axis of the lens. They are brought to a focus at point F_1 in the back focal plane. The intensity of the spot at this point depends on the phase difference between adjacent waves. Intensity maxima will be produced at points such as F_1 in the back focal plane, provided $d \sin \theta = n\lambda$ ($n = 0, 1, 2, ...$), where $d \sin \theta$ is the path difference between adjacent waves. Thus, if we place a screen in the back focal plane, we will observe a set of bright spots: the *Fraunhofer diffraction pattern* of the object. The symmetry of the object determines the symmetry of the diffraction pattern; if the object is in the form of a square grid, then the diffraction pattern will consist of a square array of spots of varying intensity. The diffraction spots are due to the constructive interference of waves leaving the object from *different points* but at the *same angle*.

In Figure 1.4, we can see that if the waves converging at the diffraction spots F_0, F_1, and F_1' are continued to the right, then the three waves shown emerging from each of the object points *A*, *B*, and *C* converge at the points *A'*, *B'*, and *C'* in the plane *I*. The bright regions that are observed at *A'*, *B'*, and *C'* are clearly the images of the object points *A*, *B*, and *C* and are due to the constructive interference of waves emerging at *different angles* from the object points *A*, *B*, and *C*, respectively. Thus, information about the object is presented in the back focal plane as the Fraunhofer diffraction pattern and in the plane *I* as an interference pattern which is an enlarged image of the object. In other words, the image arises from a double diffraction process.

Because the lens has a finite diameter, waves diffracted through angles θ greater than those shown in Figure 1.4 are not collected by the lens and therefore do not give rise to diffraction maxima in the back focal plane. Thus, some information about the object is always lost, and the image formed in the image plane can never be a perfect image of the object. This theory of image formation was first put forward by Ernst Abbe in 1873.

So far, we have used only the ideas of ray and wave optics which the reader will find in many textbooks of elementary physics, such as those by Giancoli (1984) or Resnick and Halliday (1966).

Now we must examine the physics of image formation by a lens more closely, and to do this we must introduce the ideas of Fourier optics. It would be inappropriate here to develop fully these ideas, as is done in most modern textbooks on optics, but it is important to understand clearly the fundamental concepts of Fourier optics because we shall need them when we deal specifically with electron diffraction. However, before continuing, it is necessary to digress briefly to introduce the mathematics used to describe plane and spherical waves.

The simplest type of wave in three dimensions is the *plane wave* in which, by definition, the disturbance at any instant of time has the same value at all points in any given plane that is perpendicular to the direction of propagation. Such a wave traveling along the *x* direction is described by the equation

$$\psi = \psi_0 \cos 2\pi \left(\frac{x}{\lambda} - \nu t \right) \qquad (1.3)$$

where ψ_0 is the amplitude, λ the wavelength, ν the frequency, and t time. Since

$$\cos \theta + i \sin \theta = \exp i\theta$$

Eq. (1.3) is the real part of

$$\psi = \psi_0 \exp 2\pi i \left(\frac{x}{\lambda} - \nu t \right) \qquad (1.4)$$

and it will be convenient to use the exponential notation rather than describe the wave in terms of a sine or cosine function. It will also be convenient to introduce the *wave number k*, defined by $k = 2\pi/\lambda$, and the *angular frequency* ω, defined by $\omega = 2\pi\nu$, so that Eq. (1.4) becomes

$$\psi = \psi_0 \exp i(kx - \omega t) \tag{1.5}$$

A wave that is identical to this except that it differs in phase by an angle φ is given by

$$\psi' = \psi_0 \exp i(kx - \omega t + \varphi) \tag{1.6}$$

This can be written as

$$\psi' = [\psi_0 \exp i\varphi] \exp i(kx - \omega t) \tag{1.7}$$

where the part in square brackets is the *complex amplitude*.

If Eq. (1.6) is written as

$$\psi' = \psi_0 \exp i\left[k\left(x + \frac{\varphi}{k}\right) - \omega t\right] \tag{1.8}$$

we can see that the waves ψ and ψ' are displaced along the x-axis by a distance $d = \varphi/k$ and that

$$\frac{d}{\lambda} = \frac{\varphi}{2\pi} \tag{1.9}$$

Equation (1.5) can be written in the more general form

$$\psi = \psi_0 \exp i(\mathbf{k} \cdot \mathbf{r} - \omega t) \tag{1.10}$$

where $\mathbf{k}$ is a vector, known as the *wavevector*, of magnitude $2\pi/\lambda$ and in the direction of propagation of the wave; $\mathbf{r}$ is a position vector. Sometimes the magnitude of $\mathbf{k}$ is taken as $1/\lambda$, so that Eq. (1.10) becomes

$$\psi = \psi_0 \exp 2\pi i(\mathbf{k} \cdot \mathbf{r} - \nu t) \tag{1.11}$$

For an ideal *plane wave*, the energy per unit area per unit time crossing a plane perpendicular to the direction of propagation is proportional to ψ_0^2, and this relationship does not change with distance from the source of the wave. However, for a *spherical wave* of radiation emitted by an ideal point source, this relationship does change. To understand this, consider two spherical surfaces of radii r_1 and r_2 (with $r_2 > r_1$) centered at the point source. If there is no absorption, the *total energy* that passes through the first surface in any given interval of time dt must all, at a later time, pass through the second surface in an interval of the same duration dt. Thus, the *energy per unit area* for the radiation passing through the second spherical surface during dt will be less than that for the radiation passing through the first spherical surface; so the *intensity* of the radiation

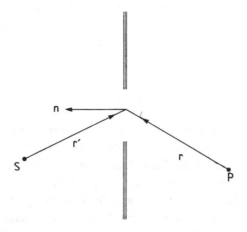

Figure 1.5. Diagram of the geometry of the Fresnel-Kirchhoff formula.

will decrease with distance r from the source. Now a spherical wave that has a very large radius should behave practically like a plane wave, for which the intensity is proportional to the square of the amplitude. Therefore, since the area of a sphere of radius r is proportional to r^2, it follows that the *amplitude* of a spherical wave will vary as $1/r$. Thus, a spherical wave is described by

$$\psi = \frac{\psi_0}{r} \exp i(kr - \omega t) \tag{1.12}$$

where $k = 2\pi/\lambda$.

1.3 Image formation by a thin lens in terms of Fourier optics*

We can use Huygens principle to explain qualitatively the essential features of diffraction, but a quantitative treatment involves casting Huygens principle into a precise mathematical form, the *Fresnel-Kirchhoff formula:*

$$U_p = -\frac{ikU_0 \exp(-i\omega t)}{4\pi} \iint_A \frac{\exp ik(r + r')}{rr'} [\cos(\mathbf{n}, \mathbf{r}) - \cos(\mathbf{n}, \mathbf{r}')] \, dA \tag{1.13}$$

The meaning of the symbols used and the physical significance of this equation can be understood with the help of Figure 1.5. Spherical waves of light from a source S are incident upon the aperture A of arbitrary

* This section is based on the account given by Fowles (1968).

shape in an opaque screen. The vector $\mathbf{r}'$ denotes the position of a point in the aperture relative to S, so the wave function at the aperture of the spherical wave (of wave number $k = 2\pi/\lambda$ and angular frequency $\omega = 2\pi\nu$) traveling outward from S is

$$U = \frac{U_0}{r'} \exp i(kr' - \omega t) \qquad (1.14)$$

The vector $\mathbf{r}$ denotes the position of the receiving point P on the opposite side of the aperture. $(\mathbf{n}, \mathbf{r})$ denotes the angle between the vector $\mathbf{r}$ and the vector $\mathbf{n}$ which is normal to the surface of integration, and similarly for $(\mathbf{n}, \mathbf{r}')$. The integration is taken only over the opening of the aperture. U_p is the "optical disturbance" at the point P. Although U_p does not accurately represent the electromagnetic field, being a scalar quantity, the square of its absolute value is a measure of the light intensity at P. U_0 in Eq. (1.13) has the same meaning. Note particularly that the factor $-i$ indicates that the diffracted wave is $\pi/2$ out-of-phase with respect to the incident wave. The term in square brackets is the *obliquity factor*. For the special case of a circular aperture with the source S symmetrically located, $\mathbf{r}'$ is constant and antiparallel to $\mathbf{n}$, so that $\cos(\mathbf{n}, \mathbf{r}') = -1$, and Eq. (1.13) becomes

$$U_p = -\frac{ik}{4\pi} \iint_A \frac{U_A}{r} \exp i(kr - \omega t)[\cos(\mathbf{n}, \mathbf{r}) + 1]\, dA \qquad (1.15)$$

where

$$U_A = \frac{U_0}{r'} \exp ikr'$$

Equation (1.15) can be interpreted in the following way. A spherical wave of complex amplitude U_A from S, incident upon the aperture, gives rise to a secondary spherical wave,

$$\Delta U = \frac{U_A}{r} \exp i(kr - \omega t)\, dA$$

from each element of area dA of the aperture. The total optical disturbance at P is the sum of the secondary waves from all such elements. This is simply a statement of Huygens principle.

If we assume, for simplicity, that the angular spread of the diffracted light is small enough for the obliquity factor to be taken as constant over the aperture, and that the variation of U_p with r and r' in the denominator of Eq. (1.13) is small compared with the exponential term, then Eq. (1.13) becomes

$$U_p = C \iint_A \exp(ikr)\,dA \tag{1.16}$$

where all the constant factors are incorporated in C. Equation (1.16) indicates that U_p is obtained simply by integrating the phase factor $\exp(ikr)$ over the area of the aperture.

Now let us consider the Fraunhofer diffraction by a two-dimensional object, which we take to be an aperture of arbitrary shape. Figure 1.6(a) shows the general geometry and the coordinates that are used. The object (the aperture in an opaque screen) lies in the xy-plane, and the diffraction pattern is formed in the XY-plane, the back focal plane of the lens. The image lies in the $x'y'$-plane.

In Section 1.2 we showed that all rays leaving a diffracting aperture in a given direction (which we specify by the unit vector $\hat{\mathbf{n}} = \alpha\hat{\mathbf{i}} + \beta\hat{\mathbf{j}} + \gamma\hat{\mathbf{k}}$) are brought to a common focus at the point $P(X, Y)$, where $X \approx f\alpha$ and $Y \approx f\beta$, and f is the focal length of the lens. Now consider two rays, one starting from the origin O and the other from $Q(x, y)$ in the xy-plane, both traveling in the direction $\hat{\mathbf{n}}$. If the position of $Q(x, y)$ relative to the origin is given by the vector $\mathbf{R} = x\hat{\mathbf{i}} + y\hat{\mathbf{j}}$, then the path difference between the rays, Figure 1.6(b), is given by

$$\delta r = \mathbf{R} \cdot \hat{\mathbf{n}} = x\alpha + y\beta$$

$$= x\frac{X}{f} + y\frac{Y}{f} \tag{1.17}$$

Therefore, from Eq. (1.16),

$$U(X, Y) = \iint_A \exp ik\,\delta r\,dA$$

$$= \iint_A \exp\left[ik\frac{xX + yY}{f}\right]dA \tag{1.18}$$

Note that, for simplicity, we have omitted the constant multiplying factor C of Eq. (1.16).

Equation (1.18) applies to a uniform aperture over which the amplitude of the diffracted wave is the same for all points (x, y). For a nonuniform aperture, we can introduce the function $g(x, y)$ which gives the amplitude of the diffracted wave originating from an element of area $dx\,dy$ of the aperture. $g(x, y)$ is usually called the aperture function, but because this nonuniform aperture is our object, we call $g(x, y)$ the *object function*.

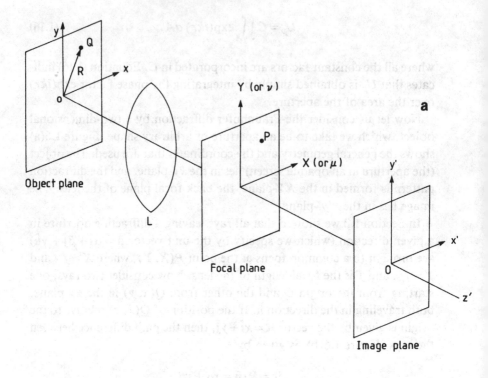

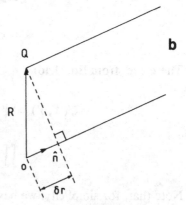

Figure 1.6. (a) Diagram showing the general geometry and coordinates used for discussing image formation by a lens in terms of Fourier optics. (b) Path difference between two parallel rays from points O and Q in the xy plane.

Now Eq. (1.18) becomes

$$U(X, Y) = \iint g(x, y) \exp\left[ik\frac{xX + yY}{f}\right] dx\,dy \qquad (1.19)$$

At this point, it is convenient to introduce the quantities μ and ν defined by

$$\mu = \frac{kX}{f} \quad \text{and} \quad \nu = \frac{kY}{f} \tag{1.20}$$

where μ and ν have the units of reciprocal length and are called *spatial frequencies*. Equation (1.19) can now be written as

$$U(\mu, \nu) = \iint_{-\infty}^{+\infty} g(x,y) \exp i(\mu x + \nu y) \, dx \, dy \tag{1.21}$$

The functions $U(\mu, \nu)$ and $g(x,y)$ are Fourier transforms of each other, and

$$g(x,y) = \frac{1}{2\pi} \iint_{-\infty}^{+\infty} U(\mu, \nu) \exp[-i(\mu x + \nu y)] \, d\mu \, d\nu \tag{1.22}$$

Thus, the diffraction pattern given by the function $U(\mu, \nu)$ is the Fourier transform of the object function $g(x,y)$.

We illustrate this relation by considering an object that is a single slit of width a, as shown in Figure 1.7(a). In one dimension, $g(x) = 1$ for $-a/2 < x < +a/2$ and $g(x) = 0$ for values of x outside this range. Thus, Eq. (1.21) becomes

$$
\begin{aligned}
U(\mu) &= \int_{-a/2}^{+a/2} \exp i\mu x \, dx \\
&= \left[\frac{1}{i\mu} \exp i\mu x \right]_{-a/2}^{+a/2} \\
&= \frac{1}{i\mu} [\cos \mu x + i \sin \mu x]_{-a/2}^{+a/2} \\
&= \frac{1}{i\mu} \left[\cos \frac{\mu a}{2} + i \sin \frac{\mu a}{2} - \cos\left(-\frac{\mu a}{2}\right) - i \sin\left(-\frac{\mu a}{2}\right) \right] \\
&= \frac{1}{i\mu} 2i \sin\left(\frac{\mu a}{2}\right)
\end{aligned}
$$

since $\cos \theta = \cos(-\theta)$ and $\sin(-\theta) = -\sin \theta$. Hence,

$$U(\mu) = a \frac{\sin(\mu a/2)}{(\mu a/2)} \tag{1.23}$$

This is the well-known expression for the Fraunhofer diffraction by a single slit, and is plotted in Figure 1.7(b).

As another example, consider the diffraction by a grating consisting of a large number N of parallel slits of width a, separated by opaque strips that are also of width a, as shown in Figure 1.8(a). The fundamental spatial period is $d = 2a$. We now make use of the Fourier theorem, which

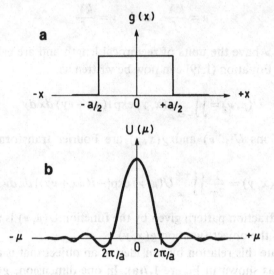

Figure 1.7. (a) Object function for a single slit and (b) its Fourier transform.

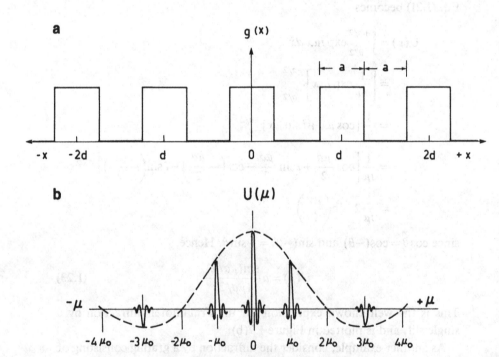

Figure 1.8. (a) Object function for a diffraction grating and
(b) its Fourier transform.

states that any periodic function $f(x)$ can be expressed as the sum of a series of sinusoidal functions which have spatial periods that are integral submultiples of the fundamental spatial period d of $f(x)$. Zero is counted as an integer, thus making the first term of the series a constant. The object function $g(x)$, graphed in Figure 1.8(a), can be written as

$$g(x) = g_0 + g_1 \cos \frac{2\pi x}{d} + g_2 \cos \frac{2\pi x}{d/2} + \cdots + g_n \cos \frac{2\pi x}{d/n} \qquad (1.24)$$

In terms of the fundamental spatial frequency $\mu_0 \, (= 2\pi/d)$, this equation becomes

$$g(x) = g_0 + g_1 \cos \mu_0 x + g_2 \cos 2\mu_0 x + \cdots + g_n \cos n\mu_0 x \qquad (1.25)$$

and substitution of this into Eq. (1.21) leads to

$$U(\mu) = \frac{a \sin(\mu a/2)}{\mu a/2} \frac{\sin(N\mu a)}{\sin(\mu a)} \qquad (1.26)$$

which is plotted in Figure 1.8(b).

A number of important aspects of Eq. (1.26) and Figure 1.8(b) must be discussed in detail. It will be seen that there are sharp spikes (usually called spectra) in $U(\mu)$ at values of μ corresponding to the spatial frequencies 0, μ_0, $3\mu_0$, $5\mu_0$, and so on. (The fact that the spectra appear only at odd multiples of μ_0 is considered later). Each spectrum is made up of a principal maximum with several, much smaller, secondary maxima on each side. The spread of the spectrum is inversely proportional to the width Nd of the grating. Also, the amplitudes of the spectra vary with μ and are indicated by the dashed envelope that is given by the first term on the right-hand side of Eq. (1.26). Note that this term is identical to Eq. (1.23), and thus it describes the diffraction from each *separate* slit of width a of the grating. The spectra (and their secondary maxima) are described by the second term of Eq. (1.26), and each spectrum arises from the interference of waves traveling in a specific direction from the slits of the grating. Thus, the interference pattern is amplitude-modulated by the diffraction from each individual slit. For this grating in which $d = 2a$, $\mu_0 = 2\pi/d$ and therefore $\mu_0 = \pi/a$. Thus, from Eq. (1.19), when $\mu = (2n)\mu_0 = 2n\pi/a$, where n is an integer, $U(\mu) = 0$ and no spectra appear at values of μ corresponding to even multiples of μ_0. If the width of the slits were different from the width of the opaque strips, these spectra would appear in the diffraction pattern.

The significant point which emerges from this discussion is that *the diffraction pattern is a display of the sinusoidal components of the Fourier* *series that represents the object function $g(x)$,* Eq. (1.25). The diffraction spectrum at $\mu = 0$ corresponds to the constant g_0 and is, as we have seen, the image of the light source. The components $g_1 \cos \mu_0 x$, $g_2 \cos 2\mu_0 x$, $g_3 \cos 3\mu_0 x$, ... correspond to the spectra at $\mu = \mu_0$, $2\mu_0$, $3\mu_0$, ..., respectively, in Figure 1.8(b). Note the essential correspondence between the angle θ at which a diffraction maximum occurs and the value of μ. We saw in Section 1.2 that a diffraction maximum occurs when $d \sin \theta = n\lambda$, where $n = 0, 1, 2, \ldots$. From Figure 1.4 it is clear that for $n = 1$, $\sin \theta = \tan \theta = X/f$ for small θ; so $1/d = X/f\lambda$ and $2\pi/d = kX/f = \mu_0$, the fundamental spatial frequency, Eq. (1.20). Thus, $2\mu_0$ corresponds to $n = 2$, $3\mu_0$ corresponds to $n = 3$, and so on. n is usually called the *order* of the diffraction maximum.

We now determine the nature of the image that is formed in the $x'y'$-plane of Figure 1.6(a). We have shown that the diffraction pattern $U(\mu, \nu)$ which occurs in the $\mu\nu$-plane (equivalent to the XY-plane) is the Fourier transform of the object function $g(x, y)$. Thus, the *image function* $g(x', y')$ that appears in the $x'y'$-plane is, in turn, simply the Fourier transform of $U(\mu, \nu)$. If *all* the spatial frequencies in the range $\mu = \pm\infty$ and $\nu = \pm\infty$ were transmitted equally through the aperture in the $\mu\nu$-plane (i.e., the back focal plane of the lens), then the image will be a faithful reproduction of the object. However, because of the finite size of the aperture in the back focal plane, some higher order spatial frequencies are inevitably excluded, making it impossible to produce a perfect image. Therefore, we introduce the *transfer function* $T(\mu, \nu)$ to describe the characteristics of the aperture; it is defined implicitly by

$$g(x', y') = \iint_{-\infty}^{+\infty} T(\mu\nu)U(\mu, \nu) \exp[-i(\mu x' + \nu y')] \, d\mu \, d\nu \quad (1.27)$$

As an example, we consider the image of an object that is a single slit of width a, specified by the object function $g(x)$ shown in Figure 1.9(a). The amplitude $U(\mu)$ in the back focal plane is shown in Figure 1.9(b), and if the edges of the aperture are defined by the points $\mu = -D/2$ and $\mu = +D/2$, as shown, then an essentially perfect image will result, Figure 1.9(c), where M is the magnification. However, suppose the width of the slit is very narrow such that the characteristic dimension $2\pi/a$ indicated in Figure 1.9(b) is large compared with $D/2$, as shown in Figure 1.9(d). If we assume that $U(\mu)$ is essentially constant, say unity, over the width of the aperture (i.e., over the range $-D/2 < \mu < +D/2$) and zero outside these limits, then the amplitude in the image plane will be as shown in Figure 1.9(e). Thus, the final image is characteristic only of the microscope, and the width of the image is determined by the size of the aperture

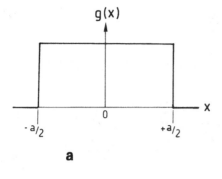

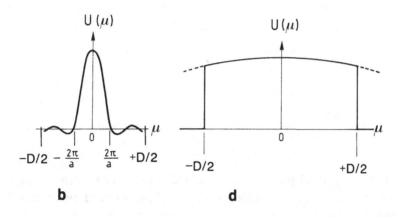

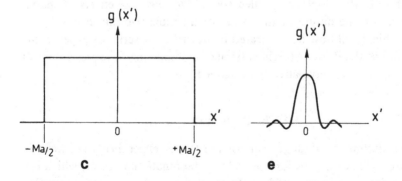

Figure 1.9. Diagrams illustrating the influence of the size of the aperture in the back focal plane on the nature of the image of a single slit. (a) Object function of single slit; (b) amplitude in back focal plane; (c) perfect image if edges defined by $\mu = \pm U(\mu)/2$; (d) amplitude when slit is very narrow and $2\pi/a$ is large compared with $U(\mu)/2$; (e) amplitude in image plane for $U(\mu)$ is constant across the aperture, as is approximately so in (d).

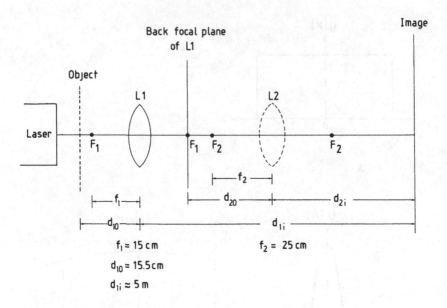

Figure 1.10. Experimental arrangement for the Porter experiments.

in the back focal plane of the lens. This situation can arise for objects such as narrow twin lamellae, subgrain and grain boundaries, and deformation lamellae; the topic is discussed in detail by McLaren et al. (1970).

The process of removing specific spatial frequencies by means of an aperture in the $\mu\nu$-plane is called *spatial filtering*. The effects of spatial filtering on the nature of the image of a simple two-dimensional periodic object will now be illustrated by describing a series of experiments (known as the Porter experiments) and by discussing the observations in terms of the concepts already developed.

1.4 The Porter experiments

The experimental arrangement for the Porter experiments is shown in Figure 1.10. A small helium–neon laser illuminates an object, which is a 3-mm diameter copper grid of the type commonly used for supporting specimens for transmission electron microscopy. The spacing d between grid bars is 125 μm.

The image of the grid formed by lens L1 only is shown in Figure 1.11(1b). The diffraction pattern of the grid is formed in the back focal plane of L1. If lens L2 is now placed in the position shown in Figure 1.10, then an

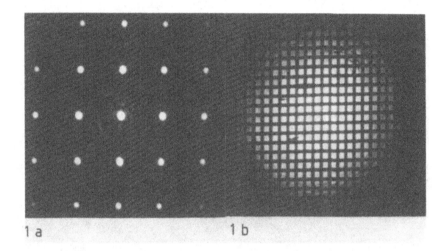

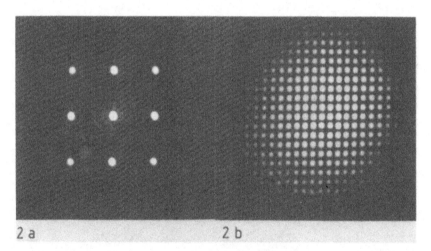

Figure 1.11. A series of ten pairs of photographs that show the effects of spatial filtering on the image of a two-dimensional grating. Each photograph shows (a) the diffraction pattern and (b) the resulting image. (*Continued, pp. 24–27*)

enlargement of the diffraction pattern will appear in the image plane, as shown in Figure 1.11(1a). Due to the finite diameter of L1, only 25 beams (including the central beam) are involved in forming the image in Figure 1.11(1b); nevertheless, the grid bars are reasonably sharply defined and the holes quite square.

By using apertures of various shapes and sizes placed in the back focal plane of L1, we can exclude specific beams and observe the effect on the nature of the image of the grid.

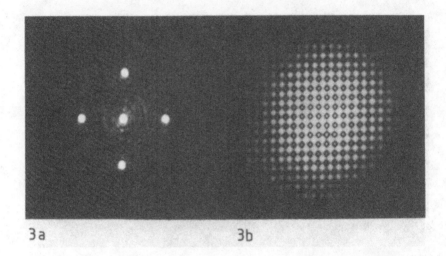

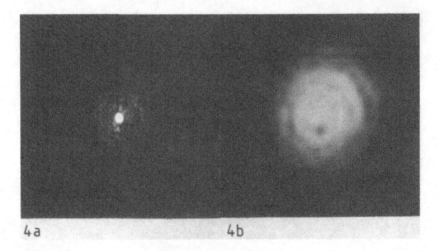

Figure 1.11. *Continued.*

Figure 1.11(2b) shows the image formed when we reduce the number of beams (including the central beam) to nine, as in Figure 1.11(2a). The grid bars in the image are now distinctly less sharp, compared with Figure 1.11(1b), and the holes clearly rounded.

If we reduce the number of beams to five, as in Figure 1.11(3a), significant changes occur in the corresponding image, Figure 1.11(3b). For example, a hole seems to have appeared at the intersection of the grid bars.

The image in Figure 1.11(4b) formed when only the central beam was used, Figure 1.11(4a). Now the grid is not resolved at all. This is, of course,

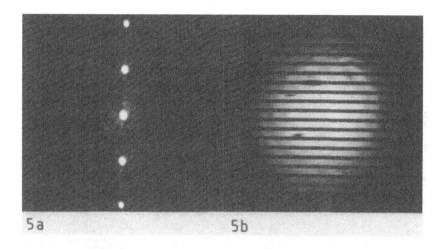

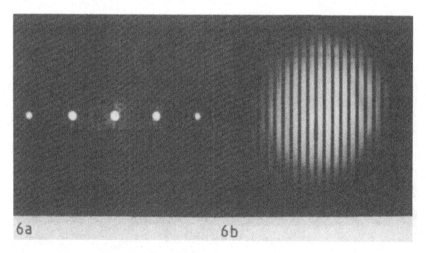

Figure 1.11. *Continued.*

not surprising since we know that the central spot of the diffraction pattern is in fact the image of the light source and contains no information about the grid. Thus, the resolution of this optical system is controlled by the size of the aperture in the back focal plane of the image-forming lens L1. The minimum radius R_c of a circular aperture placed symmetrically about the central beam in the back focal plane must be large enough to let through the first-order diffracted beams in order to produce an image that is recognizably a grid, even if the image is not a very faithful representation of the object. The angular radius of this aperture is equal to the

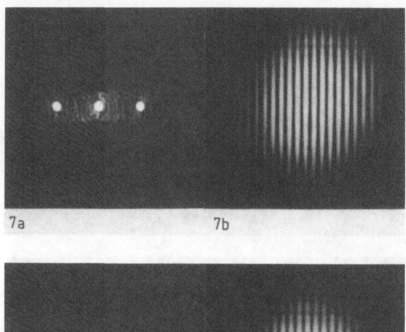

7a   7b

8a   8b

Figure 1.11. *Continued.*

angular separation θ between the central beam and the first-order diffracted beam, which (see Section 1.2) is given by $d\sin\theta = \lambda$, where d is
the spacing between the bars of the grid and λ the wavelength of the light.
Thus, since the aperture is a distance f from L1, the minimum radius
R_c of the circular aperture which is required to resolve a grid with a bar-
spacing d is given by

$$R_c = f\tan\theta = f\sin\theta = \frac{f\lambda}{d} \tag{1.28}$$

for small angles θ.

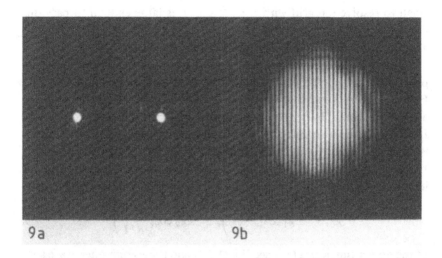

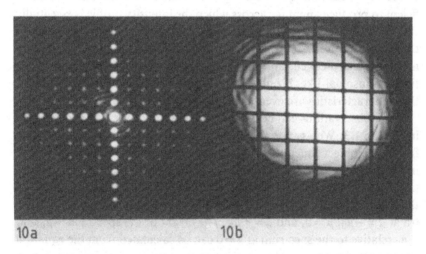

Figure 1.11. *Continued.*

Figure 1.11(10b) shows the image of a grid with a bar-spacing of 375 μm, which is three times larger than that of the grid shown in Figure 1.11(1b). The same diameter aperture was used in both cases; as a consequence, about 169 beams are used to form the image in Figure 1.11(10b), compared with 25 beams for Figure 1.11(1b), as can be seen from the associated diffraction patterns in Figures 1.11(10a) and (1a). The relative improvement in image quality that is obtained by using a larger number of beams (i.e., Fourier components) is apparent on close inspection of the images. Further, it is clear that whereas the aperture used in Figure 1.11(4a) is too small to resolve the grid with $d = 125$ μm, it is large enough to pass the first-order diffracted beams in Figure 1.11(10a); hence, the grid with $d = 375$ μm could be resolved, as expected from Eq. (1.28).

Suppose we replace the circular aperture by a slit that is wide enough to pass only a single, vertical row of diffracted beams, as in Figure 1.11(5a). Figure 1.11(5b) shows that the image formed using these beams is simply a set of horizontal bars; there is no resolution in the horizontal direction. If we orient the slit so that only a single, horizontal row of diffracted beams is used to form the image, as in Figure 1.11(6a), then the image, in Figure 1.11(6b), consists of a set of vertical bars. The five beams used to form the images in Figures 1.11(5) and (6) correspond to the Fourier components

$$ g_0, \quad \pm g_1 \cos\left(\frac{2\pi x}{d}\right), \quad \pm g_2 \cos\left(\frac{2\pi x}{d/2}\right) $$

of the object function $g(x)$, see Eq. (1.21), and these components are sufficient to produce an image across which the intensity profile is essentially that of a square wave. However, if we reduce the number of beams to three, as in Figure 1.11(7a), the grid bars in the image, Figure 1.11(7b), are no longer sharp-edged, and the intensity profile across the image is distinctly sinusoidal-like. If only two beams are used, as in Figure 1.11(8a,b), these characteristics are even more pronounced.

These observations can be understood in terms of the theory developed in Section 1.3. We assume the object is a one-dimensional grating specified by the object function $g(x)$, which is given by Eq. (1.25) and plotted in Figure 1.8(a). The amplitude distribution $U(\mu)$ of the resulting diffraction pattern is shown in Figure 1.8(b). First, we investigate the nature of the image formed using the only three beams corresponding to the spectra at $\mu = -\mu_0$, $\mu = 0$, and $\mu = +\mu_0$. The height of the spectra at $-\mu_0$ and $+\mu_0$ relative to the spectrum at $\mu = 0$ can be calculated from the equation of the dotted envelope. If the height of the spectrum at $\mu = 0$ is taken as Na, then the height of the spectra at $-\mu_0$ and $+\mu_0$ is $2Na/\pi$. Equation (1.26) now becomes

$$ U(\mu) = \frac{2Na}{\pi} \frac{\sin[N(\mu - \mu_0)a/2]}{\sin[(\mu - \mu_0)a/2]} + Na \frac{\sin(N\mu a/2)}{\sin(\mu a/2)} $$
$$ + \frac{2Na}{\pi} \frac{\sin[N(\mu + \mu_0)a/2)]}{\sin[(\mu + \mu_0)a/2)]} \tag{1.29} $$

If N is very large, then only those secondary maxima that lie in a small range of μ on each side of each principal maximum of a spectrum have heights which are comparable to that of the principal maximum itself. In this case, we can replace all the sine terms in the denominator of Eq. (1.29) by the angle itself, that is, $\sin(\mu a/2)$ becomes $(\mu a/2)$ and so on. With this approximation, we calculate $g(x')$ by substituting Eq. (1.29) into Eq. (1.27). The transfer function $T(\mu)$ has been taken care of by selecting the three spectra at $\mu = -\mu_0$, $\mu = 0$, and $\mu = +\mu_0$. Therefore, we have

$$g(x') = \int_{-\infty}^{\infty} U(\mu) \exp(-i\mu x') \, d\mu \qquad (1.30)$$

Integration leads to

$$g(x') = a\left[\frac{2}{\pi} \exp\left(\frac{-i\pi x'}{Ma}\right) + 1 + \frac{2}{\pi} \exp\left(\frac{i\pi x'}{Ma}\right)\right] \qquad (1.31a)$$

$$= a\left(1 + \frac{4}{\pi} \cos\frac{2\pi x'}{2Ma}\right) \qquad (1.31b)$$

where M is the magnification. $g(x')$ is plotted in Figure 1.12(a). However, it is the *intensity* which is actually observed; this is given by $g^2(x')$ and is plotted in Figure 1.12(b). The period is $2Ma = Md$, as expected. However, the image now shows a sinusoidal-like variation and a small, subsidiary maximum between each pair of principal maxima. Thus, the image contains spurious detail and is not a faithful representation of the object. The sinusoidal nature of the image is apparent in Figure 1.11(7b), but the subsidiary maxima are too weak to be seen.

A quite remarkable effect is observed if the central beam, corresponding to the spectrum at $\mu = 0$, is eliminated by placing a small opaque obstacle at the center of the back focal plane, and an image is formed using only the two diffracted beams, corresponding to the spectra at $\mu = -\mu_0$ and $\mu = +\mu_0$. The resulting image and its associated diffraction pattern are shown in Figure 1.11(9). The period of the grating now appears to be $d/2$ because the number of bars per unit length in the image is twice that observed in Figures 1.11(6–8). The explanation of this effect follows directly from Eq. (1.31a). Recall that the central spot of the diffraction pattern in the back focal plane is the image of the light source. Thus, by removing the central beam, we eliminate the second (constant) term of Eq. (1.31a), and $g(x')$ then becomes

$$g(x') = \frac{4a}{\pi} \cos\left(\frac{2\pi x'}{2Ma}\right) \qquad (1.32)$$

But the observed image is given by $g^2(x')$, which is plotted in Figure 1.12(c). Removing the central beam has made the heights of the subsidiary maxima equal to the heights of the principal maxima, and so the

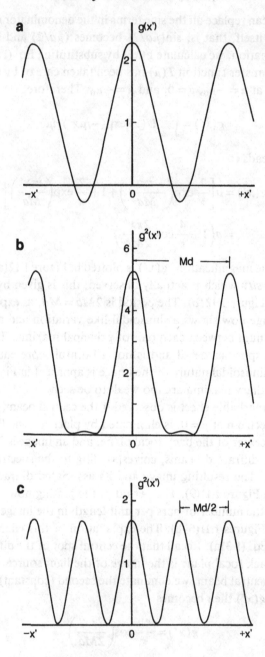

Figure 1.12. (a) Plot of $g(x')$ and (b) plot of $g^2(x')$ from Eq. (1.31b). (c) Plot of $g^2(x')$ from Eq. (1.32).

d-spacing of the bars in the image is halved. We can look at this another way. The spacing between any pair of adjacent spots in the diffraction pattern of a grating is inversely proportional to the d-spacing of the grating. If we form an image with any two (or more) of the corresponding beams, then the period in the image is simply Md, as in Figures 1.11(6-8). However, the spacing between the spots in Figure 1.11(9a) is twice the spacing in the previous three diffraction patterns, and hence the period in the image will be $Md/2$.

Finally, we consider the nature of the image formed with a single diffracted beam, say the first-order beam at $\mu = +\mu_0$. Under these conditions, the first two terms of Eq. (1.31a) are eliminated and

$$g(x') = \frac{2a}{\pi} \exp\left(\frac{ix'}{Ma}\right) \tag{1.33a}$$

To obtain the intensity distribution in the image plane, we must multiply $g(x')$ by its complex conjugate

$$g^*(x') = \frac{2a}{\pi} \exp\left(\frac{-ix'}{Ma}\right) \tag{1.33b}$$

and thus $gg^* = (2a/\pi)^2$. Hence, the image is a uniformly illuminated area and contains no information about the structure of the grating. To obtain an image that contains information about the grating, we must use at least two beams.

If the object (say a piece of grid) is smaller than the area of the illuminated field of view, and if an image is formed using the central beam (either alone or together with one or more diffracted beams), then the image of the object will be seen in a bright field. Such an image is called a *bright field* (BF) image. On the other hand, if an image is formed using one or more diffracted beams only, then the image of the object will be seen in a dark field because the illuminated region around the object itself cannot give rise to any diffracted beams. An image formed in this way is called a *dark field* (DF) image.

1.5 Imaging a defect in a periodic structure

Figure 1.13(1b) shows the image of a grid containing a defect that is coarse compared with the spacing between grid bars. In the associated diffraction pattern, Figure 1.13(1a), we see that in the neighborhood of each normal spot there is additional intensity that is clearly due to the defect.

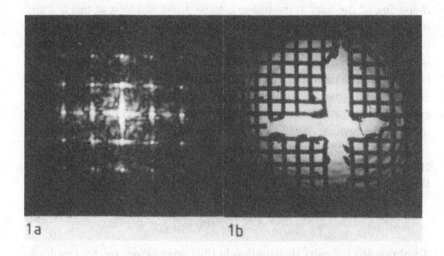

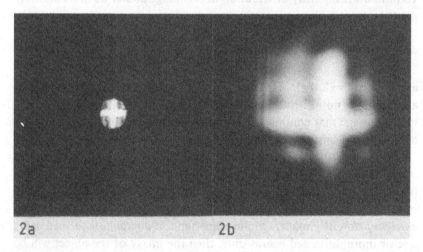

Figure 1.13. Diffraction pattern (1a) and image (1b) of a defective grid. If the aperture in the back focal plane is so small that none of the main diffracted beams can pass (2a), then the grid is not resolved but the defect is seen in the image (2b).

If a BF image, such as Figure 1.13(2b), is formed using a small aperture, as in Figure 1.13(2a), then the defect is clearly resolved but the grid itself is not. If the aperture is placed around one of the diffracted beams, then a DF image is formed which is of opposite contrast to the BF image. These modes of imaging are extensively used in transmission electron microscopy and will be discussed in detail in subsequent chapters.

1.6 Coherence

So far in this discussion of diffraction, we have assumed that the periodic object is illuminated by coherent light, such as that produced by a small laser of the type used in the Porter experiments. However, the light produced by a thermal source (e.g., a sodium vapor lamp or a heated filament coupled with a narrow bandpass filter) is never strictly monochromatic; even the sharpest spectral line has a finite width. Moreover, such a source has finite extent, and the light is emitted by many independent radiators (atoms). These two characteristics of thermal sources are directly related to what are usually referred to as *temporal* and *spatial coherence,* respectively.

In an electron microscope, the electron beam is produced by an electron gun, in which the electrons are "boiled off" a heated filament and then accelerated by a highly stabilized electrical potential difference of some hundreds of kilovolts. An electron beam produced in this way is not coherent, but its degree of coherence can be understood in terms of the concepts of optical coherence theory. Later chapters will show that the nature of electron diffraction patterns from crystals that exhibit long-period superstructures (which are not uncommon in many important rock-forming minerals) depends critically on the degree of spatial coherence of the incident electron beam. Therefore, it is important to conclude this chapter with a brief review of the basic ideas of optical coherence. A detailed account of the theory is given by Born and Wolf (1965).

Using the Fourier theorem, we can represent the quasi-monochromatic disturbance produced by a thermal source as a sum of strictly monochromatic and therefore infinitely long wave trains of slightly different frequencies. This disturbance is nearly sinusoidal, but both the amplitude and the frequency (and, therefore, the phase) vary slowly about some mean values. The complex amplitude remains constant only for a time interval Δt, known as the *coherence time,* which is roughly equal to $1/\Delta \nu$, where $\Delta \nu$ is the effective spectral width of the radiation. The corresponding length of the wave train for which the complex amplitude remains constant is often called the *coherence length* and is equal to $c \, \Delta t$, where c is the speed of propagation of the wave.

To understand spatial coherence, consider two points P_1 and P_2 on a diffracting object illuminated by radiation from an extended quasi-monochromatic source, as in Figure 1.14. If P_1 and P_2 are so close to each other that the difference $\Delta d = SP_1 - SP_2$ between the paths from all points S on the source is small compared with the mean wavelength $\bar{\lambda}$, then it is

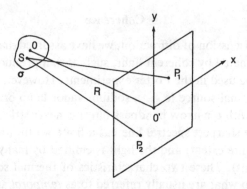

Figure 1.14. Diagram illustrating the van Cittert–Zernike theorem. An extended
source σ illuminates a diffracting object in the xy-plane, a distance R away.

reasonable to expect that the fluctuations of amplitude and phase at P_1
and P_2 are effectively the same; that is, the vibrations are highly correlated. We also expect some correlation to exist between the fluctuations
at P_1 and P_2 if the separation $P_1 P_2$ is increased, provided that for all
points S in the source, the path difference Δd does not exceed the coher-
ence length $c\,\Delta t \approx c/\Delta \nu = \bar{\lambda}^2/\Delta\lambda$. The separation $(P_1 P_2)_0$ corresponding
to this condition defines the radius of a *region of coherence* about any
point P on the diffracting object. To describe adequately the quasi-monochromatic radiation produced by a thermal source of finite size, we must
introduce some measure of the degree of correlation that exists between
the vibrations at the points P_1 and P_2. This measure is expected to be
closely related to the sharpness of the diffraction maxima that would result from the superposition of the waves originating from the two points.
When the light arriving at P_1 and P_2 comes from a very small source with
a narrow spectral range, the correlation is high and sharp diffraction maxima are produced; that is, the diffracting object is *coherently* illuminated.
However, if P_1 and P_2 each receive light from a different, independent
source, then there is no correlation and no diffraction maxima are produced; that is, the diffracting object is *incoherently* illuminated. In general, for light emitted by a source of finite size, the degree of correlation
at P_1 and P_2 is intermediate between these two extremes, and therefore
the light is *partially coherent*.

It is clear from the preceding discussion that, for a partially coherent
light source, the correlation μ_{12} between the vibrations at a fixed point P_1
and a variable point P_2 will be perfect ($\mu_{12} = 1$) when the separation $P_1 P_2$

is zero. As the separation P_1P_2 is increased, μ_{12} decreases and eventually becomes zero. The precise variation of μ_{12} with P_1P_2 and the effect of the size of the source are given by the *van Cittert–Zernike theorem*. To understand this theorem, imagine that the source is replaced by an aperture of exactly the same size and shape as the source, and that the aperture is illuminated by a plane wave of wavelength λ. The Fraunhofer diffraction pattern of this aperture appears in the plane of P_1 and P_2, a distance R from the aperture. We choose the fixed point P_1 so that it is at the central peak of the diffraction pattern, and we *normalize* the intensity by making the intensity of the central peak equal to unity. The van Cittert–Zernike theorem states that the (normalized) intensity at the point P_2 is equal to the correlation μ_{12} for the separation P_1P_2. If we decrease the size of the aperture, the diffraction pattern spreads out further; thus, for a given separation P_1P_2, the correlation μ_{12} increases as we decrease the size of the source. The smaller the size of the source, the greater the degree of coherence.

These ideas can be illustrated by considering the coherence of a circular source of diameter a. The correlation μ_{12} is equal to the normalized intensity of the Fraunhofer diffraction pattern from a circular aperture of diameter a (see Figure 1.7b). From this it follows that μ_{12} becomes zero when

$$P_1P_2 = (P_1P_2)_0 = \frac{1.22\lambda}{\alpha} \tag{1.34}$$

where $\alpha = a/R$ is the angular diameter of the source as viewed from P_1 on the diffracting object.* It is clear from the Fraunhofer diffraction pattern that μ_{12} increases when P_1P_2 is increased beyond $(P_1P_2)_0$, but the degree of coherence is always small and can be neglected in practice. If a departure of 12% from the ideal value of unity for μ_{12} is the maximum permissible departure, then the diameter Δ of the circular area that is illuminated almost completely coherently by a quasi-monochromatic uniform source of angular diameter α is given by

$$\Delta = \frac{0.32\lambda}{\alpha} \tag{1.35}$$

Δ, which defines the *coherence area,* clearly increases as the size of the source is decreased.

---

* The numerical factor 1.22 emerges from the mathematical analysis involving the integration of the elementary radiators over a circular aperture. For a single slit, the numerical factor is unity.

The effect of the size of the source on the coherence area, and hence on the visibility of diffraction maxima, can be easily demonstrated using an ordinary microscope with a tungsten filament lamp and an electron microscope specimen grid (~8 bars/mm) as an object. If we view the intensity distribution in the back focal plane of a ×4 objective lens by inserting the Bertrand lens, we will not see sharp diffraction maxima when using the normal illumination system, even with the condenser apertures decreased to their minimum size. However, sharp diffraction maxima (similar to those shown in Figure 1.11) are easily observed if we replace the condenser system by an aperture of 0.1 mm diameter placed about 100 mm from the grid. With an aperture of this size (which can be made quite easily in a piece of aluminum foil with a sharp needle), the diameter Δ of the coherence area calculated from Eq. (1.35) is 0.176 mm, taking $\bar{\lambda} =$ 550 nm. The corresponding value of $(P_1P_2)_0$ is 0.67 mm. Because both $(P_1P_2)_0$ and Δ are larger than the spacing between the bars of the grid (0.125 mm), we expect sharp diffraction maxima. Of course, if we use white light, each diffraction maximum will be a spectrum; clearly, we can increase the sharpness by using a colored filter or, better still, a quasi-monochromatic light source such as a sodium lamp. With an aperture of 1 mm diameter (for which $\Delta = 0.0176$ mm), we still observe diffraction maxima, but they are overlapping disks rather than sharp maxima. Mc-Laren and MacKenzie (1976) found that for a 100-kV electron microscope operating under normal illuminating conditions with a standard hairpin filament, Δ was about 15 nm and $(P_1P_2)_0$ about 60 nm.

As already indicated, the degree of coherence of the electron beam can have an important influence on the nature of electron diffraction patterns. However, in Chapters 3 and 4, which deal with the theories of electron diffraction, we assume, as we did earlier in this chapter, that the diameter of the coherence area is large compared with the *d*-spacings of the crystals.

2

The transmission electron microscope

2.1 Introduction

This chapter describes briefly the basic construction and characteristics of the modern transmission electron microscope and discusses its principal modes of operation. Because the electron microscope is an analogue of the optical (or light) microscope, we also consider briefly the basic features of the optical microscope; this will also provide a link with our earlier discussion of the optical principles of image formation by a lens.

2.2 The optical microscope

The basic components of an optical microscope for viewing a transparent object are shown in Figure 2.1. We need not be concerned here with the details of the illumination system, other than to note that although the object in the diagram is illuminated with parallel light, the light need not necessarily be parallel in practice. The object O is located outside the front focal plane of the objective lens that forms the first (or intermediate) image I_1 which is real and inverted; the relevant ray diagram is shown in Figure 1.1(a). The image I_1 becomes the object for the eyepiece. The eyepiece is located so that I_1 is behind the front focal plane of this lens, which forms an image I_2 that is observed by the eye; the relevant ray diagram is shown in Figure 1.1(b). I_2 is a virtual and erect image of I_1, and so the final image is inverted with respect to the object.

The objective lens ultimately determines the performance of the microscope. Any detail that is not revealed in the intermediate image I_1 formed by this lens cannot be added later by the eyepiece. The limit of resolution is set, therefore, by the effective size of the aperture in the back focal plane of the objective. From Eq. (1.28), the resolution is

$$d = \frac{\lambda}{R_c/f} = \frac{\lambda}{\sin \theta} \tag{2.1}$$

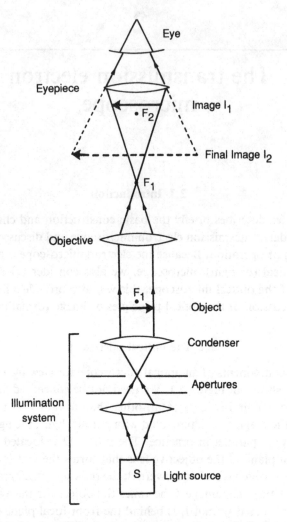

Figure 2.1. Layout of the basic components of an optical microscope
for viewing a transparent object.

R_c/f is the numerical aperture (N.A.) and is marked on all objective
lenses, together with the magnification. For example, a $\times 25$ objective with
N.A. $= 0.5$ is expected to resolve detail of the order of $2\lambda \approx 1\ \mu m$, ideally.
Equation (2.1) has been derived by considering diffraction by a coher-
ently illuminated periodic object. Rayleigh's well-known criterion of resolution, derived for a nonperiodic object incoherently illuminated, gives
nearly the same limit of resolution as that determined from the Abbe approach and need not be considered here (see, e.g., Giancoli 1984).

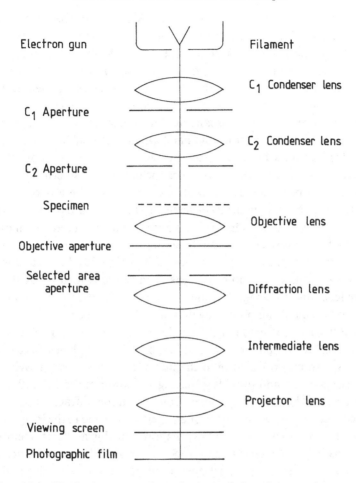

Figure 2.2. The basic construction of a transmission electron microscope.

For a more detailed account of the construction of the optical microscope and the characteristics of its various component parts, the reader is referred to Southworth (1975) and the references cited there.

2.3 General description of the transmission electron microscope

The basic construction of a modern transmission electron microscope is shown schematically in Figure 2.2. It consists of an electron gun and an assembly of electromagnetic lenses, all within a column which is evacuated to about 10^{-7} Torr $(= 2.7 \times 10^{-5}$ Pa). The beam of electrons produced by the electron gun is accelerated by a high voltage and then focused onto the thin specimen by two condenser lenses. A number of specimen holders for specific applications are available, but the minimum requirement for crystallographic studies is a double-tilt holder. With such a holder, the specimen can be tilted about two axes that are at right-angles to each other and to the optic axis of the microscope. The first condenser has a fixed aperture, and the second has a series of interchangeable apertures of different sizes. There are usually five imaging lenses (objective, diffraction, intermediate, and projectors 1 and 2), and the final image of the object is formed on a fluorescent screen, which can be viewed directly or through a pair of binoculars. To record the image, this screen is tilted out of the optical path so that the electrons are incident on a photographic film. Even though the fluorescent screen and the film are not in the same plane, the image remains in focus because of the very large depth of focus, which probably exceeds the dimensions of the microscope. Another series of interchangeable apertures of various sizes can be located in the back focal plane of the objective lens. The image is focused by the objective lens, and the magnification is controlled by the intermediate and projector lenses. A simplified ray diagram for a microscope with a single condenser lens and three magnifying lenses is shown in Figure 2.3. The electron diffraction pattern, formed in the back focal plane of the objective lens, can be projected onto the fluorescent screen by removing the objective aperture and suitably adjusting the power of the diffraction and lower lenses. The area of specimen from which the diffraction pattern is derived can be selected by means of an aperture located in the image plane of the objective. The simplified ray diagram for this mode of operation, known as *selected area diffraction* (SAD), is also shown in Figure 2.3. However, because of spherical aberration of the objective lens and because the selected area aperture is generally not located precisely in the image plane, the area defined by this aperture generally does not correspond exactly to the area of specimen from which the diffraction pattern originates. An error of about 1 μm is typical of most modern electron microscopes; so if an area of less than 1 μm^2 is selected with a small SAD aperture, it cannot be assumed that the diffracted beams seen in the diffraction pattern originate only from the area of the specimen defined by the aperture.

There is another diffraction mode, called *convergent beam electron diffraction* (CBED), in which the incident electron beam is focused to a fine spot on the specimen. If the convergence angle is appropriately chosen, the diffraction pattern consists of an array of nonoverlapping disks. For thin specimens (≈ 50 nm) the CBED disks are featureless, but

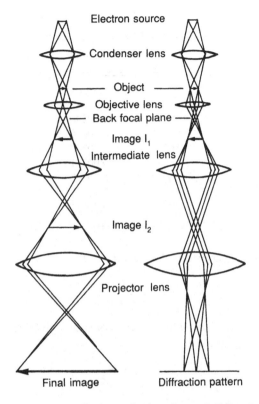

Figure 2.3. Ray diagrams for imaging and diffraction
in a transmission electron microscope.

for thicker specimens some contrast is observed in each disk, which can
provide information about crystal symmetry that is not available from a
normal SAD pattern. CBED is discussed more fully in Section 3.10.

2.4 The electron wavelength

When an electron of rest mass m_e and charge e is accelerated through a
potential difference V, the kinetic energy it acquires is

$$\tfrac{1}{2}m_e v^2 = Ve \tag{2.2}$$

where v is the velocity. The wavelength λ associated with the electron is
given by the de Broglie relation

$$\lambda = \frac{h}{m_e v} \tag{2.3}$$

Table 2.1. *Electron wavelength* λ
*for voltages V commonly used
in electron microscopy*

V (kV)	λ (nm)
50	0.00536
80	0.00418
100	0.00370
200	0.00251
300	0.00197
400	0.00164
1,000	0.00087

where h is Planck's constant. Using Eq. (2.2), we have

$$\lambda = \frac{h}{(2m_e Ve)^{1/2}} \qquad (2.4)$$

Inserting the appropriate values of the constants gives an equation that is a good approximation to Eq. (2.4),

$$\lambda = \left(\frac{1.5}{V}\right)^{1/2} \text{nm} \qquad (2.5)$$

if V is measured in volts.

However, these equations are valid only if the velocity v is small compared with the velocity of light c. This will be true for $V \leq 10,000$ volts, but for the higher voltages commonly used in electron microscopy (50 to 1,000 kV), m_e must be replaced by $m_e/(1-v^2/c^2)^{1/2}$. Values of λ for some commonly used voltages are given in Table 2.1.

For an accelerating voltage of 50 kV, the wavelength is about 10^{-5} times that of visible light. If the numerical aperture of the objective lens could be made as large as that of a good ×10 optical microscope objective (say, N.A. = 0.5), then Eq. (2.1) predicts a resolution of the order of 0.01 nm at 50 kV, which is less than the spacing between atoms in crystals. Unfortunately, due to the aberrations of magnetic lenses, this is not attainable.

2.5 Lens aberrations and the practical limit of resolution

Like their glass analogues, magnetic lenses suffer from such defects as coma, distortion, astigmatism, and chromatic and spherical aberration.

The last three are particularly important because they determine the resolution of the objective lens and, hence, of the electron microscope.

Astigmatism is due to asymmetry of the magnetic field about the axis of the lens. Fortunately, this aberration can be corrected by means of a *stigmator,* which superimposes an additional magnetic field (whose strength and direction can be controlled by the microscopist) across the gap in the pole piece of the lens.

Chromatic aberration arises because of variations of electron energy, and therefore of wavelength, that are brought about mainly by fluctuations in the accelerating voltage and energy losses within the specimen itself. The first is relatively small because, with modern high-stability power supplies, the spread ΔE in energy of electrons leaving the gun is only about 3 eV. However, the energy losses in the specimen can amount to 50 eV. Because the focal length of the lens varies with energy, chromatic aberration gives rise to a disk of confusion of radius Δr_c in the image plane, given by

$$\Delta r_c = C_c \alpha \left(\frac{\Delta E}{E} \right) \qquad (2.6)$$

where C_c is the chromatic aberration constant and α the angular radius of the aperture of the lens.

Spherical aberration in the objective lens is particularly important, and there is no convenient way of correcting it. It arises because the outer zones of the lens focus more strongly than the zones closer to the axis. As with chromatic aberration, this produces a disk of confusion of radius Δr_s, which is given by

$$\Delta r_s = C_s \alpha^3 \qquad (2.7)$$

where C_s is the spherical aberration constant. The resolution can be increased by decreasing the radius of the disk of confusion, which can be achieved by decreasing the angular radius α of the aperture of the lens. But decreasing the size of the objective aperture leads to an increase in the Abbe diffraction limit d of Eq. (2.1) and hence to a decrease in the resolution. If we assume that the optimum size of objective aperture is obtained when $d = \Delta r_s$, then from Eqs. (2.1) and (2.7),

$$C_s \alpha^3 = \frac{\lambda}{\sin \alpha} \approx \frac{\lambda}{\alpha}$$

and

$$\alpha = \left(\frac{\lambda}{C_s} \right)^{1/4} \qquad (2.8)$$

Therefore, the minimum spacing $d(\min)$ that can be resolved is

$$d(\min) = \frac{\lambda}{\alpha} = \lambda^{3/4} C_s^{1/4} \qquad (2.9)$$

As an example, consider the standard objective lens in the Philips 420 electron microscope. It has a focal length of 2.7 mm, and the aberration constants C_c and C_s are both 2 mm. Thus, at 100 kV, $\alpha = 6.6 \times 10^{-3}$ radian (corresponding to an aperture of diameter $= 2\alpha f = 36\ \mu\text{m}$) and $d(\min) = 0.56$ nm. However, we must point out that in deriving Eqs. (2.8) and (2.9), we used Eq. (1.7), which applies to the diffraction of light for which the angle θ of the first diffraction maximum is given by $d \sin \theta = \lambda$. We shall see in the next chapter that for the diffraction of electrons by crystals the corresponding expression is the Bragg equation, $2d \sin \theta = \lambda$. Thus, the crystal d-spacing that can just be resolved using a 36 μm diameter objective aperture is given by $\lambda/2\alpha = 0.28$ nm. Resolutions of this order are commonly obtained with suitably thin crystals.

2.6 Defect of focus

For reasons that will become clear later, we shall say a little here about how exact focus of the objective lens is recognized, together with the meaning and effects of underfocus and overfocus. Figure 2.4(a) is a ray diagram showing an object O and its focused real image I in the observation plane produced by a lens of focal length f. If we reduce the object distance d_o to $(d_o - \Delta d_o)$, then the Gaussian image plane will lie *below* the fixed observation plane, and an *underfocused* image will form in the observation plane, as shown in Figure 2.4(b). Figure 2.4(c) shows that if we increase the focal length of the lens by an amount Δf, while keeping the object and observation planes fixed, an underfocused image will form in the observation plane. We can achieve this increase in focal length in the electron microscope by decreasing the current through the objective lens. If we increase the current, the focal length will decrease by an amount Δf. The Gaussian image plane will lie *above* the observation plane, and an *overfocused* image will form in the observation plane, as shown in Figure 2.4(d).

In Figure 2.4(a), the arbitrary paraxial ray (dotted) from P_o making an angle β with the axis of the lens arrives at the point P_i in phase with

---

Figure 2.4. *Pages 45–46.* Ray diagrams illustrating (a) Gaussian focus, (b) underfocus by decreasing the object distance, (c) underfocus by increasing the focal length, and (d) overfocus by decreasing the focal length.

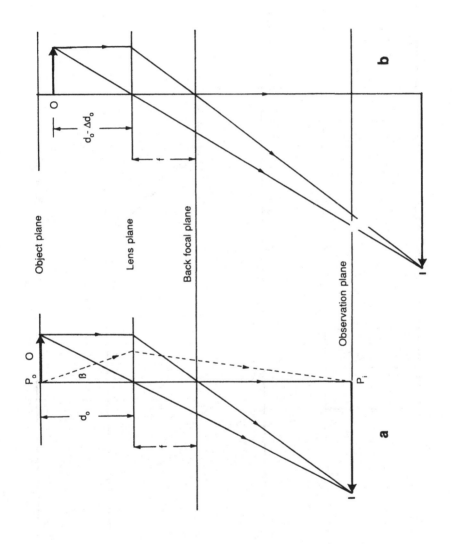

Object plane

Lens plane

Back focal plane

P_o O

β

d_o

f

P_i

I

a

Observation plane

O

$d_o - \Delta d_o$

f

I

b

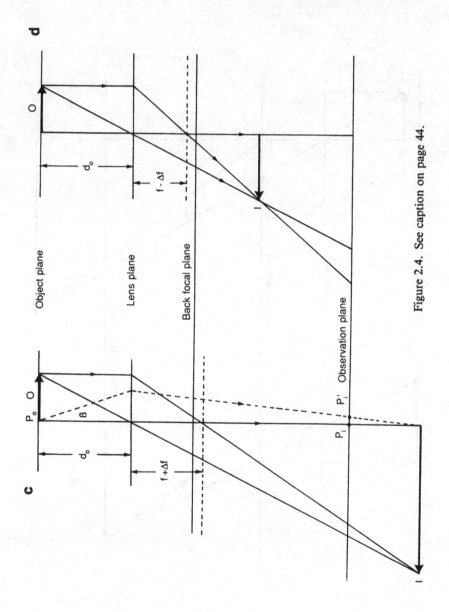

Figure 2.4. See caption on page 44.

46

the axial ray. However, if we decrease the current through the lens, the Gaussian image plane moves down, Figure 2.4(c), and the paraxial ray will pass through a point P_i' in the observation plane. If paraxial rays fill the angular range 0 to β, the defocused image (in the observation plane) of the object point P_o will be a disk of confusion of radius $P_i P_i' = M \Delta f \beta$. Paraxial rays no longer arrive at the observation plane with the same phase as the axial ray, and (since the paraxial rays and the axial ray are in phase B in the Gaussian image plane) the path difference δ is given by

$$\delta = P_i'B - P_iB = \tfrac{1}{2} \Delta f \beta^2 \tag{2.10}$$

Therefore, the phase difference is

$$\chi_f = \frac{2\pi}{\lambda}\delta = \frac{\pi}{\lambda}\Delta f \beta^2 \tag{2.11}$$

The sign of χ_f is determined by the sign of Δf, positive for underfocus and negative for overfocus, as in Figure 2.4. Unfortunately, the opposite sign convention is adopted by some authors such as Spence (1981).

The nature of the Fresnel fringes that are formed in electron microscope images of edges depends on the amount of defocus and provides a convenient way of recognizing Gaussian focus and under- and overfocus. The origin of Fresnel fringes can be understood by considering the image of an opaque straight edge. Figure 2.5(a) shows the straight edge at $z = 0$, a point source P_s of electrons at $z = -z_1$, and the viewing plane at $z = L$. Because there is a path difference between the direct beam $P_s P_1 P_2$ and the beam $P_s O P_2$ passing infinitely close to the straight edge, a Fresnel diffraction pattern forms in the viewing plane. The intensity distribution is shown in Figure 2.5(b). The maxima are located approximately at

$$x(\text{max}) = \left[\frac{L(z_1+L)}{z_1} (2n-1)\lambda \right]^{1/2} \tag{2.12}$$

where n is an integer (Heidenreich 1964). The first maximum corresponding to $n = 1$ is by far the strongest. The appearance of the Fresnel fringes in the electron microscope depends on the amount of defocus. If we adjust the current through the objective lens so that a perfectly focused image of the edge is formed in observation plane, then the conjugate image plane (i.e., the object plane) coincides with the plane of the edge and the distance $L = 0$ in Figure 2.4(a); thus, from Eq. (2.10), no fringes will appear. However, if we defocus the objective lens, the conjugate plane moves away from the plane of the edge and $L = |\Delta f|$, the amount of defocus. Since $|\Delta f|$ is usually less than z_1, the Fresnel fringes are located approximately at

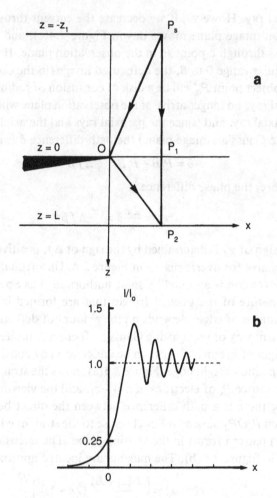

Figure 2.5. (a) Geometry for Fresnel diffraction at an opaque straight edge. (b) Intensity distribution of the Fresnel diffraction at $z = L$.

$$x_n = [|\Delta f|(2n-1)\lambda]^{1/2} \tag{2.13}$$

or at Mx_n in the observation plane, where M is the magnification.

Under normal working conditions, only the first Fresnel fringe is seen in defocused images, and its displacement $x_1 = (|\Delta f|\lambda)^{1/2}$. The phase change on going through focus greatly alters the intensities. For under-focus (lens current less than that for focus), the fringe is bright, whereas for overfocus (lens current greater than that for focus), the fringe is dark, as shown in Figure 2.6. Higher order fringes ($n = 2, 3, ...$) are seen only if

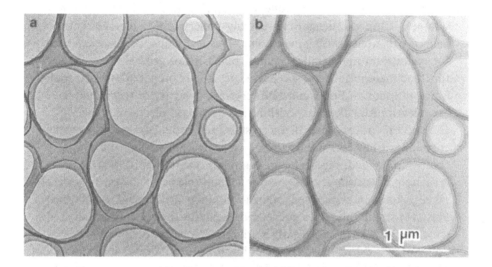

Figure 2.6. Fresnel fringes at the edge of a hole in a carbon film, observed in the electron microscope: (a) underfocus, and (b) overfocus.

the coherence of the illumination is greatly increased by using a very much smaller condenser aperture than usual. In practice in the electron microscope, electrons are transmitted through the object near the edge. This introduces an additional phase shift so that the phase relations become quite complicated, and the intensity distribution is different from the ideal case for an opaque straight edge (Figure 2.5).

The Fresnel fringe provides a satisfactory criterion for focusing if a suitable edge is available on the specimen. An alternative method of finding exact focus is mentioned in Section 6.3.7.

2.7 Specimens and specimen preparation

Because of the very strong interaction of electrons with materials (see Section 3.6), the specimen must be less than about 200 nm thick to be usefully transparent to 100-kV electrons. Observations at the limit of instrumental resolution require the specimen thickness to be about 10 nm at 100 kV. Thicker specimens can be used at higher accelerating voltages. This fact, together with the prospect of obtaining higher resolution due to the decrease in wavelength with increasing voltage (see Table 2.1), has been one of the main reasons for the development of 200-, 300-, and 400-kV microscopes for general use.

There is no universal technique for preparing thin specimens from bulk samples. Thin specimens of essentially noncrystalline biological specimens are sectioned with a microtome. This technique has also been used for metals and nonmetallic crystals, but it is not generally satisfactory for these materials. Metal specimens are usually prepared by electropolishing techniques. For nonmetallic crystals, a variety of techniques have been developed. Chemical polishing has been used successfully for MgO, Ge, and Si, for example. If the crystals are brittle, the regions near the edges of crushed fracture fragments are often thin enough and extensive enough to be used, but the crystallographic orientation of the fragments is random unless the crystal possesses definite cleavage planes. This technique is commonly employed for high-resolution lattice imaging, for which only very small, thin areas are needed due to the high magnification used ($> \times 500,000$). Most of the early work on silicates and other minerals was done using crushed fragments, but ion (or atom) bombardment of 25–30-μm thin sections (previously prepared by standard petrological methods) is now the preferred technique (Barber 1970; Gillespie et al. 1971). This technique is suitable for both single-crystal and polycrystalline (rock) samples. In the latter, preferential thinning at the grain boundaries can occur if undercutting has already occurred there during preparation of the petrological thin section. Ion-bombardment thinning always leaves a radiation-damaged surface layer on the thinned specimen, but for most materials it is not visible in the electron microscope, except perhaps in the ultrathin regions near edges. In ion-thinned sulphides, the damage is visible throughout the transparent areas but can be dissolved chemically.

2.8 Modes of operation

The electron microscope can be operated to form on the fluorescent screen either (i) a diffraction pattern of the specimen or (ii) one of several types of image, depending on which beams are allowed to pass through the aperture in the back focal plane of the objective lens. The geometry of the electron diffraction pattern will be discussed in detail in Chapter 3.

The most commonly employed imaging mode is *bright field* (BF), which uses only the central beam, usually called the *transmitted* beam. Image contrast is observed if, for whatever reason, the diffracted beams originating from different parts of the specimen vary in intensity. If a crystalline specimen is so oriented with respect to the incident beam that many strong diffracted beams are excited, then an extremely complex BF image

can be formed, the details of which may not be readily interpreted. Therefore, it is usual, especially when observing images due to crystal defects,
to orient the specimen to produce two-beam conditions, that is, only one
strongly diffracted beam in addition to the transmitted beam. *Dark-field*
(DF) images formed with this strongly diffracted beam are commonly
used, often to complement the BF image. If the DF image is produced
simply by displacing the objective aperture, then a poor-quality image
will be obtained because the diffracted beam does not travel along the
optic axis and the lens aberrations (especially spherical aberration and
astigmatism) will be greatly increased. To overcome this difficulty, the incident beam is tilted by means of an electromagnetic device in the illumination system so that the diffracted beam travels along the optic axis.
This mode of imaging is usually called *centered dark-field* (CDF).

A variant of CDF imaging is the *weak-beam* (WB) mode. In this mode,
the specimen is tilted away from the orientation for CDF imaging until the
centered diffracted beam is of low intensity, but care must be taken to ensure that no other diffracted beams are strongly excited. WB images of dislocations are much narrower than strong-beam BF or CDF images. Therefore, this type of imaging is particularly useful for studying regions of
high dislocation density and the dissociation of dislocations into partials.

It is clear, of course, that since these imaging modes use only the transmitted beam or a single diffracted beam, the resulting images contain only
information on a scale that is coarse compared with the spacing between
crystallographic planes. To produce images corresponding to the planes
in a crystal, we must allow at least two beams to pass through the aperture
in the back focal plane of the objective. The conditions that are necessary
for the formation of one- and two-dimensional lattice images of crystals
in the electron microscope are, in principle, the same as those required to
produce an image of a grid by a thin lens (see Chapter 1), but the details
of the images are critically dependent on the thickness of the specimen
and the defect of focus.

The detailed interpretation of electron microscope images produced
using any of the operating modes discussed in this chapter requires as
complete an understanding as possible of the diffraction process. The next
two chapters develop and explain as simply as possible the current theories of electron diffraction by crystals in order to provide a basis for the
interpretation of images of crystal defects (such as dislocations, stacking
faults, and twins) and of lattice images.

3

Kinematical theory of
electron diffraction

3.1 Introduction

We now consider the diffraction of electrons by a single crystal in terms
of the so-called kinematical theory. Although this theory has serious limitations, it is useful in practice under certain conditions, and it also provides an introduction to the more satisfactory dynamical theory, which
we develop in Chapter 4.

In the kinematical theory, we consider the diffraction of a plane wave
(of wavelength λ) incident upon a three-dimensional lattice array of identical scattering points, each of which consists of a group of atoms and
acts as the center of a spherical scattered wave. Our problem is to find the
combined effect of the scattered waves at a point outside the crystal, at
a distance from the crystal that is large compared with its linear dimensions. In developing the theory, we make several important assumptions:

1. There is no attenuation of the incident wave in the crystal so that the
   incident wave has the same amplitude at each scattering point. This is
   equivalent to neglecting any interaction between the incident wave in
   the crystal and the scattered waves.
2. Each scattered wave travels through the crystal without being rescattered by other scattering points.
3. There is no absorption of either the incident or the scattered waves in
   the crystal.

Since the theory makes no assumptions about the nature of the wave or
about the detailed mechanism of the interaction of the wave with the scattering points, it is applicable to x-rays, electrons, and neutrons.

3.2 Derivation of the Laue equations

We begin by considering the scattering of a plane wave of wavevector $\mathbf{K}_0$
(of magnitude $1/\lambda$) by two identical scattering points P_1 and P_2 at the

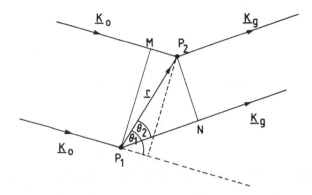

Figure 3.1. Diagram used for the calculation of the phase difference of the waves scattered by two lattice points.

lattice points of a space lattice, as in Figure 3.1. The position of P_2 relative to P_1 is given by the vector $\mathbf{r}$, and the wavevector of the scattered wave is $\mathbf{K_g}$.

We want the phase difference between the two scattered waves at Q, a point at a distance R ($\gg P_1 P_2$) from P_1 and P_2. The path difference δ is given by

$$\delta = P_1 N - P_2 M$$

$$= P_1 P_2 \cos \theta_2 - P_1 P_2 \cos \theta_1$$

$$= \frac{\mathbf{r} \cdot \mathbf{K_g}}{|\mathbf{K_g}|} - \frac{\mathbf{r} \cdot \mathbf{K_0}}{|\mathbf{K_0}|}$$

$$= \lambda \mathbf{r} \cdot (\mathbf{K_g} - \mathbf{K_0}) \tag{3.1}$$

Putting

$$\mathbf{K_g} - \mathbf{K_0} = \mathbf{g} \tag{3.2}$$

we have

$$\delta = \lambda (\mathbf{r} \cdot \mathbf{g}) \tag{3.3}$$

The significance of the vector $\mathbf{g}$ can be understood with the aid of Figure 3.2. *OP* represents the incident wavevector $\mathbf{K_0}$ and *OP'* the scattered wavevector $\mathbf{K_g}$. P is a scattering point. The direction *PP'* is given by the vector

$$\mathbf{g} = \mathbf{K_g} - \mathbf{K_0}$$

and as there is no absorption,

$$|\mathbf{K_0}| = |\mathbf{K_g}| = \frac{1}{\lambda} \tag{3.4}$$

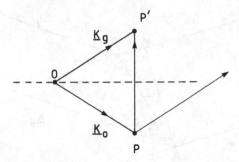

Figure 3.2. Construction of the normal to the reflecting plane.

Thus, **g** is normal to a plane that would reflect the incident wave into the scattering direction. The plane is, therefore, called the *reflecting plane*.

If θ is the glancing angle of incidence, then the angle between the incident and the scattered wavevectors is 2θ, which is called the *scattering angle*.

From Figure 3.2,

$$PP' = |\mathbf{K}_g - \mathbf{K}_0|$$

$$= |\mathbf{g}|$$

$$= 2\left(\frac{1}{\lambda}\right)\sin\theta$$

and hence,

$$2\left(\frac{1}{|\mathbf{g}|}\right)\sin\theta = \lambda \tag{3.5}$$

Now the phase difference φ at the point Q, between the waves scattered by P_1 and P_2 in Figure 3.1, is given by

$$\varphi = \frac{2\pi}{\lambda} \times \text{path difference}$$

$$= \frac{2\pi}{\lambda} \times \lambda(\mathbf{r} \cdot \mathbf{g})$$

$$= 2\pi(\mathbf{r} \cdot \mathbf{g}) \tag{3.6}$$

The amplitude of the scattered wave is a maximum in a direction such that the contributions from each lattice point differ in phase by integral multiples of 2π. For this condition, the separate scattered amplitudes add up constructively, making the intensity of the scattered wave a maximum.

Now if **a**, **b**, **c** are the primitive translation vectors that define the unit cell of the three-dimensional array of scattering points, then we have, using Eq. (3.6), the following conditions for diffraction maxima:

$$\varphi_a = 2\pi(\mathbf{a} \cdot \mathbf{g}) = 2\pi h$$
$$\varphi_b = 2\pi(\mathbf{b} \cdot \mathbf{g}) = 2\pi k \qquad (3.7)$$
$$\varphi_c = 2\pi(\mathbf{c} \cdot \mathbf{g}) = 2\pi l$$

where h, k, and l are integers. If we define α, β, and γ as the cosines of the angles between **g** and **a**, **g** and **b**, and **g** and **c**, respectively, then from Eq. (3.5) we have

$$\mathbf{a} \cdot \mathbf{g} = a\left[2\left(\frac{1}{\lambda}\right)\sin\theta\right]\alpha$$
$$\mathbf{b} \cdot \mathbf{g} = b\left[2\left(\frac{1}{\lambda}\right)\sin\theta\right]\beta \qquad (3.8)$$
$$\mathbf{c} \cdot \mathbf{g} = c\left[2\left(\frac{1}{\lambda}\right)\sin\theta\right]\gamma$$

Hence, from Eqs. (3.7) and (3.8), we have

$$2a\alpha \sin\theta = h\lambda$$
$$2b\beta \sin\theta = k\lambda \qquad (3.9)$$
$$2c\gamma \sin\theta = l\lambda$$

These are the Laue equations. Note there are solutions only for special values of θ and λ. It will now be shown that these equations are equivalent to the Bragg law.

From Eq. (3.9) we see that the direction cosines of **g** are given by

$$\alpha = \left(\frac{h}{a}\right)\left(\frac{\lambda}{2}\sin\theta\right)$$
$$\beta = \left(\frac{k}{b}\right)\left(\frac{\lambda}{2}\sin\theta\right) \qquad (3.10)$$
$$\gamma = \left(\frac{l}{c}\right)\left(\frac{\lambda}{2}\sin\theta\right)$$

Now consider a set of planes (hkl), spaced a distance $d(hkl)$ apart, as shown in two dimensions in Figure 3.3. By the definition of Miller indices, adjacent planes (hkl) intersect the crystallographic axes **a**, **b**, **c** at intervals of a/h, b/k, and c/l, and with the aid of Figure 3.3 we can see that

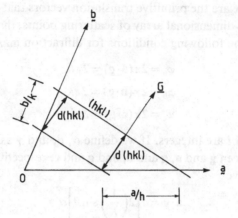

Figure 3.3. Diagram used in deriving Eqs. (3.11).

$$\frac{d(hkl)}{a/h} = \alpha'$$

$$\frac{d(hkl)}{b/k} = \beta' \qquad (3.11)$$

$$\frac{d(hkl)}{c/l} = \gamma'$$

where α', β', and γ' are the cosines of the angles between the normal **G** to the planes (hkl) and **a**, **b**, and **c**, respectively. If the planes (hkl) are the reflecting planes, then **G** is parallel to **g** and $\alpha' = \alpha$, $\beta' = \beta$, and $\gamma' = \gamma$; it therefore follows from Eqs. (3.10) and (3.11) that

$$2d(hkl) \sin\theta = \lambda \qquad (3.12)$$

which is the Bragg law.

If we compare Eq. (3.12) with Eq. (3.5), we can see that

$$|\mathbf{g}| = \frac{1}{d(hkl)} \qquad (3.13)$$

Thus, **g** is normal to the reflecting plane (hkl) and has a magnitude equal to the reciprocal of the spacing between the reflecting planes.

The integers h, k, l in the Laue equations may not be identical with the Miller indices because the h, k, l of the Laue equations may contain a common integral factor n which has been eliminated from the Miller indices. Thus, the Bragg law can be written as

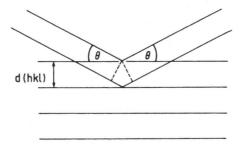

Figure 3.4. Diagram used for the derivation of the Bragg law, $2d \sin \theta = \lambda$.

$$2d \sin \theta = n\lambda \tag{3.14}$$

where d is the spacing between adjacent planes with Miller indices $(h/n, k/n, l/n)$. The integer n is the order of the reflection.

The Bragg law can also be derived in a simple manner. Suppose a plane wave is incident upon a crystal at a grazing angle θ to a set of atomic planes (hkl) which behave like partially reflecting mirrors spaced a distance $d(hkl)$ apart, as shown in Figure 3.4. It is clear that the path difference between the two waves shown is $2d(hkl) \sin \theta$. Constructive interference between waves reflected from successive planes occurs when this path difference is an integral number of wavelengths. Thus,

$$2d(hkl) \sin \theta = n\lambda$$

In deriving the Bragg law in this way, we have assumed that if any phase change occurs on reflection, it is the same for all planes. Further we have assumed there is no refraction; that is, the refractive index of the crystal is unity.

For electrons accelerated through 100 kV, $\lambda = 0.0037$ nm; hence, for d-spacings of about 0.2 nm, θ is of the order of 10^{-2} radian.

3.3 Reciprocal lattice

It is often easier and more elegant to express the conditions for diffraction in terms of a mathematical transformation known as the reciprocal lattice. The reciprocal lattice vectors $\mathbf{a^*}, \mathbf{b^*}, \mathbf{c^*}$ are defined in terms of the real lattice vectors $\mathbf{a}, \mathbf{b}, \mathbf{c}$ by the relations:

$$(1) \quad \mathbf{a^*} \cdot \mathbf{b} = \mathbf{a^*} \cdot \mathbf{c} = \mathbf{b^*} \cdot \mathbf{a} = \mathbf{b^*} \cdot \mathbf{c} = \mathbf{c^*} \cdot \mathbf{a} = \mathbf{c^*} \cdot \mathbf{b} = 0 \tag{3.15}$$

that is,

a* is normal to both **b** and **c**,
b* is normal to both **a** and **c**, and
c* is normal to both **a** and **b**.

$$(2) \quad \mathbf{a^*} \cdot \mathbf{a} = \mathbf{b^*} \cdot \mathbf{b} = \mathbf{c^*} \cdot \mathbf{c} = 1 \tag{3.16}$$

that is,

$$|\mathbf{a^*}| = \frac{1}{\text{spacing of the planes containing } \mathbf{b} \text{ and } \mathbf{c} \text{ in real space}}$$

and similarly for $|\mathbf{b^*}|$ and $|\mathbf{c^*}|$. Since **a*** is normal to **b** and **c**, we can write

$$\mathbf{a^*} = \beta(\mathbf{b} \times \mathbf{c})$$

and similarly for **b*** and **c***. It can be shown that β is equal to $1/V_c$, where V_c is the volume of the real-space unit cell. Thus,

$$\mathbf{a^*} = \frac{\mathbf{b} \times \mathbf{c}}{\mathbf{a} \cdot (\mathbf{b} \times \mathbf{c})} \tag{3.17}$$

and similarly for **b*** and **c***.

Two properties of the reciprocal lattice make it of value in diffraction theory. We state these without proof.

1   The vector $\mathbf{r^*}(hkl)$ from the origin to the point h, k, l of the reciprocal lattice is normal to the plane (hkl) of the real lattice.
2   $|\mathbf{r^*}(hkl)| = 1/d(hkl)$.

As an example, we use these properties to construct the reciprocal lattice corresponding to a face-centered cubic (fcc) real lattice. In the direct fcc lattice, the distance between lattice planes in each cube edge direction is $a/2$, where a is the lattice constant. Thus, there are points along the x, y, and z axes in the reciprocal lattice at $2/a$ from the origin. There is only one other set of lattice planes in the real lattice for which the separation between adjacent planes is greater than $a/2$. These are the (111) planes, which are separated by $a/\sqrt{3}$. These planes are, therefore, represented by a point in the reciprocal lattice in the [111] direction at a distance of $\sqrt{3}/a$ from the origin. This distance is one-half of the length of the body diagonal of the cubic unit cell in the reciprocal lattice. Thus, an fcc unit cell of a lattice constant a in real space becomes a body-centered cubic (bcc) unit cell of lattice constant $2/a$ in reciprocal space. These two unit cells are shown in Figure 3.5.

3.4 Ewald sphere construction in reciprocal space

The Ewald sphere construction is an elegant method for determining the diffracted wavevectors from a crystal, and will be used a great deal in this

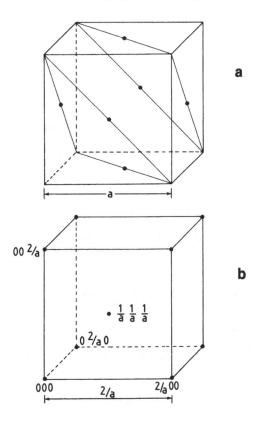

Figure 3.5. (a) Face-centered cubic unit cell in real space. (b) The corresponding body-centered cubic unit cell in reciprocal space.

and the following chapters. Having constructed the reciprocal lattice of the crystal being studied, we adopt the following procedure (see Figure 3.6). Draw a vector LO of length $1/\lambda$ in the direction of the incident wave and terminating at the origin O of the reciprocal lattice. This incident wavevector is designated $\mathbf{K}_0$ and the point L is called the Laue point. Now draw a sphere of radius $|\mathbf{K}_0| = 1/\lambda$ about L as center. The possible directions of diffracted waves are determined by the intersections of the sphere with reciprocal lattice points. In Figure 3.6, the point $G(hkl)$ intersects the sphere, so there will be a diffracted wave in the direction from L to G, the corresponding wavevector being designated $\mathbf{K}_g$.

We prove this result by showing that it is consistent with the Bragg law. The vector OG, which is designated $\mathbf{g}$, is normal to a set of lattice planes (hkl) and is of magnitude $1/d(hkl)$. From Figure 3.6, it is clear that

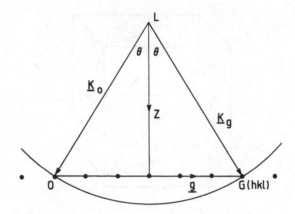

Figure 3.6. Ewald sphere construction in reciprocal space.

$$|\mathbf{g}| = 2\left(\frac{1}{\lambda}\right)\sin\theta = 2|\mathbf{K_0}|\sin\theta = 2|\mathbf{K_g}|\sin\theta \qquad (3.18)$$

from which it follows that

$$2d(hkl)\sin\theta = \lambda$$

Also note that

$$\mathbf{K_0} + \mathbf{g} = \mathbf{K_g} \qquad (3.19)$$

$\mathbf{g}$ is called the reciprocal lattice vector of the operating reflection.

The wavelength λ associated with electrons accelerated through 100 kV is 0.0037 nm, giving the Ewald sphere a radius of 270 nm^{-1}. Crystal d-spacings are typically of the order of 0.2 nm, so the magnitude of the corresponding reciprocal lattice vectors is about 5 nm^{-1}. Thus, the Ewald sphere is nearly a plane in the neighborhood of the reciprocal lattice points O and $G(hkl)$; so, to a first approximation, the electron diffraction pattern represents a plane section through the reciprocal lattice normal to the incident wavevector (see also Section 3.7).

3.5 The relative intensities of the waves diffracted by a unit cell: the structure factor $F(hkl)$

So far, we have been concerned with the directions of the waves diffracted by a crystal. We must now inquire into their relative intensities.

First consider a plane wave

$$\psi = \psi_0 \exp 2\pi i (\mathbf{k_0}\cdot\mathbf{r} - \nu t) \qquad (3.20)$$

incident upon a scattering point. In general, the scattered wave has a different amplitude, wavevector, and phase. Thus, the scattered wave can be represented by

$$\psi = f\psi_0 \exp 2\pi i(\mathbf{k_g} \cdot \mathbf{r} - \nu t - \varphi) \tag{3.21}$$

This may be rewritten as

$$\psi = [f \exp 2\pi(-i\varphi)] \psi_0 \exp 2\pi(\mathbf{k_g} \cdot \mathbf{r} - \nu t) \tag{3.22}$$

Since we are interested in determining relative intensities, we can ignore the time-dependent part of Eq. (3.22) and consider only the part in square brackets, which is called the *complex scattering amplitude. f* is the *scattering factor* and φ the *scattering phase shift.*

We now consider the relative intensity of a wave diffracted by a unit cell situated at a point in the space lattice. Each atom in the unit cell scatters incident radiation, and the contribution of the unit cell as a whole is the resultant of these separate waves. This resultant amplitude is called the *structure factor, F_g* or $F(hkl)$. Referring to Figure 3.1, let the scattering point P_1 be at the origin of the unit cell and the scattering point P_2 at a position $\mathbf{r}$ given by

$$\mathbf{r} = u\mathbf{a} + v\mathbf{b} + w\mathbf{c} \tag{3.23}$$

where $\mathbf{a}$, $\mathbf{b}$, and $\mathbf{c}$ define the unit cell. The phase difference between the waves scattered by P_1 and P_2 is $\varphi = 2\pi(\mathbf{r} \cdot \mathbf{g})$, which is Eq. (3.6); and with Eq. (3.23), we have

$$\varphi = 2\pi(hu + kv + lw) \tag{3.24}$$

Thus, for a unit cell containing n atoms that scatter with amplitudes $f_1, f_2, ..., f_n$ and phases $\varphi_1, \varphi_2, ..., \varphi_n$, the resultant diffracted amplitude is given by

$$F(hkl) = \sum_n f_n \exp[-2\pi i(hu_n + kv_n + lw_n)] \tag{3.25}$$

where u_n, v_n, and w_n are the fractional coordinates of the nth atom. The diffracted intensity $I(hkl) = FF^*$, where F^* is the complex conjugate of F.

In vector notation, Eq. (3.25) becomes

$$F_g = \sum_n f_n \exp[-2\pi i \mathbf{g} \cdot \mathbf{r}_n] \tag{3.26}$$

In terms of sines and cosines, we can write

$$F(hkl) = \left\{ \sum_n f_n \cos 2\pi(hu_n + kv_n + lw_n) \right\}$$
$$- i \left\{ \sum_n f_n \sin 2\pi(hu_n + kv_n + lw_n) \right\} \tag{3.27}$$

If the crystal has a center of symmetry, the sine terms vanish.

As an example, we calculate $F(hkl)$ for a body-centered cubic crystal containing two atoms (e.g., CsCl) that are located in the unit cell at $0, 0, 0$ and $\frac{1}{2}, \frac{1}{2}, \frac{1}{2}$, respectively. Equation (3.27) now becomes

$$
\begin{aligned}
F(hkl) = f_1 &[\cos 2\pi(0+0+0) - i \sin 2\pi(0+0+0)] \\
+ f_2 &[\cos 2\pi(\tfrac{1}{2}h + \tfrac{1}{2}k + \tfrac{1}{2}l) - i \sin 2\pi(\tfrac{1}{2}h + \tfrac{1}{2}k + \tfrac{1}{2}l)] \\
= f_1 &+ f_2 \cos \pi(h+k+l)
\end{aligned} \tag{3.28}
$$

We consider two special cases:

(a) If $h+k+l$ is even, Eq. (3.28) becomes

$$F(hkl) = f_1 + f_2 \quad \text{and} \quad I(hkl) = (f_1 + f_2)^2$$

(b) If $h+k+l$ is odd, Eq. (3.28) becomes

$$F(hkl) = f_1 - f_2 \quad \text{and} \quad I(hkl) = (f_1 - f_2)^2$$

Clearly, if f_1 and f_2 approach each other, then $I(hkl)$ tends to zero for case (b). If $f_1 = f_2 = f$, which is true if the crystal consists of atoms of only one kind (e.g., iron), then $I(hkl) = 4f^2$ when $(h+k+l)$ is even, and $I(hkl) = 0$ when $(h+k+l)$ is odd.

Thus, for CsCl, the reflections from (100) planes are very weak, but there are no 100 reflections for iron. On the other hand, the 200 and 110 reflections, for example, are strong in both crystals.

For an fcc crystal (such as aluminum) in which all atoms are the same, the atoms are at 000, $\frac{1}{2}0\frac{1}{2}$, $0\frac{1}{2}\frac{1}{2}$, $\frac{1}{2}\frac{1}{2}0$, and $I(hkl) = 16f^2$ when h, k, l are all even or all odd. For all other reflections $I(hkl) = 0$.

For a given crystal structure the diffracted intensity is zero at certain Bragg angles because of the destructive interference of the waves scattered by the atoms in directions corresponding to these Bragg angles. Similar behavior is observed in the optical diffraction by a grating (Section 1.3 and Figure 1.8).

Occasionally, reflections that are forbidden by the structure factor are observed in a diffraction pattern. These forbidden reflections are due to *double diffraction,* which occurs when a strong diffracted beam in the crystal acts as an incident beam for further diffraction by the crystal. These extra spots can be found by translating the diffraction pattern, without rotation, so that the 000 spot coincides successively with all the strong diffraction spots of the pattern. All new spots introduced by this procedure are geometrically possible double-diffraction spots. If $h_1 k_1 l_1$ and $h_2 k_2 l_2$ are the indices of any two allowed primary diffraction spots, then all spots with indices of the form h_1+h_2, k_1+k_2, l_1+l_2 are possible double-diffraction spots. In quartz, for example, the 0003 is an allowed reflection, but the forbidden reflections 0001 and 0002 are often observed in sections cut normal to an **a**-axis. This is presumably due to double diffraction involving strong pairs of reflections such as $10\bar{1}1$ and $\bar{1}010$, and $10\bar{1}2$ and $\bar{1}010$, respectively.

3.6 The atomic scattering factor

Equation (3.25), which was derived for the intensity of the radiation scattered by a unit cell, contains the atomic scattering factor f_n. This factor depends principally on the nature of the radiation (and therefore the scattering mechanism), the nature of the scattering point, and the scattering angle.

It will be helpful to discuss the scattering of x-rays by a single atom before considering the atomic scattering factor for electrons.

X-rays, which are electromagnetic waves, are scattered by an atom because of their interaction with the electronic charge of the atom. The incident electromagnetic wave causes the charge to oscillate at the same frequency as the wave, and the accelerating charge radiates electromagnetic waves. The atomic scattering factor $f(x)$ is a measure of the amplitude scattered by an atom when x-radiation of a given amplitude falls on it, and is expressed in terms of the amplitude scattered by a single electron under the same conditions.

To derive an expression for $f(x)$ using classical arguments, we make the following assumptions:

1   The scattering atom contains electrons that are distributed throughout a volume comparable to atomic dimensions and to the wavelength of the radiation.
2   Each electron is loosely bound in the atom so that it scatters as a free electron. This means that if the frequency ν of the incident radiation is very large compared with the natural frequency of oscillation (ν_0), then the scattered wave will be exactly π radians out of phase with the incident wave for all the electrons.
3   The orbital motion of the electrons in the atom is so slow that there is no appreciable change in the configuration of the electrons during a time corresponding to many oscillations of the incident wave.

These assumptions have several immediate consequences. In the forward direction there is no phase difference between waves scattered by

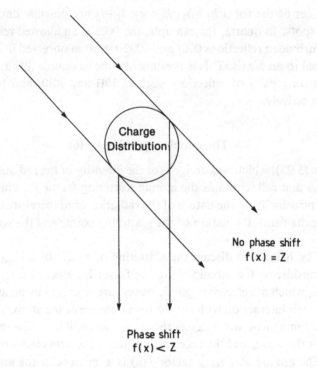

Figure 3.7. Diagram illustrating the origin of the phase difference between the waves scattered from opposite sides of an atom when the scattering angle is different from zero, and the corresponding decrease in the atomic scattering factor.

electrons in different parts of the atom. Therefore, because any change of phase on scattering is the same for all the electrons, the amplitude scattered by the atom in the forward direction is simply Z times that due to a single electron, where Z is the atomic number (number of electrons in the atom). On the other hand, for radiation scattered in a direction making a finite angle with the direction of the incident radiation (Figure 3.7), there will be a path difference between waves scattered from electrons in different parts of the atom. Therefore, interference will produce a scattered amplitude smaller than the scattered amplitude in the forward direction. The phase differences and hence the scattered amplitude will depend on the scattering angle 2θ, the wavelength, and the effective volume of the atom. Thus, for small scattering angles, $f(x)$ will approach Z and will decrease with increasing scattering angles.

If we consider a space lattice in which every lattice point is occupied by an atom of the same kind, the electrons of the atom then form a group of scattering points associated with each lattice point. Initially, we assume that each atom contains only one electron whose position with respect to the corresponding lattice point is given by a vector **r**. The phase difference φ (at a distance large compared with atomic dimensions) between a wave scattered by this electron and a wave scattered by an electron at the lattice point is given by Eq. (3.6):

$$\varphi = 2\pi \mathbf{g} \cdot \mathbf{r}$$

where **g** is normal to the reflecting planes. The amplitude of the wave scattered by the electron at **r** is $\exp(2\pi i \mathbf{g} \cdot \mathbf{r})$ times the amplitude f_0 of the wave scattered by an electron at the lattice point.

Now consider an atom containing a number of electrons and imagine that each electron is spread out into a diffuse cloud of negative charge, characterized by a charge density ρ, expressed in units of the charge on a single electron. Thus $\rho \, dV$ is the ratio of the charge contained in an element of volume dV to the charge on a single electron. The amplitude scattered by an element of charge $\rho \, dV$ at a position **r** is f_{dv} given by

$$f_{dv} = (f_0 \exp 2\pi i \mathbf{g} \cdot \mathbf{r}) \rho \, dV \tag{3.29}$$

To obtain the amplitude scattered by an atom containing *a single electron* (whose charge is spread diffusely throughout the volume of the atom), we must integrate Eq. (3.29) over the volume of the atom. Hence,

$$f_v = f_0 \int_v \exp 2\pi i \mathbf{g} \cdot \mathbf{r} \, dV \tag{3.30}$$

We define the atomic scattering factor f_e for an atom containing one electron by

$$f_e = \int_v \exp 2\pi i \mathbf{g} \cdot \mathbf{r} \, dV \tag{3.31}$$

Now let **r** make an angle α with the direction **g** as shown in Figure 3.8. Therefore, because $2d(hkl) \sin\theta = \lambda$, we have

$$\varphi = 2\pi \mathbf{g} \cdot \mathbf{r} = \frac{2\pi r \cos\alpha}{d(hkl)}$$

$$= \frac{4\pi r}{\lambda} \sin\theta \cos\alpha$$

$$= ur \cos\alpha \tag{3.32}$$

where $u = (4\pi \sin\theta)/\lambda$.

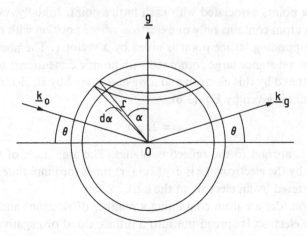

Figure 3.8. Diagram used in the calculation of the atomic scattering factor.

Suppose that ρ is a function of the magnitude of $\mathbf{r}$ only and not of its direction so that the charge distribution is spherically symmetrical. We take as a convenient element of volume dV an annulus of a spherical shell of radius $\mathbf{r}$, thickness dr, and width dx, with $\mathbf{g}$ as axis, as shown in Figure 3.8. Thus,

$$dV = 2\pi r^2 \sin\alpha \, d\alpha \, dr \qquad (3.33)$$

and Eq. (3.31) becomes

$$f_e = \int_v 2\pi r^2 \rho(r) \exp(iur\cos\alpha) \sin\alpha \, d\alpha \, dr$$

$$= \int_0^\infty 2\pi r^2 \rho(r) \, dr \int_0^\pi \exp(iur\cos\alpha) \sin\alpha \, d\alpha \qquad (3.34)$$

If we put $(ur\cos\alpha) = x$, then $dx = -ur\sin\alpha \, d\alpha$ and Eq. (3.34) becomes

$$f_e = \int_0^\infty 2\pi r^2 \rho(r) \, dr \int_{-ur}^{ur} \frac{\exp(ix)\,dx}{ur}$$

$$= \int_0^\infty 4\pi r^2 \rho(r) \frac{\sin ur}{ur} \, dr \qquad (3.35)$$

For an atom containing n electrons, the amplitude scattered by the atom is the sum of the amplitudes scattered by each electron, that is,

$$f(x) = \sum_n f_{en} = \sum_n 4\pi r^2 \rho_n(r) \frac{\sin ur}{ur} \, dr \qquad (3.36)$$

$f(x)$ is called the atomic scattering factor for x-rays and is clearly the amplitude scattered by the atom expressed in units of the amplitude scattered by a single electron at the lattice point on which the atom is centered. To compute $f(x)$, we need to know the radial dependence of the electron density $\sum_n \rho_n(r)$ in the atom. Since $f(x)$ is a function of u, it is a function of $(\sin\theta)/\lambda$. For $\theta = 0$, $u = 0$ and $(\sin ur)/ur = 1$; hence,

$$f(x) = \int_0^\infty 4\pi r^2 \rho(r)\, dr = Z \qquad (3.37)$$

where Z is the number of electrons in the atom.

Tables of $f(x)$ as a function of $(\sin\theta)/\lambda$ are given for all atoms by MacGillavry and Rieck (1983). $f(x)$ for Ni is graphed in Figure 3.9(a).

By definition, the atomic scattering factor $f(x)$ is given in terms of the amplitude scattered by a single electron at the lattice point. It is useful, however, to have the scattered amplitude A in terms of the incident amplitude A_0. From classical electromagnetic theory, it follows that if a wave of amplitude A_0 is incident on a free electron, the amplitude A of the radiation emitted in the forward direction, at a distance R (meters) from the electron, is given by

$$A = \frac{A_0}{R}\left(\frac{e^2}{4\pi\epsilon_0 mc^2}\right) \qquad (3.38)$$

using SI units.* Thus, at a distance of a few centimeters from the electron, the ratio of the intensities $(A/A_0)^2$ is of the order of 10^{-26}. For scattering by an atom,

$$\frac{A}{A_0} = \frac{f(x)}{R} \times 2.84 \times 10^{-15} \qquad (3.39)$$

The scattering of electrons by atoms is discussed in various textbooks (such as Hirsch et al. 1965), where it is shown that the atomic scattering factor $f(e)$, analogous to $f(x)$ for x-rays, is given by the expression

$$f(e) = \frac{me^2}{(4\pi\epsilon_0)2h^2}\left\{Z - f(x)\left(\frac{\lambda}{\sin\theta}\right)^2\right\} \qquad (3.40)$$

where $me^2/(4\pi\epsilon_0)2h^2 = 2.38 \times 10^4\ \mathrm{m}^{-1}$.

Thus $f(e)$ has the units of meters. If we express $f(e)$ in centimeters and λ in angstroms (Å), Eq. (3.40) becomes

$$f(e) = 2.38 \times 10^{-10}\{Z - f(x)\}\left(\frac{\lambda}{\sin\theta}\right)^2 \qquad (3.41)$$

---

* ϵ_0, the permittivity of free space, has a value of 8.85415×10^{-12} F/m.

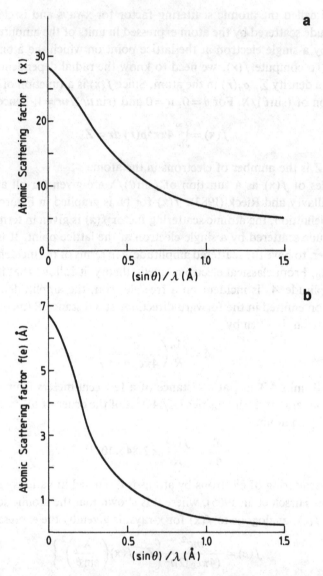

Figure 3.9. (a) The x-ray and (b) the electron atomic scattering factors
as a function of $(\sin\theta)/\lambda$ for Ni.

We can see from Eq. (3.40) that two mechanisms are involved in the
scattering of an electron by an atom. The first is Rutherford scattering by
the nucleus [note that $f(e)$ is proportional to Ze^2], and the second is due
to scattering by the electron cloud [note the presence of the x-ray atomic

scattering factor $f(x)$ in the equation]. Also, $f(e)$, like $f(x)$, decreases rapidly with increasing $(\sin\theta)/\lambda$. Values of $f(e)$ as a function of $(\sin\theta)/\lambda$ for all atoms are given by Hirsch et al. (1965) and by MacGillavry and Rieck (1983). Values for Ni are plotted in Figure 3.9(b).

The ratio of the scattered amplitude A to the amplitude A_0 of the incident electron beam is given by

$$\frac{A}{A_0} = \frac{1}{R} f(e) \tag{3.42}$$

It is instructive to compare the values of A/A_0 for x-rays and electrons scattered by an atom. Consider a Ni atom as an example. From the tables of atomic scattering factors* we have $f(x) = 24.9$ and $f(e) = 5.27$ Å for $(\sin\theta)/\lambda = 0.1$ Å^{-1}. Thus, from Eqs. (3.39) and (3.42) we have, for R equal to one centimeter,

$$\left(\frac{A}{A_0}\right)^2_x = (24.9 \times 2.84 \times 10^{-13})^2 = 5 \times 10^{-23}$$

$$\left(\frac{A}{A_0}\right)^2_e = (5.27 \times 10^{-8})^2 = 28 \times 10^{-16}$$

Therefore, the scattered intensity for electrons is about 5×10^7 times that for x-rays.

For $(\sin\theta)/\lambda = 1.0$ Å^{-1}, we have $(A/A_0)^2_x = 0.4 \times 10^{-23}$ and $(A/A_0)^2_e = 0.25 \times 10^{-16}$. Note that by increasing $(\sin\theta)/\lambda$ from 0.1 Å^{-1} to 1.0 Å^{-1}, the x-ray scattered intensity has been reduced by a factor of 15 and the electron scattered intensity has been reduced by a factor of more than 100.

It is also instructive to compare the electron scattering from a light element such as C ($Z = 6$) with that from Ni ($Z = 28$). For $(\sin\theta)/\lambda = 0.1$ Å^{-1}, $f(e)$ for C is 2.09 Å, so

$$\left(\frac{A}{A_0}\right)^2_e = 4.4 \times 10^{-16}$$

which is about $\frac{1}{6}$ of that for Ni. In general, the intensity of the scattered radiation increases with Z.

3.7 Intensity of the wave diffracted from a perfect crystal

We now consider the amplitude of the wave diffracted from a perfect crystal in directions that differ slightly from the exact Bragg angle. This

---

* Since wavelengths are given in angstroms in the *International Tables for X-ray Crystallography* (MacGillavry and Rieck 1983), these units are used here.

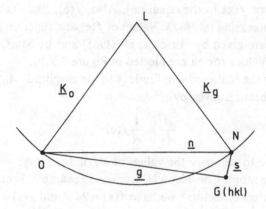

Figure 3.10. The Ewald sphere construction in reciprocal space when there is a slight deviation from the exact Bragg angle.

deviation can be described by a vector s in reciprocal space that gives the distance from the reciprocal lattice point $G(hkl)$ to a point N on the Ewald sphere, as in Figure 3.10. In this diagram, $G(hkl)$ is outside the Ewald sphere, which indicates that the angle between $\mathbf{K}_0$ and the reflecting planes (which are normal to $\mathbf{g}$) is slightly smaller than the exact Bragg angle. Therefore, the vector s is negative. If $G(hkl)$ were inside the Ewald sphere, s would be positive.

From Figure 3.10, we have

$$\mathbf{n} = \mathbf{K}_g - \mathbf{K}_0 = \mathbf{g} + \mathbf{s} \tag{3.43}$$

Therefore, for one unit cell, the diffracted amplitude becomes

$$F_g(s) = \sum_n f_n \exp[-2\pi i(\mathbf{g}+\mathbf{s})\cdot\mathbf{r}_n] \tag{3.44}$$

Here $\mathbf{r}_n$ is the position of the nth atom in the unit cell with respect to some arbitrarily chosen origin. We now express $\mathbf{r}_n$ as the sum of two vectors

$$\mathbf{r}_n = \mathbf{r}_j + \mathbf{r}_k \tag{3.45}$$

where $\mathbf{r}_j$ is the position of the origin of the jth unit cell, and $\mathbf{r}_k$ is the position of the kth atom in the unit cell with respect to the origin of the unit cell. Therefore, the amplitude of the wave diffracted by the *crystal* is given by

$$A(s) = \sum_{j,k} f_k \exp[-2\pi i(\mathbf{g}+\mathbf{s})\cdot(\mathbf{r}_k+\mathbf{r}_j)]$$

$$= \sum_{k} f_k \exp[-2\pi i(\mathbf{g}+\mathbf{s})\cdot\mathbf{r}_k] \times \sum_{j} \exp[-2\pi i(\mathbf{g}+\mathbf{s})\cdot\mathbf{r}_j]$$

$$= F_g(s) \sum_{j} [\exp(-2\pi i\mathbf{g}\cdot r_j)][\exp(-2\pi i\mathbf{s}\cdot\mathbf{r}_j)] \qquad (3.46)$$

Because $\mathbf{g}$ is a reciprocal lattice vector and $\mathbf{r}_j$ is a lattice vector in real space, $\mathbf{g}\cdot\mathbf{r}_j$ is an integer; hence, Eq. (3.46) becomes

$$A(s) = F_g(s) \sum_{j} \exp(-2\pi i\mathbf{s}\cdot\mathbf{r}_j) \qquad (3.47)$$

Consider an electron beam traveling in the z direction and incident normally on the top surface of a thin crystal plate of thickness t. Now consider specifically the diffraction by a column of unit cells parallel to the diffracted wave. Since the Bragg angles are typically of the order of 10^{-2} radian, the column can be taken as perpendicular to the crystal plate, that is, parallel to z. Now s is small, and if we assume that $\mathbf{s}\cdot\mathbf{r}_j$ varies slowly from one unit cell to another, the summation in Eq. (3.47) can be replaced by an integral. If we choose the origin at the center of the crystal plate, Eq. (3.47) becomes

$$A(s,z) = \frac{1}{V_c}F_g \int_{-t/2}^{t/2} \exp(-2\pi i s z)\, dz \qquad (3.48)$$

where s is the component of $\mathbf{s}$ in the z direction. On integrating, we find that

$$A(s,z) = \frac{1}{V_c}F_g\left(\frac{\sin \pi t s}{\pi s}\right) \qquad (3.49)$$

If we now assume that the whole crystal is made up of such columns and that the columns diffract independently of each other, the intensity of the diffracted beam leaving the bottom surface of the crystal plate is given by

$$I_g(s,z) \sim \frac{\sin^2 \pi t s}{(\pi s)^2} \qquad (3.50)$$

I_g is plotted as a function of s in Figure 3.11. In the other directions (x and y), the integration leads to expressions for $I_g(s,x)$ and $I_g(s,y)$ of the same form as Eq. (3.50). Thus, in three dimensions, for a crystal in the form of a parallelepiped with edges t_1, t_2, t_3 along the directions x, y, z, we have

$$A(s) = \left(\frac{\sin \pi t_1 s_x}{\pi s_x}\right)\left(\frac{\sin \pi t_2 s_y}{\pi s_y}\right)\left(\frac{\sin \pi t_3 s_z}{\pi s_z}\right) \qquad (3.51)$$

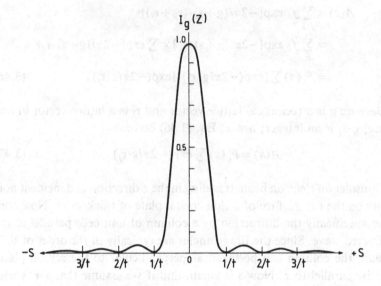

Figure 3.11. The diffracted intensity I_g as a function of the deviation from the exact Bragg angle calculated from the kinematical theory.

where s_x, s_y, and s_z are the components of **s**. Points to be noted in Eq. (3.51) are (i) that the *intensity distribution* is independent of *hkl* and so is the same about each reciprocal lattice point and (ii) that the intensity distribution depends on the *shape* of the crystal. If the crystal is platelike with both t_1 and $t_2 \gg t_3$, then the first two terms in Eq. (3.51) tend to unity and

$$I_g(s, z) \sim \frac{\sin^2 \pi t_1 s_z}{(\pi s_z)^2}$$

which is identical to Eq. (3.50).

This intensity distribution means that each reciprocal lattice point is effectively extended into a rod that is normal to the thin crystal specimen. Due to this extension of the reciprocal lattice points and to the fact that the radius of the Ewald sphere is large compared with the spacing between reciprocal lattice points, the Ewald sphere usually intersects many reciprocal lattice rods, as shown in Figure 3.12(a). Thus, many diffracted

Figure 3.12. *Facing page.* (a) Diagram showing the Ewald sphere cutting the reciprocal lattice rods of zero and higher order Laue zones and (b) a schematic diagram of the corresponding diffraction pattern.

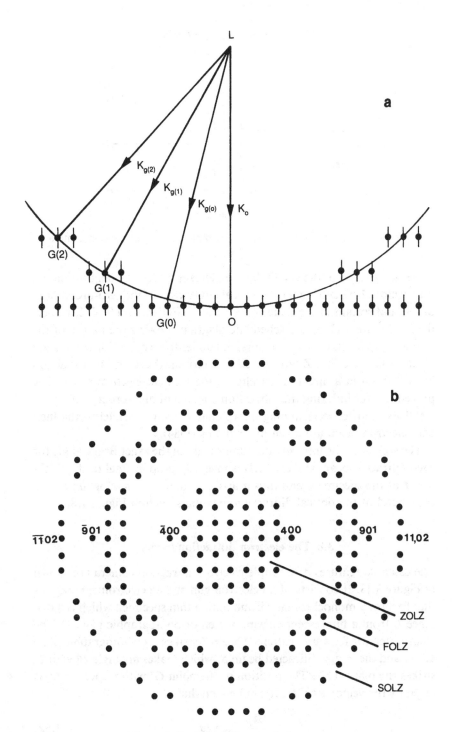

a

$K_{g(2)}$

$K_{g(1)}$

$K_{g(o)}$ K_o

G(2)

G(1)

G(0)   O

b

$\overline{11}02$ $\overline{9}01$ $\overline{4}00$ 400   901   11,02

ZOLZ

FOLZ

SOLZ

Figure 3.12. See caption on facing page.

73

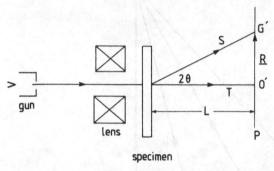

Figure 3.13. The essential features of a simple electron diffraction camera.

waves are excited in the crystal by the incident wave. The diagram shows the Ewald sphere cutting the reciprocal lattice rods of the zero-order, first-order, and second-order Laue zones (ZOLZ, FOLZ, and SOLZ, respectively). Figure 3.12(b) is a schematic diagram showing the nature of the diffraction pattern that will be observed under these conditions. The radii of the FOLZ and SOLZ can be used to determine the interplanar spacings in the crystal in a direction parallel to the incident electron beam. The procedures for indexing the diffraction maxima of high-order Laue zones (HOLZs) and for determining interplanar spacings parallel to the incident electron beam are given by Williams (1984).

The spread of the intensity distribution about the exact Bragg angle for each diffracted beam (Figure 3.11) is inversely proportional to the thickness t of the specimen, and thus it arises in an exactly analogous way to the spread of the optical diffraction maxima shown in Figure 1.8.

3.8 The electron diffraction camera

The essential features of a simple electron diffraction camera are shown in Figure 3.13. It consists of an electron gun and an electromagnetic lens that focuses a monochromatic beam onto a thin specimen which is a distance L from a fluorescent viewing screen or photographic plate P. The transmitted (or forward-scattered) beam T strikes the photographic plate at O', and the Bragg diffracted beam S, which makes an angle 2θ with T, strikes the plate at G'. The position of the point G' with respect to O' is given by the vector $\mathbf{R}(hkl)$. It can be seen that

$$\frac{|\mathbf{R}|}{L} = \tan 2\theta \tag{3.52}$$

But $2d(hkl) \sin\theta = \lambda$; so for small θ, $|\mathbf{R}|/L = 2\theta = 2\lambda/2d(hkl)$ and

$$|\mathbf{R}| = \frac{\lambda L}{d(hkl)} \qquad (3.53)$$

However, $1/d(hkl) = |\mathbf{g}|$; so Eq. (3.53) can be written as

$$|\mathbf{R}| = (\lambda L)|\mathbf{g}| \qquad (3.54)$$

Since $\theta \approx 10^{-2}$ radian, $\mathbf{R}$ must be very nearly parallel to $\mathbf{g}$; hence,

$$\mathbf{R} = (\lambda L)\mathbf{g} \qquad (3.55)$$

We see, therefore, a direct relation between $\mathbf{R}$ and the reciprocal lattice vector $\mathbf{g}$ of the operating reflection, which is normal to the reflecting planes. (λL) is called the *camera constant*.

In a transmission electron microscope, the electron diffraction pattern is formed in the back focal plane of the objective lens, and we can produce an enlargement of this diffraction pattern on the viewing screen by suitable adjustments of the currents through the intermediate and projector lenses. Clearly, the value of L in Eq. (3.54) is not simply the geometrical distance between the crystal specimen and the viewing screen. However, (λL) can be determined experimentally by measuring the radii R of the diffraction rings corresponding to known d-spacings for a polycrystalline specimen, such as an evaporated thin silver foil. Once λL is known, the spots in any diffraction pattern obtained from a known single crystal can, in principle, be indexed using Eq. (3.55) and the orientation of the specimen with respect to the incident electron beam determined.

3.9 Kikuchi lines and bands

Electrons of the energies usually used in transmission electron microscopes (e.g., 100–300 keV) are scattered essentially *elastically* (i.e., without any appreciable loss of energy) by crystal specimens of the order of 100 nm thick or less. The SAD patterns of such specimens consist of sharp, bright spots on a uniformly black background. However, for thicker regions of crystal, the diffraction spots are superimposed on a diffuse background produced by electrons that have lost a few (< 50) eV of energy, that is, by electrons which have been *inelastically* scattered. This diffuse background is usually crossed by a network of bright and dark lines, called *Kikuchi lines* (Figure 3.14). Kikuchi lines provide a very convenient method of determining with high accuracy the crystallographic orientation of the specimen and, in particular, the magnitude and sign of s_g, the deviation from the exact Bragg angle, for any operating reflection $\mathbf{g}$.

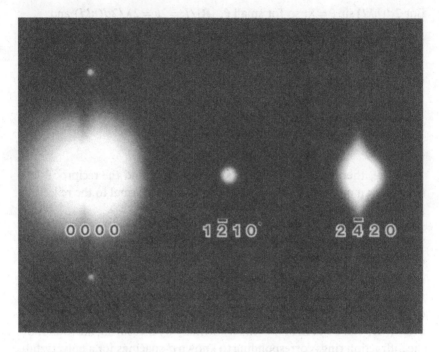

Figure 3.14. Kikuchi lines in an electron diffraction pattern of quartz. $\mathbf{g}_2 = 2\bar{4}20$ is close to the exact Bragg angle. Compare with Figure 3.16(e, f).

The geometry of Kikuchi lines can be understood in terms of a simple theory (first given by Kikuchi in 1928), in which inelastically scattered electrons are subsequently elastically scattered. Consider a primary beam of electrons I_0 incident on a crystal, as shown in Figure 3.15(a). Suppose that, at some point P_o just below the surface of the crystal, these electrons are inelastically scattered diffusely as indicated. Now consider two rays a and b of this diffuse fan of electrons that are inclined at the Bragg angle to a set of atomic planes, also indicated. The beams a and b are diffracted (elastically) and become the beams b' and a', respectively. a' is parallel to a, and b' is parallel to b. Thus, the beams a and b are interchanged by diffraction. If these beams were of the same intensity, this interchange would produce no observable effect. However, the angular distribution of the intensity of the diffuse scattering has a maximum in the direction of the incident beam I_0. Thus, the beams a and b' are more intense than the beams b and a'.

Now the inelastically scattered beams from P_o fan out in three dimensions. All the beams, such as a, and the corresponding Bragg diffracted beams, such as b', lie on the surface of a cone (called a Kossel cone) whose axis is normal to the reflecting plane at P_a and whose semiangle is $\frac{1}{2}(180° - 2\theta) = (90° - \theta)$, as shown in Figure 3.15(b). Similarly, all the beams, such as b and a', lie on the surface of a cone of the same semiangle and whose axis is normal to the reflecting plane at P_b. The loci of the points D and E on the photographic plate, which is placed some distance L away from the specimen and normal to the incident beam, are the intersections of the Kossel cones with the photographic plate. Since θ is small, the Kossel cones have semiangles that are only slightly less than $90°$; so the intersections of the Kossel cones with the photographic plate are two parallel straight lines: the Kikuchi lines. It is evident from Figure 3.15(a) that the angular separation between the beams a' and b' is 2θ, so the distance x between the pair of Kikuchi lines on the photographic plate is $2\theta L$, where L is the camera length. For small θ, the Bragg law becomes $2d\theta = \lambda$, so

$$xd = \lambda L \tag{3.56}$$

The Kikuchi line D is darker than background and is therefore called the *defect line*. This line is always closer to the point at which the incident beam strikes the photographic plate than the Kikuchi line E, which is brighter than background and hence is called the *excess line*. Also note that the reflecting plane through P_o intersects the photographic plate along a line that is midway between the two Kikuchi lines. Thus, if we set the crystal at the exact Bragg angle for a reflection $\mathbf{g}$, then the defect line will pass through the central spot O, and the excess line will pass through the diffracted spot $\mathbf{g}$. If the crystal is now tilted about an axis normal to the beam (as in a standard tilting specimen holder), the spots of the SAD pattern will not move (the individual spots will change in intensity, some will disappear and others appear in different positions), but the Kikuchi lines will move as though they were rigidly attached to the crystal. Thus, the direction and magnitude of the movement of the Kikuchi lines indicate with high accuracy the change of orientation of the crystal.

In practical electron microscopy, it is often important to be able to determine accurately the deviation $\Delta\theta$ from the exact Bragg angle θ. We now consider in some detail how we can use Kikuchi lines for doing this.

Figure 3.16(a) shows the Ewald sphere construction for a systematic row of reflections when the first-order reflection $\mathbf{g}$ satisfies the exact Bragg condition. The corresponding reciprocal lattice point G_1 is thus on the Ewald sphere and $s_g = 0$. The reciprocal lattice point G_2 (corresponding to the second-order reflection $2\mathbf{g}$) is outside the sphere, and so s_{2g} is negative.

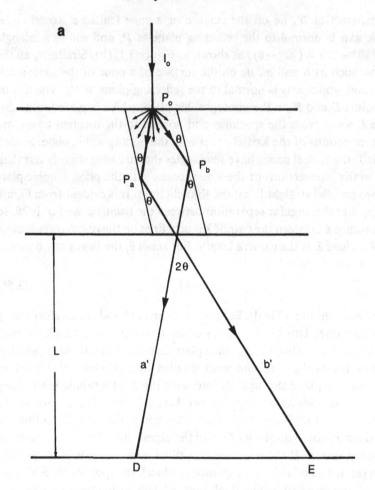

Figure 3.15. (a) Diagram showing the origin of Kikuchi lines.

The corresponding SAD pattern of spots and Kikuchi lines is shown in Figure 3.16(b). The spacing between adjacent spots of the systematic row is x, and the Kikuchi lines D_1 and E_1 pass through the diffraction spots O and g, respectively. The second-order Kikuchi lines D_2 and E_2 pass midway between $-g$ and O and between g and $2g$, respectively, and hence are a distance $2x$ apart.

Suppose we tilt the crystal through an angle $\Delta\theta_1$ about an axis normal to the systematic row (i.e., normal to the direction of **g**) so that G_1 now lies inside the Ewald sphere, as shown in Figure 3.16(c). The diffraction error s_g is positive, and its magnitude is given by

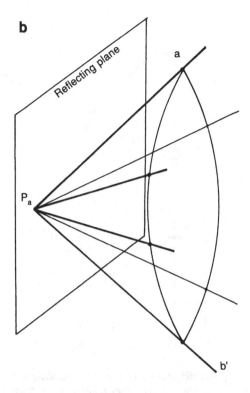

Figure 3.15. *Continued.* (b) Kossel cone of rays such as *a* and *b'* whose axis is normal to the reflecting plane at P_a.

$$s_g = |\mathbf{g}| \Delta\theta_1 \qquad (3.57)$$

This tilting causes the Kikuchi lines to move to the right: E_1 is displaced from the spot g by a distance Δx_1 and E_2 is displaced from the spot $2g$ by a distance Δx_2. This is shown in Figure 3.16(d), and it is clear that

$$\Delta x_1 + \Delta x_2 = \frac{x}{2} \qquad (3.58)$$

A rotation of 2θ would cause the Kikuchi lines to move through a distance x. Thus, Δx_1 and $\Delta\theta_1$ are related by the equation

$$\frac{\Delta\theta_1}{2\theta} = \frac{\Delta x_1}{x} \qquad (3.59)$$

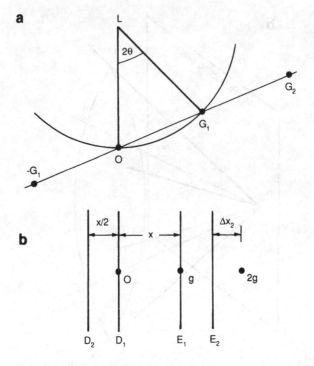

Figure 3.16. Ewald sphere diagrams and the corresponding diffraction patterns showing the positions of the Kikuchi lines relative to the main Bragg beam. (a, b) $s_g = 0$; (c, d) $s_g < 0$; and (e, f) $s_{2g} = 0$.

For small angles, the Bragg law can be written as $2\,d\theta = \lambda$, and so from Eqs. (3.57) and (3.59),

$$s_g = |g|^2 \lambda \left(\frac{\Delta x_1}{x} \right) \tag{3.60}$$

If we tilt the crystal until G_2 lies on the Ewald sphere (i.e., $s_{2g} = 0$), then the Kikuchi lines will be displaced so that $\Delta x_2 = 0$, as shown in Figure 3.16(e, f). Thus, from Eq. (3.58), $\Delta x_1 = x/2$ and

$$s_g = |g|^2 \frac{\lambda}{2} \tag{3.61}$$

By drawing similar diagrams for higher-order reflections (3g, 4g, etc.) satisfying the Bragg condition, we can easily see that if a reciprocal lattice point G_n of the systematic row lies on the Ewald sphere (i.e., $s_{ng} = 0$), then the magnitude of the diffraction error s_g of the first-order reflection will be given by

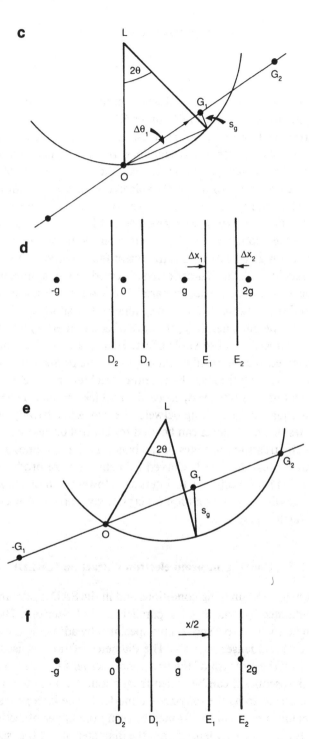

Figure 3.16. *Continued.*

$$s_g = \frac{(n-1)}{2} |g|^2 \lambda \qquad (3.62)$$

This equation is particularly useful when setting up for weak-beam dark field imaging, to be discussed in Chapter 5.

It is instructive to estimate the smallest tilt angle $\Delta\theta$ which can be measured from displacements of Kikuchi lines. If we assume that a displacement of $\Delta x = 0.1x$ can just be measured, then from Eq. (3.59), $\Delta\theta = (0.1)2\theta$, which is equal to about 0.05 degree with $d = 0.5$ nm and $\lambda = 0.004$ nm. Thus, the accuracy with which an orientation can be determined from a diffraction pattern is greatly increased if Kikuchi lines are present.

Although the geometry of Kikuchi lines can be understood from the simple treatment given earlier, this treatment fails to account for two features of Kikuchi patterns that are often observed. For example, when the primary beam is oriented exactly parallel to a set of crystal planes, the beams a and b in Figure 3.15(a) are symmetrically situated with respect to the primary beam. Therefore, they are of equal intensity and their interchange produces no observable effect. Hence, a pair of Kikuchi lines that are oriented symmetrically with respect to the primary beam cannot exist. Under these conditions, a band (either brighter or darker than background) is commonly observed, especially in thick crystals. The width of the band is equal to the spacing between the two geometrically predicted Kikuchi lines. Kikuchi bands can be used for control or determination of orientation by treating the edges of the bands as Kikuchi lines and using the procedures just discussed. Curved Kikuchi lines are produced when straight Kikuchi lines attempt to intersect. However, these curved lines provide no additional information on crystal geometry and need not be discussed further here.

3.10 Convergent beam electron diffraction (CBED)

Under normal TEM imaging conditions and in the SAD mode, the specimen is illuminated by a more or less parallel beam of electrons. The illumination can be focused to a spot on the specimen by adjusting the strength of the second condenser lens C_2. The diameter of this focused spot is determined by the strength of the first condenser lens C_1, and in modern electron microscopes it can be varied from about 2 μm to 40 nm. This is generally referred to as the *microprobe* mode. In the *nanoprobe* mode, an even smaller probe can be formed by using the upper objective polepiece as a third condenser lens. Again the diameter of the focused probe is controlled by the strength of C_1 and can be varied from about 0.1 μm to 2 nm. In either mode, the focused beam is highly convergent, and the convergence is determined by the diameter of the C_2 aperture. Having focused the probe onto the specimen and made sure the objective and SAD apertures are retracted, we can view the convergent beam diffraction pattern in the back focal plane of the objective lens on the fluorescent screen by simply switching to the diffraction mode.

Whereas an SAD pattern consists of an array of sharp spots, the convergent beam electron diffraction (CBED) pattern consists of an array of disks whose diameter depends on the convergence of the incident beam and hence on the diameter of the C_2 aperture. If the crystal is less than about 50 nm thick, the disks are featureless. A CBED pattern of small (nonoverlapping) featureless disks contains the same two-dimensional crystallographic information as a normal SAD pattern, and because it is derived from a much smaller area of crystal, it is usually called a *microdiffraction* pattern. With increasing crystal thickness, it is found that contrast (associated with the dynamical interaction of the many electron beams excited in the crystal; see Chapter 4) appears in the disks as well as Kikuchi lines. The contrast within the disks contains important crystallographic information that is not present in normal SAD patterns. To see the detail within the disks as clearly as possible, it is usual to select a C_2 aperture such that the disks become as large as possible without overlapping. CBED patterns of nonoverlapping disks in which contrast is visible are called *Kossel–Mollenstedt* (K–M) patterns. A CBED pattern in which the disks overlap is often called a *Kossel* pattern. A K–M CBED pattern is shown in Figure 3.17.

The volume of specimen contributing to a CBED pattern is much smaller than that contributing to an SAD pattern. Thus, there is less likelihood that the effects of strain, specimen bending, or crystal defects will influence the nature of a CBED pattern. Consequently, Kikuchi lines are usually more often observed and are usually clearer in CBED patterns than in SAD patterns.

The nature and origin of Kikuchi lines that arise from planes of the ZOLZ were discussed in Section 3.9. Kikuchi lines can also arise from HOLZ planes and are observed outside the diffraction disks of a CBED pattern. However, within the disks there are the so-called HOLZ lines, which are continuous with the HOLZ Kikuchi lines.

K–M and Kossel patterns are extremely sensitive to crystal orientation, and are therefore particularly useful for tilting a crystal into a precise orientation. However, their most important use is in obtaining three-dimensional crystal symmetry information, including the complete determination of the point group and space group of a crystal. Extensive reviews of

Figure 3.17. A convergent beam electron diffraction pattern (CBED) from a specimen of $Tl_2U_xO_{11}$. Note that the ZOLZ suggests six-fold symmetry, but the HOLZ's indicate that the symmetry is only three-fold.
(Courtesy of Dr R. L. Withers.)

the techniques for acquiring the various types of CBED patterns and their interpretations have been given by Williams (1984) and Steeds (1979).

CBED has not been used to any appreciable extent in minerals research. However, because of its many advantages over the SAD mode and the ease with which CBED patterns can be formed in modern microscopes, it is likely to be applied to many mineralogical problems in the near future.

3.11 Image contrast in terms of the kinematical theory of diffraction

It will be recalled from Chapter 2 that a dark field (DF) image is obtained by allowing a diffracted beam to pass through the aperture in the back focal plane of the objective lens. The DF image is determined by I_g and reveals those parts of the specimen that are giving rise to that beam. The *bright field* (BF) image is obtained by allowing only the central, transmitted beam to pass through the objective aperture. The BF image is determined by I_0. If there is no absorption (as required by the kinematical theory), then the value of I_0 in any part of the crystal will be given by the intensity of the incident beam minus the intensities of the diffracted beams from that part of the crystal that do not pass through the objective aperture. If we arrange the orientation of the crystal so that there is only one strong diffracted beam I_g (i.e., $|s|$ is small for only one reciprocal lattice point), then

$$I_0 = 1 - I_g \tag{3.63}$$

and the BF and DF images will be complementary.

We can see from Figure 3.11 that $I_g = 0$ when

$$s = \frac{m}{t} \tag{3.64}$$

where m is an integer. Thus, for t constant, I_g varies sinusoidally with s. In a bent crystal (where s varies slowly and continually), we expect to observe in DF a fringe pattern (called a *bend contour*), across which the intensity profile is as shown in Figure 3.11. We also expect the BF image to be complementary. The BF image and the two DF images, together with the associated selected area electron diffraction pattern observed in a bent crystal of constant thickness, are shown in Figure 3.18. In the diffraction pattern, Figure 3.18(a), it will be seen that on either side of the central spot there are two spots (designated $+g$ and $-g$), that are significantly brighter than all the other diffraction spots in the pattern. The DF images observed using $+g$ and $-g$, Figure 3.18(b, d), both show a bright central band on either side of which there are weak fringes. The intensity profile across each DF image is at least qualitatively similar to that expected from Figure 3.11. However, the BF image, Figure 3.18(c), is noticeably different from the DF images in several respects. By careful comparison of the three images, we see that the bright regions *AB* and *CD* on the sides of the BF image correspond to the bend contours in the DF images observed with $+g$ and $-g$, respectively. It is clear that the BF and DF images of region *AB* (or *CD*) of the crystal are certainly not complementary, as we would expect from the kinematical theory.

From Eq. (3.64), we also expect I_g to vary sinusoidally with t for constant s ($\neq 0$). Thus, in a crystal of varying thickness, we expect to see bright and dark fringes that correspond to regions of equal thickness.

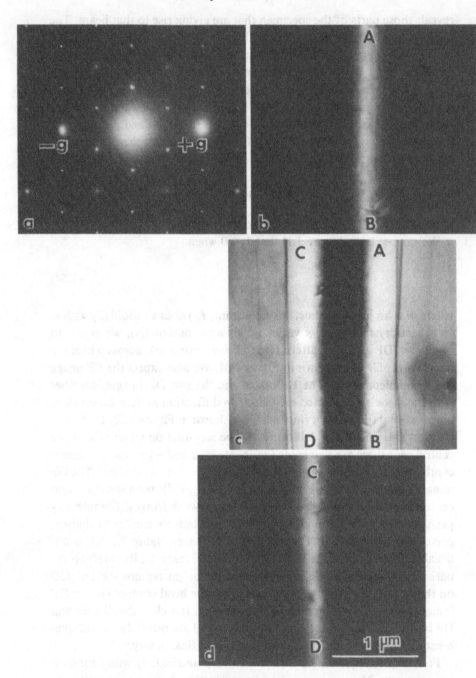

Figure 3.18. See caption on facing page.

Such *thickness fringes* are observed. In DF images we expect a dark fringe, when t is a multiple of $1/s$, and a complementary image in BF. Equation (3.64) also predicts that as the magnitude of s is decreased, the spacing between adjacent dark (or bright) fringes will increase until at $s = 0$ the spacing is infinitely large and no fringes will be observed. An increase in fringe spacing is observed as s is decreased; but at $s = 0$, fringes of high contrast are still observed, as shown in Figure 3.19. This complete break-down of the kinematical theory at $s = 0$ means that the foundations of the theory must be critically reexamined.

3.12 Reexamination of the foundations of the kinematical theory

The assumptions of the kinematical theory – that the incident wave is monochromatic and plane and that there is no absorption of either the transmitted wave or the scattered waves – are reasonable assumptions with which to begin a theory. However, the other assumptions – that a scattered wave is never rescattered and that there is no interaction between the transmitted and scattered waves – are gross oversimplifications of the physical situation. For example, consider a plane monochromatic wave incident upon a crystal plate at such an angle that the Bragg law is satisfied for a set of planes approximately normal to the crystal plate, as shown in Figure 3.20. It is clear from this diagram that at A the diffracted wave S_1 in the crystal is rediffracted so that it travels in the same direction as the transmitted wave T. This rediffracted wave is denoted S_2. There is no reason why S_2 should not be also rediffracted at B to produce the wave S_3, as shown. Thus, the assumption that the diffracted wave, once produced in the crystal, is never rediffracted, is clearly unacceptable.

Figure 3.18. *Facing page.* Bright field (BF) and dark field (DF) images of a bend in a thin crystal of cadmium iodide. The different images were recorded from the same area of specimen without changing its orientation with respect to the incident electron beam. The corresponding selected area diffraction (SAD) pattern is shown in (a). The BF image (c) and the two DF images (b and d) were obtained by locating the objective aperture around the central beam 000, and the two strong diffracted beams $+g$ and $-g$, respectively. The BF bend contour consists of a band of closely spaced fringes, on each side of which is a bright region, AB and CD. Using the dislocations near B as markers, we see that the DF bend contour for $+g$ is located along AB, whereas the DF contour for $-g$ is located along CD. It is clear that whereas the BF image is asymmetrical about both $s_{+g} = 0$ and $s_{-g} = 0$, the DF images are symmetrical.

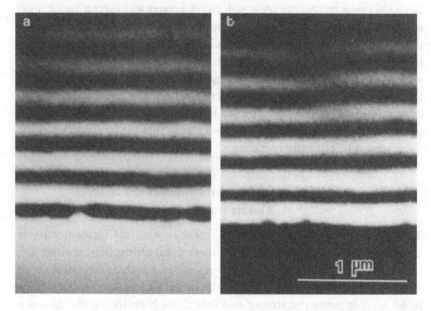

Figure 3.19. Thickness fringes in a wedge-shaped crystal of quartz oriented at the exact Bragg angle ($s = 0$) for $\mathbf{g} = 10\bar{1}1$. (a) Bright field. (b) Dark field.

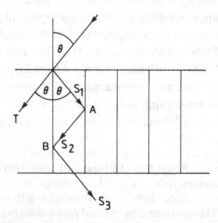

Figure 3.20. Diagram illustrating multiple scattering in a thick crystal.

Moreover, interaction between S_2 and a transmitted wave T is also probable since both are traveling in the same direction. To understand this possible interaction, we must obtain some information about the phase relationship between S_2 and T. As a direct consequence of the Huygens-

Fresnel–Kirchhoff theory of diffraction discussed in Section 1.3, the diffracted wave is $\pi/2$ out of phase with the incident wave. Thus, the twice-diffracted wave S_2 is π out of phase with T, and the two waves interfere destructively. Consequently, in a perfect crystal we should expect the intensities of both the transmitted and the diffracted waves to decrease very rapidly as they penetrate the crystal. This phenomenon is observed and is known as *primary extinction*. The degree of primary extinction is clearly related to the thickness of the crystal and to the crystal perfection.

In x-ray diffraction, primary extinction is rarely encountered with powder specimens because the individual crystals are very small. However, primary extinction can become pronounced for single crystals of the order of 1 mm thick. For such crystals, the diffracted intensity is critically dependent on the crystal perfection. In a distorted crystal, the singly diffracted wave S_1 may not be at the exact Bragg angle at A. This wave may, therefore, pass through the crystal without having its intensity significantly reduced by interference with a transmitted wave T.

In developing the kinematical theory, we made use in Section 3.7 of the *column approximation*. In view of the preceding discussion, it is important to estimate the crystal thickness t for which the column approximation is valid. If kinematical conditions are to apply within a single column of unit cells of width x, and adjacent columns are to diffract independently of each other, then $2\theta t < x$. Thus, for $x \approx 0.5$ nm, the crystal thickness must be less than about 25 nm.

Thus, the kinematical theory is valid only for very thin crystals that are not diffracting strongly. To interpret the details of the images observed from the much thicker crystals commonly used in TEM (particularly when $s = 0$), we require a theory that includes the dynamical interaction of the many beams excited in such crystals. The dynamical theory, which is developed in the next chapter, overcomes the critical limiations of the kinematical theory and provides the basis for the interpretation of the images due to crystal defects, which are discussed in detail in Chapter 5.

4

Dynamical theory of electron diffraction

4.1 Introduction

In developing the kinematical theory of diffraction in Chapter 3, we assumed that a beam of electrons has wavelike characteristics and can be described by an equation of the form

$$\psi = \psi_0 \exp 2\pi i (\mathbf{k} \cdot \mathbf{r} - \nu t) \tag{4.1}$$

and that the intensity is obtained simply by multiplying ψ by its complex conjugate. Except for a discussion of the atomic scattering factor for x-rays and electrons in Section 3.6, we were not concerned with the nature of the radiation or with the detailed mechanism by which this radiation was diffracted by the crystal, which consisted of a three-dimensional periodic arrangement of scattering centers (atoms).

The behavior of a beam of electrons, each of total energy eE, moving in an electric field in which the potential $V(\mathbf{r})$ at any point specified by a vector $\mathbf{r}$, is given by the Schrödinger equation

$$\nabla^2 \psi(\mathbf{r}, t) - \frac{4\pi m}{ih} \frac{\partial}{\partial t} \psi(\mathbf{r}, t) + \frac{8\pi^2 me}{h^2} [E + V(\mathbf{r})] \psi(\mathbf{r}, t) = 0 \tag{4.2}$$

Because we are concerned with intensities of the various waves under steady-state conditions, we can omit the time-dependent term, so Eq. (4.2) becomes

$$\nabla^2 \psi(\mathbf{r}) + \frac{8\pi me}{h^2} [E + V(\mathbf{r})] \psi(\mathbf{r}) = 0 \tag{4.3}$$

where

$$\nabla^2 = \frac{\partial^2}{\partial x^2} + \frac{\partial^2}{\partial y^2} + \frac{\partial^2}{\partial z^2}$$

The physical significance of ψ is that the value of $\psi \psi^* \, d\tau$ is the probability of finding an electron in any element of volume $d\tau$. Thus, the intensity of the wave associated with a beam of electrons is effectively $\psi \psi^*$.

To solve the Schrödinger equation for a beam of electrons moving inside a crystal, we must obtain an expression for the potential $V(\mathbf{r})$ that suitably represents the crystal. We begin by representing the crystal as a continuous electron density function $\rho(\mathbf{r})$ expressed in electrons per unit volume and including all the electrons in the unit cell. In this representation, atoms show up as local concentrations or peaks in the continuous electron density. Since the crystal consists of a three-dimensional periodic arrangement of atoms, $\rho(\mathbf{r})$ is triply periodic and can be expressed as a triple Fourier series,

$$\rho(\mathbf{r}) = \frac{1}{V_c} \sum_g C_g \exp(2\pi i \mathbf{g} \cdot \mathbf{r}) \tag{4.4}$$

where V_c is the volume of the unit cell. $\mathbf{g}$ is a reciprocal lattice vector given by

$$\mathbf{g} = h\mathbf{a}^* + k\mathbf{b}^* + l\mathbf{c}^*$$

where $\mathbf{a}^*, \mathbf{b}^*, \mathbf{c}^*$ are the reciprocal lattice vectors defining the unit cell in reciprocal space and h, k, l are the Miller indices of the reflecting plane corresponding to the reciprocal lattice point specified by $\mathbf{g}$. The summation is taken over all $\mathbf{g}$ and signifies all possible values of h, k, l, including negative values.

By taking the Fourier transform of Eq. (4.4), we can show that

$$C_g = \sum_n f_n(\mathbf{x}) \exp(-2\pi i \mathbf{g} \cdot \mathbf{r}) \tag{4.5}$$

By comparing this equation with Eq. (3.26), we see that the coefficients C_g of the Fourier series, Eq. (4.4), turn out to be the structure factors $F_g(\mathbf{x})$ for x-ray diffraction. Thus, *the diffraction pattern is essentially a display of the components of the Fourier series that represents the crystal.*

The charge $\rho(\mathbf{r})$ and the potential $V(\mathbf{r})$ are related by the Poisson equation

$$\nabla^2 V(\mathbf{r}) = -\frac{1}{\epsilon_0} \rho(\mathbf{r}) \tag{4.6}$$

Since $\rho(\mathbf{r})$ can be expressed as a triple Fourier series, so can $V(\mathbf{r})$. Thus, we write

$$V(\mathbf{r}) = \sum_g V_g \exp(2\pi i \mathbf{g} \cdot \mathbf{r}) \tag{4.7}$$

If we substitute the expressions for $V(\mathbf{r})$ and $\rho(\mathbf{r})$ in Eq. (4.6), we find that

$$V_g = \frac{e}{4\pi\epsilon_0 |g|^2 \pi V_c} \sum_n \{Z_n - f_n(x)\} \exp(-2\pi i g \cdot r_n)$$

$$= \frac{1}{4\pi\epsilon_0} \left(\frac{e}{\pi V_c}\right) \left(\frac{\lambda}{2\sin\theta}\right)^2 \left(\frac{2h^2}{me^2}\right)$$

$$\times \sum_n \frac{me^2}{2h^2} \{Z_n - f_n(x)\} \exp(-2\pi i g \cdot r_n)$$

$$= \frac{h^2}{2\pi m e V_c} \sum_n \frac{1}{4\pi\epsilon_0} \left(\frac{me^2}{2h^2}\right) \{Z_n - f_n(x)\} \left(\frac{\lambda}{\sin\theta}\right)^2 \exp(-2\pi i g \cdot r_n)$$

$$= \frac{h^2}{2\pi m e V_c} \sum_n f_n(e) \exp(-2\pi i g \cdot r_n) \tag{4.8}$$

Hence

$$V_g = \frac{0.48 \times 10^{-18}}{V_c} F_g(e) \tag{4.9}$$

It is clear from Eq. (4.9) that the components V_g of the Fourier series that represents the crystal are essentially the structure factors. If we express $f_n(e)$ in angstroms (Å) and V_c in Å^3 then Eq. (4.9) becomes

$$V_g = \frac{48}{V_c} F_g(e) \tag{4.10}$$

As an example, we calculate the value of V_g for the 200 reflection in chromium. Chromium is body-centered cubic; thus, for reflections with $(h + k + l)$ even, $F_g = 2f$. Because $a = 2.88$ Å, $d_{200} = 1.44$ Å, and thus $(\sin\theta)/\lambda = 1/2.88 = 0.35$. From the tables* of $f_n(e)$, we find that $f = 2.06$, and hence from Eq. (4.10),

$$V_g = \frac{48}{2.88^3} \times 2 \times 2.06 = 8.28 \text{ volts}$$

4.2 Solution of the Schrödinger equation

Before attempting to solve the Schrödinger equation with $V(r)$ as given by Eq. (4.7), we consider solutions for two simple cases, namely $V(r) = 0$ and $V(r) = V_0$, a constant.

For an electron beam in vacuum, $V(r) = 0$ and

$$\psi(r) = \psi_0 \exp 2\pi i k \cdot r \tag{4.11}$$

---

* MacGillavry and Rieck (1983).

is a solution of the Schrödinger equation provided

$$k^2 = \frac{2meE}{h^2} \tag{4.12}$$

as can be verified by direct substitution of Eq. (4.11) into Eq. (4.3). Similarly, if $V(\mathbf{r}) = V_0$, then

$$\psi(\mathbf{r}) = \psi_0 \exp(2\pi i \mathbf{K} \cdot \mathbf{r}) \tag{4.13}$$

is a solution of the Schrödinger equation provided

$$K^2 = \frac{2me}{h^2}(E + V_0)$$

$$= \frac{2meE}{h^2} + U_0 \tag{4.14}$$

where $U_0 = 2meV_0/h^2$. It follows from Eq. (4.14) and Eq. (4.12) that

$$K^2 = k^2 + U_0 \tag{4.15}$$

If $\mathbf{k}$ is the wavevector of the incident wave in vacuum, then K is the magnitude of $\mathbf{k}$ after correction for the mean inner potential U_0 of the crystal. The fact that K is different from k means that the crystal has a refractive index, and it is clear that the mean refractive index n must be related to U_0. Direct measurements show that n is of the order of 1×10^{-4}.

In the dynamical theory of electron diffraction, which we develop in this chapter, we formally make all the Fourier components V_g complex quantities, that is, we replace V_g by $V_g' + iV_g''$. The full physical significance of the procedure will become clear in due course, but for the moment it will be helpful to consider the consequences of making V_0 complex. If V_0 is complex, then the mean refractive index n must also be complex, and so we write $n = n' + in''$.

A wave traveling through a medium of mean refractive index n *without diffraction* in the x direction can be written as

$$\psi = \psi_0 \exp 2\pi i (nkx - \nu t)$$

$$= \psi_0 \exp 2\pi i [(n' + in'')kx - \nu t]$$

$$= \psi_0 [\exp(-2\pi n''kx)] \exp 2\pi i (n'kx - \nu t) \tag{4.16}$$

It is clear from Eq. (4.16) that n'' is a measure of the attenuation of the wave as it passes through the medium, and the *linear absorption coefficient for intensity* μ is equal to $4\pi kn''$. We could equally well have written $nk = K$ and made K complex, in which case the imaginary part K'' would have described attenuation, and the linear absorption coefficient would have been $4\pi K''$. If there is no absorption, then $n'' = 0$ and $n = n'$, which is the normal refractive index. Thus, in making all the components V_g complex, we are introducing a term V_g'' for each diffracted wave $\mathbf{K}_g$ that is directly related to absorption of the wave.

We are now in a position to begin the development of the dynamical theory.* $V(\mathbf{r})$ is expressed as a Fourier series,

$$V(r) = \sum_g V_g \exp(2\pi i g \cdot r)$$

$$= \frac{h^2}{2me} \sum_g U_g \exp(2\pi i g \cdot r) \tag{4.17}$$

and we look for a solution of the Schrödinger equation of the form

$$\psi(\mathbf{r}) = \sum_g A_g \exp[2\pi i(\mathbf{K}_g \cdot \mathbf{r})] \tag{4.18}$$

where $\mathbf{K}_g = \mathbf{K}_0 + \mathbf{g}$, which is equivalent to the Bragg equation (see Section 3.4).

The expressions for $\psi(\mathbf{r})$ and $V(\mathbf{r})$ are now substituted into the Schrödinger equation, Eq. (4.3). This leads to an expression stating that an infinite sum of plane waves is zero. Thus, the coefficient multiplying each plane wave can be set equal to zero. This gives an infinite set of linear homogeneous equations from which the ratios $A_1/A_0, A_2/A_0, \ldots$ can, in principle, be determined.

However, we now make the simplifying assumption that as we approach the condition for Bragg scattering, only *two* of the amplitudes A_g are simultaneously large and all the other A_g are negligibly small. The large A_g correspond to $g = 0$ and $g = 1$. This is known as the *two-beam approximation*. For convenience we write these two amplitudes as A_0 and A_g. The infinite set of equations now becomes

$$(K^2 - \mathbf{K}_0 \cdot \mathbf{K}_0)A_0 + U_{-g}A_g = 0 \tag{4.19a}$$

$$U_g A_0 + (K^2 - \mathbf{K}_g \cdot \mathbf{K}_g)A_g = 0 \tag{4.19b}$$

These are called the *dispersion equations* and can be conveniently written as

$$\begin{vmatrix} (K^2 - \mathbf{K}_0 \cdot \mathbf{K}_0) & U_{-g} \\ U_g & (K^2 - \mathbf{K}_g \cdot \mathbf{K}_g) \end{vmatrix} \begin{matrix} A_0 \\ A_g \end{matrix} = 0 \tag{4.20}$$

* The following account of the dynamical theory of electron diffraction is based on the treatment of the formally similar theory of x-ray diffraction given by Batterman and Cole (1964).

which has a solution only if the determinant multiplying the amplitude coefficients vanishes. This restricts the permitted values of the wavevectors.

Note that the Fourier coefficient for $-g$ (i.e., $\overline{hkl}$) appears, even though we assumed the reflection khl only was operative. This is reasonable because the wave with wavevector $\mathbf{K_g}$ may be scattered from the backside of the (kkl) planes back into the $\mathbf{K_0}$ direction.

The first and fourth terms in the dispersion equations represent the difference between the square of the wavevectors $\mathbf{K_0}$ and $\mathbf{K_g}$ inside the crystal and the square K, which is the vacuum value k corrected for the mean inner potential. If there is no difference, there is no unique solution. Thus, the refractive index for $\mathbf{K_0}$ and $\mathbf{K_g}$ waves must be different from the average refractive index. *This is the crux of the dynamical theory.*

Now from Eq. (4.20),

$$(K^2 - \mathbf{K_0} \cdot \mathbf{K_0})(K^2 - \mathbf{K_g} \cdot \mathbf{K_g}) = U_g U_{-g} \qquad (4.21)$$

This can be written as

$$(K + K_0)(K - K_0)(K + K_g)(K - K_g) = U_g U_{-g}$$

But k, K_0, and K_g differ only by about one part in 10^5, therefore,

$$K + K_0 = K + K_g = 2K$$

Thus, Eq. (4.21) becomes

$$(K - K_0)(K - K_g) = \frac{U_g U_{-g}}{4K^2} \qquad (4.22)$$

which can be written as

$$\xi_0 \xi_g = \frac{U_g U_{-g}}{4K^2} \qquad (4.23)$$

where ξ_0 and ξ_g are defined by

$$\xi_0 = (K - K_0) \qquad (4.24a)$$

$$\xi_g = (K - K_g) \qquad (4.24b)$$

Equation (4.23) is the fundamental equation of the *dispersion surface*, which we now investigate in detail.

4.3 The dispersion surface

In the usual Ewald sphere construction shown in Figure 4.1, the center L of the sphere is determined such that the magnitude of the wavevectors to

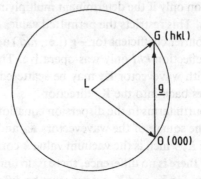

Figure 4.1. The usual Ewald sphere construction in which **LO** is the incident wavevector and **LG** the diffracted wavevector, both of magnitude $1/\lambda$. **OG** is the reciprocal lattice vector of the operating reflection and is of magnitude $1/d(hkl)$.

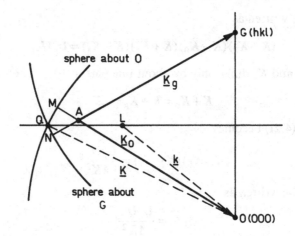

Figure 4.2. The modified Ewald sphere construction.

the origin $O(000)$ and to the reciprocal lattice point $G(hkl)$ is the vacuum value $k = 1/\lambda$.

If the construction is now performed taking into account the mean inner potential of the crystal, then the center Q of the sphere is determined such that the magnitude of the wavevectors to $O(000)$ and to $G(khl)$ is K. It can be seen from Eq. (4.15) that $K > k$, so the point Q is somewhere to the left of the Laue point L in Figure 4.1. The Ewald sphere construction with Q as center and radius K is shown in Figure 4.2. The diagram also includes the point L, but the distance between Q and L has been greatly exaggerated, as the following considerations show. From Figure 4.2, we can see that $QL \approx K - k = k(K/k - 1)$. But K/k is the mean refractive index which, as we have seen, is of the order of 1×10^{-4}; hence QL will be of the order of $10^{-4}k$.

Now the magnitudes of the wavevectors $\mathbf{K}_0$ and $\mathbf{K}_g$ are different from k and K, but the differences are not great. Therefore, we assume that the correct point A, from which $\mathbf{K}_0$ and $\mathbf{K}_g$ are to be drawn, is somewhere in the neighborhood of L and Q. Now draw spheres of radius K about the points O and G. These will, of course, intersect at Q, and in the neighborhood of Q the spheres can be represented as planes. We can see in Figure 4.2 that

$$MA = MO - AO = K - k_0 = \xi_0$$

$$NA = NG - AG = K - k_g = \xi_g$$

But ξ_0 and ξ_g are related by Eq. (4.23). Therefore, the selection of the point A must satisfy this equation. Because the spheres about O and G can be represented as planes in the neighborhood of Q, the locus of the points A are hyperbolic sheets with these planes as asymptotes. These hyperbolic sheets are called the *dispersion surfaces*. A more detailed view of the neighborhood around Q is shown in Figure 4.3. The two branches of the dispersion surface are called α and β, the one closer to the L point being α. A point on the dispersion surface is called a *tie point*. The arbitrarily selected tie points (A_1 and A_2) and the directions of their associated wavevectors are shown. Note that for the α branch, ξ_0 and ξ_g are both positive, but for the β branch, they are both negative.

For the point A_3, which lies on the line joining L and Q, we can see that $\xi_0 = \xi_g$, and thus

$$\xi_0 = \xi_g = \frac{|U_g|}{2K} \tag{4.25}$$

from the equation of the dispersion surface, Eq. (4.23). Because the angle between $\mathbf{K}_0$ and $\mathbf{K}_g$ is 2θ, it follows from Figure 4.3 and Eq. (4.25) that the diameter D of the hyperbola is given by

$$D = 2QA_3 = 2\xi_0 \sec\theta = \left(\frac{|U_g|}{K} \right) \sec\theta \tag{4.26}$$

Later we find that t_g, defined by

$$t_g = \frac{1}{D} = \frac{K\cos\theta}{|U_g|} \tag{4.27}$$

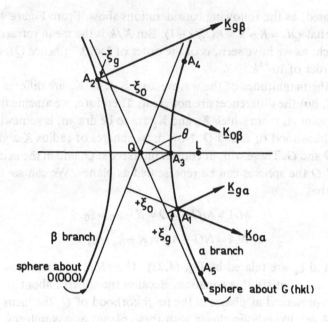

Figure 4.3. Detailed view of the dispersion surfaces
in the neighborhood of the point Q.

is an important parameter. By putting $U_g = -U_g = |U_g|$, we have assumed that the crystal has a center of symmetry.

We now show that a tie point on the dispersion surface also characterizes the ratio of the amplitudes of the waves *inside* the crystal associated with the tie point. From the first dispersion equation, Eq. (4.19a), we have

$$R_g = \frac{A_g}{A_0} = \frac{-(K^2 - K_0^2)}{U_{-g}}$$

$$= \frac{-(K + K_0)(K - K_0)}{U_{-g}}$$

$$= -\frac{2K}{U_{-g}}\xi_0 \qquad (4.28a)$$

Similarly, from the second dispersion equation, Eq. (4.19b), we have

$$R_g = -\frac{U_g}{2K}\left(\frac{1}{\xi_g}\right) \qquad (4.28b)$$

Now it can be seen from Figure 4.3 that for tie points such as A_4, ξ_0 tends to zero. Therefore, from Eq. (4.28a), A_g also tends to zero; so there is effectively only a transmitted wave. Similarly, for tie points such as A_5, ξ_g tends to zero. Therefore, from Eq. (4.28b), A_g becomes infinitely large and A_0 tends to zero. This means that no incident wave can excite this tie point.

Note that because the Fourier coefficients U_g can be complex, so can ξ_0 and ξ_g. However, only the real parts are plotted in reciprocal space.

4.4 Selection of active tie points: boundary conditions at entrance face of the crystal

In the discussion of the dispersion surface in Section 4.3, tie points on the dispersion surface were arbitrarily selected in order to show how the tie points characterize the allowed waves in the crystal. We must now examine how the tie points are selected in an actual experiment, but first we consider the boundary conditions that exist at the boundary between vacuum and the entrance face of the crystal.

Consider a wave

$$\psi_i = A \exp 2\pi i (\mathbf{k}_i \cdot \mathbf{r} - \nu t)$$

incident on a plane face of a crystal. In general, there will be a reflected wave ψ_r and a refracted (or transmitted) wave ψ_p. In transmission electron diffraction, the electron beam is usually close to normal incidence and because the effective refractive index of the crystal is very close to unity, we are justified in making the assumption that there is no reflected wave ψ_r. Therefore, the transmitted wave is

$$\psi_p = A \exp 2\pi i (\mathbf{k}_p \cdot \mathbf{r} - \nu t)$$

Now the solution of the Schrödinger equation in vacuum must fit smoothly onto the solution within the crystal, or, in other words, the wave function must be *continuous* across the boundary. This means that for *all points in the boundary*

$$\psi_i = \psi_p$$

Because we are concerned with the value of the wave function only at points within the boundary, we may replace $\mathbf{r}$ in the expressions for ψ_i and ψ_p by a vector τ in the boundary. The origin is also taken in the boundary. Therefore, for continuity of the wave function we have

$$\mathbf{k}_i \cdot \tau = \mathbf{k}_p \cdot \tau$$

Now let us write

$$\mathbf{k}_i = \mathbf{k}_p + \mathbf{n}$$

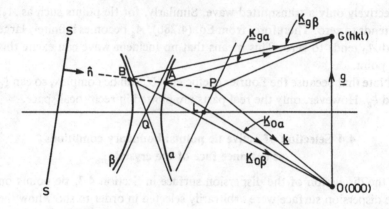

Figure 4.4. Construction in reciprocal space showing the selection of tie points *A*
and *B* on the two branches of the dispersion surface.

Then $(k_p + n) \cdot \tau$ is equivalent to $k_p \cdot \tau$. This can be so only if

$$n \cdot \tau = 0 \qquad (4.29)$$

that is, if **n** is normal to the boundary.

We see therefore that the wavevectors of the two waves can differ only
by a vector normal to the boundary. In other words, *the tangential components of the wavevectors must be equal.* There is no such restriction on
the normal components; in fact, they are not equal. In any particular experiment, the tie points are selected by the geometry of the experiment,
that is, by the angle of incidence to the crystal face and the orientation of
the diffracting planes to this face. The tie points for a given geometry can
be found by a simple construction in reciprocal space. We now describe
this construction and justify it by showing that it is consistent with the
boundary conditions just discussed.

Figure 4.4 shows the origin *O*, the *Q* point, and the two branches of
the dispersion surface. The *L* point lies on a sphere with center *O* and
radius *k*. In the neighborhood of *L*, this sphere approximates to a plane,
as shown. The wavevector **k** of the *outside* incident wave starts at some
point *P* on this sphere and ends at the origin *O*. If it started at *L*, then the
Bragg law would be satisfied exactly. Thus, *LP* is a measure of the angular deviation $\pm \Delta\theta$ from the exact Bragg angle θ. As shown in Figure 4.4,
the point *P* indicates that the angle between the incident wavevector **k**
and the reflecting planes (which are normal to **g**) is $\theta + \Delta\theta$. If *P* were on
the other side of *L*, then the grazing angle of incidence would be $\theta - \Delta\theta$.

The incident crystal surface is represented by SS, and $\hat{n}$ is a unit vector normal to this surface.

To select the tie points, draw a vector through P parallel to $\hat{n}$. It cuts the dispersion surface at two points, A on the α branch and B on the β branch; A and B are the required tie points. We see, therefore, that the *outside* incident wave gives rise to *four* waves *inside* the crystal: $\mathbf{K}_{0\alpha}$ and $\mathbf{K}_{0\beta}$ in the forward direction (transmitted waves) and two strongly diffracted waves $\mathbf{K}_{g\alpha}$ and $\mathbf{K}_{g\beta}$. This is precisely what we would expect from the splitting of the energy at the Brillouin zone boundary.

We now show that this construction is consistent with the boundary conditions at the entrance face. From Figure 4.4,

$$\mathbf{K}_{0\alpha} = \mathbf{AP} + \mathbf{k} = q\hat{n} + \mathbf{k} \qquad (4.30)$$

and there are corresponding expressions for the other inside wavevectors. But $\mathbf{AP}$ is parallel to $\hat{n}$, so the inside wavevectors differ from the outside incident wavevectors only by a vector normal to the crystal surface. That is, the tangential components of the wavevectors of the outside and inside waves are equal. We see, therefore, that our construction is consistent with the condition for the matching of the waves on either side of the crystal surface, Eq. (4.29).

Figure 4.4 has been drawn for the *Laue case,* in which the diffracted waves are directed *into* the crystal. This is the appropriate case for transmission electron diffraction in which $\hat{n}$ is usually nearly normal to $\mathbf{g}$.

For the *Bragg case,* SS is more nearly parallel to $\mathbf{k}$, and the surface normal through P cuts only the α or β branch, or passes between them. If it passes between them, total reflection occurs and there is only an exponentially attenuated wave in the crystal. It follows that the angular width of the total reflection is directly related to the diameter of the hyperbola.

4.5 The exit waves

Having found the waves that are excited in the crystal by an incident wave, we must now determine the wavevectors of the waves which pass out of the crystal at the exit surface. These wavevectors must originate at tie points on spheres (about O and G) that pass through the L point. The construction for finding the tie points of the exit waves is similar to that used for determining the tie points for the wavevectors inside the crystal, and is shown in Figure 4.5. To select the tie points, draw the normal to the exit surface through the active tie points on the dispersion surface. For simplicity, the figure shows only one tie point A on the α branch of the dispersion surface. It is clear that the exit wavevector $\mathbf{k}_{0e}$ differs from

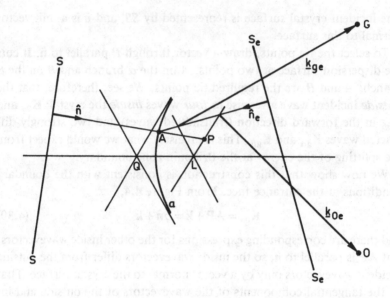

Figure 4.5. Construction for the selection of exit wavevectors k_{0e} and k_{ge}. For simplicity, only one tie point A on the α branch is shown. SS and S_eS_e refer to the entrance and exit faces of the crystal, respectively.

the inside wavevector $\mathbf{K}_{0\alpha}$ only by a vector $p\hat{\mathbf{n}}_e$ that is normal to the exit face, as required by the boundary conditions discussed in Section 4.4. The same applies to the wavevectors $\mathbf{k}_{ge}$ and $\mathbf{K}_{g\alpha}$, as well as to the wavevectors originating on the β branch. For a parallel-sided crystal, the incident wavevector $\mathbf{k}$ and the exit transmitted wavevector $\mathbf{k}_{0e}$ are parallel.

4.6 Calculation of the intensities of the transmitted and diffracted waves without absorption

The wavevectors to O from points in the neighborhood of the dispersion surface are parallel to within a few seconds of arc, as are the wavevectors to $G(hkl)$. We can, therefore, define the two unit vectors $\hat{\mathbf{s}}_0$ and $\hat{\mathbf{s}}_g$ whose directions are those of the transmitted and diffracted waves, respectively. Therefore, from Figure 4.4,

$$K_0 = PO + AP \cdot \hat{\mathbf{s}}_0$$
$$= k + q\hat{\mathbf{n}} \cdot \hat{\mathbf{s}}_0$$
$$= k + q\gamma_0 \qquad (4.31)$$

and

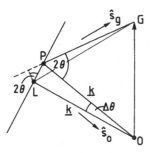

Figure 4.6. Detail of the Ewald sphere construction used in deriving Eq. (4.32).

$$K_g = PG + \mathbf{AP} \cdot \hat{\mathbf{s}}_g$$
$$= PG + q\hat{\mathbf{n}} \cdot \hat{\mathbf{s}}_g$$

With the aid of Figure 4.6 we see that $PG = LG - LP \sin 2\theta$; so

$$K_g = LG - LP \sin 2\theta + q\hat{\mathbf{n}} \cdot \hat{\mathbf{s}}_g$$
$$= k - k \Delta\theta \sin 2\theta + q\gamma_g \qquad (4.32)$$

Therefore, from Eqs. (4.31) and (4.32), and from the definitions of ξ_0 and ξ_g in Eq. (4.24),

$$\xi_0 = K - k - q\gamma_0 \qquad (4.33a)$$
$$\xi_g = K - k + k \Delta\theta \sin 2\theta - q\gamma_g \qquad (4.33b)$$

Solving these equations for q, we have

$$q = \frac{1}{\gamma_0}(K - k - \xi_0) = \frac{1}{\gamma_g}(K - k + k \Delta\theta \sin\theta - \xi_g) \qquad (4.34)$$

and hence,

$$\xi_g = -\frac{\gamma_g}{\gamma_0}(K - k - \xi_0) + K - k + k \Delta\theta \sin 2\theta$$
$$= (K - k)\left(1 - \frac{\gamma_g}{\gamma_0}\right) + k \Delta\theta \sin 2\theta + \frac{\gamma_g}{\gamma_0}\xi_0 \qquad (4.35)$$

If we assume the symmetrical Laue case, in which the incident wave is normal to the entrance face, then $\hat{\mathbf{n}}$ is normal to $\mathbf{g}$ and $\gamma_g/\gamma_0 = 1$, as can be seen with the help of Figure 4.7. Therefore, Eq. (4.35) becomes

$$\xi_g = k \Delta\theta \sin 2\theta + \xi_0 \qquad (4.36)$$

If we now substitute ξ_g from Eq. (4.36) into the equation of the dispersion surface, Eq. (4.23), we have

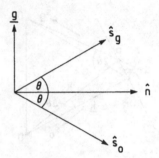

Figure 4.7. Vector diagram showing the directions
of $\hat{n}$, g, $\hat{s}_0$, and $\hat{s}_g$ for the symmetrical Laue case.

$$\xi_0^2 + (k\,\Delta\theta\sin2\theta)\xi_0 - \frac{U_g U_{-g}}{4K^2} = 0 \tag{4.37a}$$

and if we put $U_g = U_{-g} = |U_g|$, this becomes

$$\xi_0^2 + (k\,\Delta\theta\sin2\theta)\xi_0 - \frac{\cos^2\theta}{4t_g^2} = 0 \tag{4.37b}$$

where $t_g = (K\cos\theta)/|U_g|$, previously defined in Eq. (4.27). Solving Eq. (4.37b) for ξ_0, we have

$$\xi_0 = \frac{1}{2}\left\{-k\,\Delta\theta\sin2\theta \pm \left[(k\,\Delta\theta\sin2\theta)^2 + \frac{\cos^2\theta}{t_e^2}\right]^{1/2}\right\}$$

$$= \frac{1}{2}\left\{-k\,\Delta\theta\cdot2\sin\theta\cos\theta \pm \cos\theta\left[\left(\frac{k\,\Delta\theta\cdot2\sin\theta\cos\theta}{\cos\theta}\right)^2 + \frac{1}{t_g^2}\right]^{1/2}\right\}$$

$$= \frac{1}{2}\left\{-2k\,\Delta\theta\sin\theta \pm \left[(2k\,\Delta\theta\sin\theta)^2 + \frac{1}{t_g^2}\right]^{1/2}\right\} \tag{4.38}$$

by putting $\cos\theta = 1$, for small θ. If we define

$$s = 2k\,\Delta\theta\sin\theta = |g|\,\Delta\theta \tag{4.39}$$

then Eq. (4.38) becomes

$$\xi_0 = \frac{1}{2}\left\{-s \pm \left(s^2 + \frac{1}{t_g^2}\right)^{1/2}\right\}$$

$$= \frac{1}{2}\left\{-s \pm \frac{1}{t_g}(1 + s^2 t_g^2)^{1/2}\right\}$$

$$= \tfrac{1}{2}(-s \pm \sigma) \tag{4.40}$$

where σ is defined by

$$\sigma = \frac{1}{t_g}(1+s^2t_g^2)^{1/2} \tag{4.41}$$

s is proportional to the angular deviation from the exact Bragg angle. When $s=0$, $\sigma = 1/t_g$.

The two solutions of Eq. (4.40) correspond to $\xi_{0\alpha}$ and $\xi_{0\beta}$. Since $\xi_{0\alpha}$ is positive and $\xi_{0\beta}$ is negative (see Figure 4.3), we have

$$\xi_{0\alpha} = \tfrac{1}{2}(-s+\sigma) \tag{4.42a}$$

$$\xi_{0\beta} = \tfrac{1}{2}(-s-\sigma) \tag{4.42b}$$

Hence, from Eq. (4.24),

$$K_{0\alpha} = K - \xi_{0\alpha}$$
$$= K + \frac{s}{2} - \frac{\sigma}{2} \tag{4.43a}$$

and

$$K_{0\beta} = K - \xi_{0\beta}$$
$$= K + \frac{s}{2} + \frac{\sigma}{2} \tag{4.43b}$$

Thus,

$$K_{0\alpha} + K_{0\beta} = 2K + s \tag{4.44a}$$

$$K_{0\alpha} - K_{0\beta} = -\sigma \tag{4.44b}$$

In vector form these equations become

$$\mathbf{K}_{0\alpha} + \mathbf{K}_{0\beta} = (2K+s)\hat{\mathbf{s}}_0 \tag{4.45a}$$

$$\mathbf{K}_{0\alpha} - \mathbf{K}_{0\beta} = -\sigma\hat{\mathbf{n}} \tag{4.45b}$$

Now

$$\mathbf{K}_{g\alpha} = \mathbf{K}_{0\alpha} + \mathbf{g}$$
$$\mathbf{K}_{g\beta} = \mathbf{K}_{0\beta} + \mathbf{g}$$

So, with Eq. (4.45),

$$\mathbf{K}_{g\alpha} + \mathbf{k}_{g\beta} = \mathbf{K}_{0\alpha} + \mathbf{K}_{0\beta} + 2\mathbf{g}$$
$$= (2K+s)\hat{\mathbf{s}}_0 + 2\mathbf{g} \tag{4.46a}$$

and

$$\mathbf{K}_{g\alpha} - \mathbf{K}_{g\beta} = \mathbf{K}_{0\alpha} - \mathbf{K}_{0\beta} = -\sigma\hat{\mathbf{n}} \tag{4.46b}$$

We now turn to calculating the amplitudes $A_{0\alpha}$, $A_{0\beta}$, $A_{g\alpha}$, and $A_{g\beta}$ in terms of s, t_g, and σ. From Eq. (4.28a),

$$R_{g\alpha} = \frac{A_{g\alpha}}{A_{0\alpha}} = -\frac{2K}{U_{-g}}\xi_0$$

$$= -\frac{2K}{U_{-g}}\frac{(-s+\sigma)}{2}$$

$$= -t_g(-s+\sigma) \tag{4.47a}$$

from Eqs. (4.42a) and (4.27) with $\cos\theta = 1$.

Similarly, using Eqs. (4.28a), (4.42b), and (4.27), we have

$$R_{g\beta} = \frac{A_{g\beta}}{A_{0\alpha}} = t_g(s+\sigma) \tag{4.47b}$$

At the entrance face of the crystal there is an incident (or transmitted) beam with amplitude A_0, but no diffracted beam. Hence,

$$A_{0\alpha} + A_{0\beta} = 1 \tag{4.48a}$$

$$A_{g\alpha} + A_{g\beta} = 0 \tag{4.48b}$$

Using the definitions of $R_{g\alpha}$ and $R_{g\beta}$, Eq. (4.47) and Eq. (4.48) can be written as

$$A_{0\alpha}R_{g\alpha} + A_{0\beta}R_{g\beta} = 0 \tag{4.48c}$$

By substituting $A_{0\beta} = 1 - A_{0\alpha}$ into Eq. (4.48c), we obtain

$$A_{0\alpha} = \frac{R_{g\beta}}{R_{g\beta} - R_{g\alpha}}$$

which, with Eq. (4.47), gives

$$A_{0\alpha} = \frac{1}{2}\left(1 + \frac{s}{\sigma}\right) \tag{4.49a}$$

Similarly,

$$A_{0\beta} = \frac{1}{2}\left(1 - \frac{s}{\sigma}\right) \tag{4.49b}$$

Also, from the definitions of $R_{g\alpha}$ and $R_{g\beta}$, we have $A_{g\alpha} = A_{0\alpha}R_{g\alpha}$ and $A_{g\beta} = A_{0\beta}R_{g\beta}$. Hence, from Eqs. (4.49) and (4.47),

$$A_{g\alpha} = -\frac{1}{2}t_g\left(1 + \frac{s}{\sigma}\right)(-s+\sigma)$$

$$= -\tfrac{1}{2}(t_g\sigma)^{-1} \tag{4.50a}$$

$$A_{g\beta} = \frac{1}{2}t_g\left(1 - \frac{s}{\sigma}\right)(s+\sigma)$$

$$= \tfrac{1}{2}(t_g\sigma)^{-1} \tag{4.50b}$$

The total crystal wave function is

$$\Psi = [A_{0\alpha} \exp(2\pi i \mathbf{K}_{0\alpha}\cdot\mathbf{r}) + A_{0\beta} \exp(2\pi i \mathbf{K}_{0\beta}\cdot\mathbf{r})]$$
$$+ [A_{g\alpha} \exp(2\pi i \mathbf{K}_{g\alpha}\cdot\mathbf{r}) + A_{g\beta} \exp(2\pi i \mathbf{K}_{g\beta} + \mathbf{r})]$$
$$= \psi_0 + \psi_g \tag{4.51}$$

where ψ_0 is the wave diffracted in the forward direction (transmitted wave) and ψ_g is the strongly diffracted wave. Now,

$$\psi_0 = A_{0\alpha} \exp(2\pi i \mathbf{K}_{0\alpha}\cdot\mathbf{r}) + A_{0\beta} \exp(2\pi i \mathbf{K}_{0\beta}\cdot\mathbf{r})$$
$$= \exp[\pi i(\mathbf{K}_{0\alpha}+\mathbf{K}_{0\beta})\cdot\mathbf{r}]\{A_{0\alpha} \exp[\pi i(\mathbf{K}_{0\alpha}-\mathbf{K}_{0\beta})\cdot\mathbf{r}]$$
$$+ A_{0\beta} \exp[-\pi i(\mathbf{K}_{0\alpha}-\mathbf{K}_{0\beta})\cdot\mathbf{r}]\}$$
$$= \exp[\pi i(2K+s)\hat{\mathbf{s}}_0\cdot\mathbf{r}]\{A_{0\alpha} \exp[-\pi i\sigma\hat{\mathbf{n}}\cdot\mathbf{r}] + A_{0\beta} \exp[\pi i\sigma\hat{\mathbf{n}}\cdot\mathbf{r}]\} \tag{4.52}$$

where $\hat{\mathbf{s}}_0$ is the unit vector in the transmitted beam direction and $\hat{\mathbf{n}}$ the unit vector normal to the incident surface (and therefore normal to $\mathbf{g}$). In electron diffraction, because the Bragg angles are small, $\hat{\mathbf{s}}_0$ is approximately parallel to $\hat{\mathbf{n}}$ so that

$$\hat{\mathbf{n}}\cdot\mathbf{r} = \hat{\mathbf{s}}_0\cdot\mathbf{r} = z \tag{4.53}$$

where z is the depth in the crystal measured from the incident surface where $z = 0$. Therefore, with the values of the amplitudes and wavevectors just determined, Eq. (4.52) becomes

$$\psi_0 = \exp 2\pi i \mathbf{K}\cdot\mathbf{r} \exp \pi i s z \left\{ \frac{1}{2}\left(1+\frac{s}{\sigma}\right)\exp[-\pi i \sigma z] \right.$$
$$\left. + \frac{1}{2}\left(1-\frac{s}{\sigma}\right)\exp[\pi i \sigma z] \right\}$$
$$= \exp 2\pi i \mathbf{K}\cdot\mathbf{r} \exp \pi i s z \left\{ \frac{1}{2}\left(1+\frac{s}{\sigma}\right)[\cos \pi\sigma z - i \sin \pi\sigma z] \right.$$
$$\left. + \frac{1}{2}\left(1-\frac{s}{\sigma}\right)[\cos \pi\sigma z + i \sin \pi\sigma z] \right\}$$
$$= \exp 2\pi i \mathbf{K}\cdot\mathbf{r} \exp \pi i s z \left\{ \cos \pi\sigma z - \left(\frac{is}{\sigma}\right)\sin \pi\sigma z \right\} \tag{4.54}$$

The intensity is found by multiplying ψ_0 by its complex conjugate, thus,

$$I_0 = \psi_0 \psi_0^* = \cos^2 \pi\sigma z + \left(\frac{s}{\sigma}\right)^2 \sin^2 \pi\sigma z \tag{4.55}$$

Similarly for the diffracted beam,

$$\psi_g = A_{g\alpha} \exp(2\pi i K_{g\alpha} \cdot r) + A_{g\beta} \exp(2\pi i K_{g\beta} \cdot r)$$

$$= -\tfrac{1}{2}(t_g \sigma)^{-1} \exp(2\pi i K_{g\alpha} \cdot r) + \tfrac{1}{2}(t_g \sigma)^{-1} \exp(2\pi i K_{g\beta} \cdot r)$$

$$= \exp[\pi i (K_{g\alpha} + K_{g\beta}) \cdot r] \{ -\tfrac{1}{2}(t_g \sigma)^{-1} \exp[\pi i (K_{g\alpha} - K_{g\beta}) \cdot r]$$

$$+ \tfrac{1}{2}(t_g \sigma)^{-1} \exp[-\pi i (K_{g\alpha} - K_{g\beta}) \cdot r] \}$$

$$= \exp[2\pi i (K+g) \cdot r] \exp \pi i s \hat{s}_0 \cdot r \times \{ -\tfrac{1}{2}(t_g \sigma)^{-1} \exp[-\pi i \sigma \hat{n} \cdot r]$$

$$+ \tfrac{1}{2}(t_g \sigma)^{-1} \exp[\pi i \sigma \hat{n} \cdot r] \} \qquad (4.56)$$

As before, $\hat{s}_0 \cdot r = \hat{n} \cdot r = z$, and therefore Eq. (4.56) becomes

$$\psi_g = \exp[2\pi i (K+g) \cdot r] \exp \pi i s z$$

$$\times \{ -\tfrac{1}{2}(t_g \sigma)^{-1} \exp[-\pi i \sigma z] + \tfrac{1}{2}(t_g \sigma)^{-1} \exp[\pi i \sigma z] \}$$

$$= \exp 2\pi i (K+g) \cdot r \exp \pi i s z$$

$$\times \{ \tfrac{1}{2}(t_g \sigma)^{-1} [-\cos \pi \sigma z + i \sin \pi \sigma z + \cos \pi \sigma z + i \sin \pi \sigma z] \}$$

$$= \exp 2\pi i (K+g) \cdot r \exp \pi i s z \times \{ (t_g \sigma)^{-1} i \sin \pi \sigma z \} \qquad (4.57)$$

and the intensity is

$$I_g = \psi_g \psi_g^* = (t_g \sigma)^{-2} \sin^2 \pi \sigma z \qquad (4.58)$$

Summarizing we have, for no absorption,

$$I_0 = \cos^2 \pi \sigma z + \left(\frac{s}{\sigma} \right)^2 \sin^2 \pi \sigma z$$

$$I_g = (t_g \sigma)^{-2} \sin^2 \pi \sigma z$$

It will be noted that

$$I_0 + I_g = \cos^2 \pi \sigma z + \left[\left(\frac{s}{\sigma} \right)^2 + (t_g \sigma)^{-2} \right] \sin^2 \pi \sigma z$$

$$= \cos^2 \pi \sigma z + \frac{1}{\sigma^2} \left(s^2 + \frac{1}{t_g^2} \right) \sin^2 \pi \sigma z$$

$$= 1$$

so the bright and dark field images are *complementary* for *no absorption*.

4.7 Discussion of I_0 and I_g for no absorption

From Eq. (4.58), we see that for $s =$ constant, I_g is zero when $z = m/\sigma$ (where $m = 0, 1, 2, ...$) and has a maximum value when $z(m+\tfrac{1}{2})/\sigma$. We see, therefore, that I_g varies sinusoidally with z, the period of oscillation being $1/\sigma$. Also, since I_0 and I_g are complementary, I_0 also varies sinusoidally with z with the same period. Recall from Eq. (4.46b) that

$$\sigma = \mathbf{K}_{0\beta} - \mathbf{K}_{0\alpha} = \mathbf{K}_{g\beta} - \mathbf{K}_{g\alpha}$$

Therefore, the sinusoidal variation of I_0 and I_g with z can be considered as the beating of the two waves whose wavevectors originate on different branches of the dispersion surface. From Figure 4.4, we see that $\sigma = BA$, which is the distance between the two active tie points (A on the α branch and B on the β branch) that are selected by the incident wave conditions. For the symmetrical Laue case with $s = 0$ (i.e., $PL = 0$), A and B lie on the line QL; BA now has its *minimum* value and thus the period of oscillation ($1/\sigma$) has its *maximum* value. As we move away from the exact Bragg angle, BA increases, and thus the period of oscillation of I_0 and I_g with z decreases. When $s = 0$, $\sigma = 1/t_g$ and

$$I_0 = \cos^2 \frac{\pi z}{t_g} \tag{4.59a}$$

$$I_g = \sin^2 \frac{\pi z}{t_g} \tag{4.59b}$$

The period of oscillation of I_0 and I_g is now a maximum value and equal to t_g. In Figure 4.8, I_0 and I_g for $s = 0$ are plotted as a function z. Note that I_0 and I_g are π out of phase, so the electron energy passes backward and forward between the transmitted and diffracted waves as they pass down through the crystal. This is called the pendullösung effect. Figure 4.8 also shows that if the thickness of the crystal is t_1, then at the exit face, $I_0 = 1$ and $I_g = 0$, and there is no diffracted beam. This is always the case when the thickness of the crystal is a multiple of t_g. Similarly, if the crystal thickness is t_3, then $I_0 = 0$ and $I_g = 1$, and there is no transmitted beam. When $z = t_2$, then $I_0 = I_g$.

It follows that the image of a wedge-shaped crystal shows *thickness fringes* parallel to the edge. When the thickness of the crystal is a multiple of $1/\sigma$, I_0 has a maximum value and I_g a minimum. Thus, in the bright field image there is a bright fringe when the thickness equals a multiple of $1/\sigma$. The dark field image is complementary. When $s = 0$, $1/\sigma = t_g$ (the maximum value), and the fringe spacing is a maximum (see Figure 4.9). The fringe spacing decreases as s departs from zero. $1/\sigma$ is called the *extinction distance*. When the Bragg condition is exactly satisfied ($s = 0$), the extinction distance is t_g, given by Eq. (4.27),

$$t_g = \frac{K \cos \theta}{|U_g|}$$

$$= \frac{K \cos \theta}{(2me/h^2)|V_g|}$$

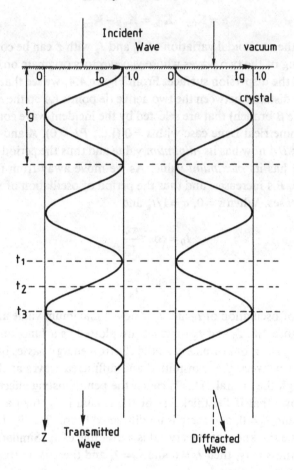

Figure 4.8. I_0 and I_g plotted as a function of z, the depth in the crystal, for $s = 0$ without absorption.

From Eq. (4.8),

$$V_g = \left(\frac{h^2}{2\pi m e V_c} \right) F_g(e)$$

and therefore

$$t_g = \frac{\pi V_c \cos \theta}{\lambda F_g(e)} \tag{4.60}$$

As an example, let us consider the 200 reflection for chromium, using 100-kV electrons, for which $\lambda = 0.037$ Å. The lattice parameter $a = 2.88$ Å, and so $d_{200} = 1.44$ Å. Hence $(\sin \theta)/\lambda = 1/2.88 = 0.35$ and the atomic

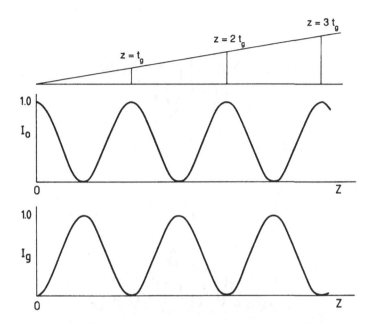

Figure 4.9. Graphs of I_0 and I_g as a function of z for a wedge-shaped crystal, for $s = 0$ without absorption.

scattering factor $f(e) = 2.06$, from the tables* of $f(e)$ as a function of $(\sin \theta)/\lambda$. Since Cr is bcc, $F_{200} = 2f(e)$. Therefore, since $\cos \theta$ is very close to unity, we have

$$t_g = \frac{\pi \times (2.88)^3 \times 1}{0.037 \times 2 \times 2.06} = 49.2 \text{ nm}$$

Consider now a crystal plate of constant thickness, which we take to be $3t_g$. Eqs. (4.55) and (4.58) then become

$$I_0 = \cos^2 3\pi (1 + s^2 t_g^2)^{1/2} + \left(\frac{s}{\sigma}\right)^2 \sin^2 3\pi (1 + s^2 t_g^2)^{1/2}$$

$$I_g = \frac{1}{1 + s^2 t_g^2} \sin^2 3\pi (1 + s^2 t_g^2)^{1/2}$$

I_0 and I_g are plotted as a function of st_g in Figure 4.10. This situation is achieved experimentally in a bent crystal, and the variation of s across the bend gives rise to a fringe pattern known as a *bend contour*. If we

* MacGillavry & Rieck (1983).

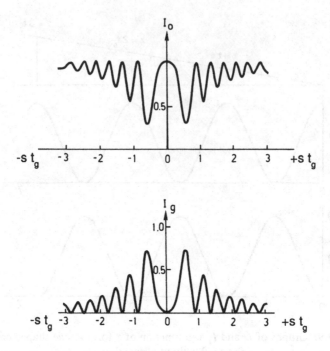

Figure 4.10. I_0 and I_g plotted as a function of st_g without absorption for a bent crystal of thickness $z = 3t_g$. Note that when z is a multiple of t_g, $I_0 = 1$ and $I_g = 0$ at $s = 0$. (From Amelinckx 1964.)

gradually tilt the crystal with respect to the incident electron beam, then the bend contour will sweep across the field of view.

The two-beam dynamical theory without absorption, unlike the kinematical theory, predicts thickness fringes at the exact Bragg angle, as observed in Figure 3.19. However, in these micrographs, we see that the fringe contrast decreases with increasing thickness z, which is not in agreement with Eq. (4.59). Furthermore, the theory predicts (see Figure 4.10) that BF and DF bend contours are complementary images, which is contrary to the observations shown in Figure 3.18. These images, particularly the thickness fringes, suggest that absorption should be included in the two-beam dynamical theory. This is carried out in the following section.

4.8 Absorption

In view of the failure of the dynamical theory without absorption to account for several important features of bend contours and thickness fringes, we now modify the theory to include absorption by making all the Fourier coefficients V_g complex (see Section 4.2). We begin by returning to Eq. (4.37a):

$$\xi_0^2 + (k\,\Delta\theta\sin 2\theta)\xi_0 - \frac{U_g U_{-g}}{4K^2} = 0$$

Solving for ξ_0, we have

$$\xi_0 = \frac{1}{2}\left\{-(k\,\delta\theta\sin 2\theta) \pm \left[(k\,\Delta\theta\sin 2\theta)^2 + \frac{U_g U_{-g}}{K^2}\right]^{1/2}\right\} \tag{4.61}$$

For the exact Bragg angle, $\Delta\theta = 0$ and

$$\xi_0 = \pm\frac{1}{2}\left(\frac{U_g U_{-g}}{K^2}\right)^{1/2} \tag{4.62}$$

Let us assume that the crystal is centrosymmetric so that $U_g = U_{-g} = |U_g|$ and Eq. (4.62) becomes

$$\xi_0 = \pm\frac{|U_g|}{2K} \tag{4.63}$$

If absorption is to be included, then we must make $|U_g|$ complex. It follows from Eq. (4.15) that K will also be complex. Thus, ξ_0 is given by

$$\xi_0 = \xi_0' + i\xi_0'' = \pm\frac{1}{2}\left[\frac{U_g' + iU_g''}{K' + iK''}\right]$$

$$= \pm\frac{1}{2}\left[K'\left(1 + \frac{iK''}{K'}\right)\right]^{-1}(U_g' + iU_g'') \tag{4.64}$$

If we assume $K''/K' = 0$, then

$$\xi_0'' = \frac{\pm U_g''}{2K'} \tag{4.65}$$

ξ_0'' is positive for the α branch and negative for the β branch of the dispersion surface.

Now it can be seen from Eq. (4.61) that at $\Delta\theta = 0$, ξ_0 is proportional to the square root of the complex constant $(U_g U_{-g}/K^2)$. As $\Delta\theta$ increases from zero (either positively or negatively), a rapidly increasing real part is added to the complex constant before taking its square root. Thus, ξ_0'' (the imaginary part of ξ_0) becomes increasingly less important, until it is essentially zero well away from the exact Bragg angle. The variation of ξ_0'' with $\Delta\theta$ is sketched in Figure 4.11.

By definition, Eq. (4.24), $\xi_0 = K - K_0$, so

$$\xi_0' + i\xi_0'' = K' + iK'' - (K_0' + iK_0'') = (K' - K_0') + i(K'' - K_0'') \tag{4.66}$$

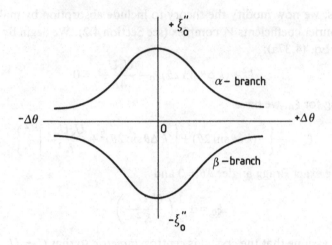

Figure 4.11. Diagram showing the form of the variation of ξ_0'' with $\Delta\theta$, the deviation from the exact Bragg angle.

and by comparing imaginary parts, we obtain

$$K_0'' = K'' - \xi_0'' \tag{4.67}$$

K'' represents the attenuation due to the imaginary part of the mean inner potential. As there is always attenuation and not amplification, $K'' > \xi_0''$. The attenuation of the wave with wavevector $\mathbf{K}_0$ is given by K_0''.

For the α branch ξ_0'' is positive, and for the β branch ξ_0'' is negative, therefore

$$K_{0\alpha}'' = K'' - \xi_0'' \tag{4.68a}$$

$$K_{0\beta}'' = K'' - (-\xi_0'') \tag{4.68b}$$

Hence $K_{0\beta}'' > K_{0\alpha}''$ in the neighborhood of $\Delta\theta = 0$.

We now look at this semiquantitative result in more detail, starting with Eq. (4.15):

$$K^2 = k^2 + U_0$$

If attenuation is included,

$$(K' + iK'')^2 = k^2 + U_0' + iU_0'' \tag{4.69}$$

and by comparing imaginary parts, we have

$$K'' = \frac{U_0''}{2K'} \tag{4.70}$$

Hence, from Eqs. (4.68a) and (4.65),

$$K''_{0\alpha} = \frac{1}{2K'}(U''_0 - U''_g) \tag{4.71a}$$

We expect U''_0 and U''_g to be of comparable magnitude, and therefore $K''_{0\alpha} \approx 0$, that is, there is no attenuation for waves with wavevectors originating on the α branch. On the other hand, for wavevectors originating on the β branch, we have from Eqs. (4.68b) and (4.65),

$$K''_{0\beta} = \frac{1}{2K'}(U''_0 + U''_g) \tag{4.71b}$$

which gives a positive attenuation.

We see, therefore, that the waves whose wavevectors originate on the α branch of the dispersion surface suffer little or no attenuation, whereas the waves whose wavevectors originate on the β branch are attenuated. Well away from the exact Bragg angle, both sets of waves suffer equal attenuation. Note that this analysis provides no information on the mechanism of the attenuation. However, we can obtain some insight into this mechanism by calculating the amplitudes of the wave functions associated with waves whose wavevectors originate on the α and β branches of the dispersion surface.

To simplify the discussion, consider the wave functions ψ_α and ψ_β for a nonabsorbing crystal of the simplest structure set at the exact Bragg angle. The total wave function Ψ is given by Eq. (4.51), and we define the wave functions ψ_α and ψ_β of the waves associated with the α and β branches, respectively, by

$$\psi_\alpha = A_{0\alpha}\exp(2\pi i \mathbf{K}_{0\alpha}\cdot\mathbf{r}) + A_{g\alpha}\exp(2\pi i \mathbf{K}_{g\alpha}\cdot\mathbf{r}) \tag{4.72a}$$

$$\psi_\beta = A_{0\beta}\exp(2\pi i \mathbf{K}_{0\alpha}\cdot\mathbf{r}) + A_{g\beta}\exp(2\pi i \mathbf{K}_{g\beta}\cdot\mathbf{r}) \tag{4.72b}$$

At $s = 0$, we see from Eqs. (4.49) and (4.50) that $A_{0\alpha} = A_{0\beta} = A_{g\beta} = \frac{1}{2}$ and $A_{g\alpha} = -\frac{1}{2}$. Recalling that $\mathbf{K}_g = \mathbf{K}_0 + \mathbf{g}$, we have

$$\psi_\alpha = \tfrac{1}{2}\exp(2\pi i \mathbf{K}_{0\alpha}\cdot\mathbf{r}) - \tfrac{1}{2}\exp[2\pi i(\mathbf{K}_{0\alpha}+\mathbf{g})\cdot\mathbf{r}]$$
$$= \tfrac{1}{2}\exp(2\pi i \mathbf{K}_{0\alpha}\cdot\mathbf{r})\{1 - \exp 2\pi i \mathbf{g}\cdot\mathbf{r}\}$$

Therefore,

$$\psi_\alpha\psi_\alpha^* = \tfrac{1}{4}(1 - \exp 2\pi i \mathbf{g}\cdot\mathbf{r})[1 - \exp(-2\pi i \mathbf{g}\cdot\mathbf{r})]$$
$$= \tfrac{1}{2}(1 - \cos 2\pi \mathbf{g}\cdot\mathbf{r})$$
$$= \sin^2 \pi \mathbf{g}\cdot\mathbf{r} \tag{4.73a}$$

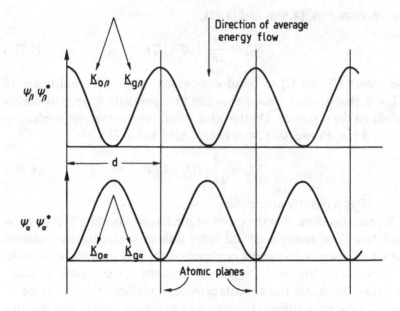

Figure 4.12. Diagram showing the standing waves associated with the wavevectors originating on the α and β branches of the dispersion surface.

Similarly,

$$\psi_\beta = \tfrac{1}{2}\exp(2\pi i \mathbf{K}_{0\beta}\cdot\mathbf{r}) + \tfrac{1}{2}\exp[2\pi i(\mathbf{K}_{0\beta}+\mathbf{g})\cdot\mathbf{r}]$$
$$= \tfrac{1}{2}\exp(2\pi i \mathbf{K}_{0\beta}\cdot\mathbf{r})\{1 + \exp 2\pi i \mathbf{g}\cdot\mathbf{r}\}$$

and, therefore,

$$\psi_\beta\psi_\beta^* = \tfrac{1}{4}(1 + \exp 2\pi i \mathbf{g}\cdot\mathbf{r})[1 + \exp(-2\pi i \mathbf{g}\cdot\mathbf{r})]$$
$$= \cos^2 \pi\mathbf{g}\cdot\mathbf{r} \qquad\qquad (4.73\text{b})$$

In a transmission electron diffraction experiment (which is essentially a symmetrical Laue case), the electron flow (current) is on the average parallel to the atomic (reflecting) planes. However, it can be seen from the expressions for $\psi_\alpha\psi_\alpha^*$ and $\psi_\beta\psi_\beta^*$ which have just been derived that the probability density (intensity) of each of these wave fields is modulated normal to the reflecting planes. This is illustrated in Figure 4.12. Note that when $r=0$ or nd, where d is the spacing between the reflecting planes and n is an integer, $\psi_\alpha\psi_\alpha^* = 0$ and $\psi_\beta\psi_\beta^* = 1$. That is, the antinodes of the standing wave $\psi_\alpha\psi_\alpha^*$ are midway between the atomic planes, whereas the antinodes of the standing wave $\psi_\beta\psi_\beta^*$ are coincident with the atomic planes.

We might expect, therefore, that the waves whose wavevectors originate on the β branch would be absorbed due to the production of x-rays, whereas the waves whose wavevectors originate on the α branch would pass through the crystal unattenuated when the crystal is set at the exact Bragg angle.

This analysis is formally similar to the generally accepted explanation of the Bormann effect, which is the enhanced transmission of x-rays through an absorbing crystal when it is oriented at the exact Bragg angle (Batterman and Cole, 1964). It is clear from Figure 4.12 that, for electrons, the corresponding "anomolous transmission" of the transmitted beam is associated with the waves whose wavevectors originate on the α branch of the dispersion surface.

We now return to the consideration of absorbing crystals. From the definition of t_g, Eq. (4.27), we have

$$\frac{1}{t_g} = \frac{|U_g|}{K\cos\theta}$$

It follows that if $|U_g|$ is complex, then so is $1/t_g$. Therefore,

$$\frac{1}{t_g} = \frac{1}{t_g'} + \frac{i}{t_g''} \tag{4.74}$$

t_g'' is called the *absorption length*.

As defined by Eq. (4.41),

$$\sigma = \frac{1}{t_g}\sqrt{1 + (st_g)^2}$$

and if $1/t_g$ is complex, then so is σ. Therefore, we write

$$\sigma = \sigma' + i\sigma'' \tag{4.75}$$

and from the definition of σ, together with Eqs. (4.74) and (4.75), we have

$$\sigma^2 = \sigma'^2 + 2i\sigma'\sigma'' - \sigma''^2 = \left(\frac{1}{t_g'} + \frac{i}{t_g''}\right)^2 + s^2$$

Comparing imaginary parts gives

$$\frac{1}{\sigma''} = \sigma' t_g' t_g'' \tag{4.76}$$

Comparing real parts gives

$$\sigma'^2 - \sigma''^2 = \left(\frac{1}{t_g'}\right)^2 - \left(\frac{1}{t_g''}\right)^2 + s^2$$

and, with Eq. (4.76), this becomes

$$\sigma'^2 - \left(\frac{1}{\sigma' t'_g t''_g}\right)^2 = \left(\frac{1}{t'_g}\right)^2 - \left(\frac{1}{t''_g}\right)^2 + s^2$$

Because we expect $1/t''_g$ to be small compared with $1/t'_g$, we can put $(1/t''_g)^2 = 0$ so that

$$\sigma'^2 \approx \left(\frac{1}{t'_g}\right)^2 + s^2 = \left(\frac{1}{t'_g}\right)^2 [1 + (s t'_g)^2]$$

Thus, σ' is approximately equal to the value of σ for a nonabsorbing crystal, as can be seen by comparing this equation with the definition of σ, Eq. (4.41).

The selective attenuation of waves originating on different branches of the dispersion surface can thus be incorporated into the expressions for I_0 and I_g, Eqs. (4.55) and (4.58), by making $1/t_g$ and σ complex. However, a uniform absorption associated with the mean inner potential U_0 (discussed in Section 4.2) must also be included. This absorption is described by K'' which, from Eq. (4.70), is given by

$$K'' = \frac{U''_0}{2K'}$$

Hence, the linear absorption coefficient μ for intensity is given by

$$\mu = 2\pi K'' = \frac{2\pi U''_0}{K'} \tag{4.77}$$

As in Eq. (4.74), we introduce an absorption length t''_0 so that

$$\mu = \frac{2\pi}{t''_0} \tag{4.78}$$

The final expressions for $I_0(a)$ and $I_g(a)$ become

$$I_0(a) = \exp\left(-\frac{2\pi z}{t''_0}\right)\left\{\left[\cosh u + \left(\frac{s}{\sigma'}\right)\sinh u\right]^2 \right.$$

$$\left. - (\sigma' t'_g)^{-2} \sin^2 \pi\sigma' z\right\} \tag{4.79a}$$

$$I_g(a) = \exp\left(-\frac{2\sigma z}{t''_0}\right)\left\{(\sigma' t'_g)^{-2}[\sinh^2 u + \sin^2 \pi\sigma' z]\right\} \tag{4.79b}$$

where $u = \pi\sigma'' z$.

When there is no absorption, $t''_0 \to \infty$, $u = 0$, $t''_g \to \infty$, $\cosh u = 1$, $\sinh u = 0$, $t'_g = t_g$, $\sigma' = \sigma$, and therefore Eqs. (4.79) become

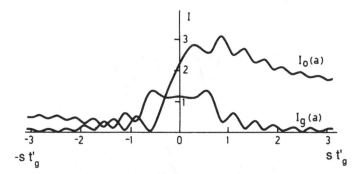

Figure 4.13. Calculated curves showing $I_0(a)$ and $I_g(a)$ as a function of st'_g. These curves correspond to the intensity profiles across a bend contour observed in BF and DF, respectively. Note the asymmetrical nature of $I_0(a)$ and the symmetrical nature of $I_g(a)$ about $s = 0$. The crystal thickness was taken as $3t'_g$. $t'_g/t''_g = 0.1$ and $t''_0 = t''_g$. (From Amelinckx 1964.) Compare with Figure 3.18.

$$I_0 = \cos^2 \pi \sigma z + \left(\frac{s}{\sigma}\right)^2 \sin^2 \pi \sigma z$$

$$I_g = (\sigma t_g)^{-2} \sin^2 \pi \sigma z$$

which are, as they should be, identical to expressions calculated in Section 4.6.

4.9 Discussion of $I_0(a)$ and $I_g(a)$

The first point to be noted about the expressions for $I_0(a)$ and $I_g(a)$ calculated in the previous section is that $I_0(a)$ is asymmetrical with respect to s, whereas $I_g(a)$ is symmetrical; that is

$$I_0(a)(+s) \neq I_0(a)(-s)$$

but

$$I_g(a)(+s) = I_g(a)(-s)$$

This is because in $I_g(a)$, s always appears as s^2, but in $I_0(a)$, s appears as s in the product term $2(s/\sigma')(\cosh u \sinh u)$. Thus, we expect the contrast at a bend contour to be symmetrical about $s = 0$ in a DF image and asymmetrical in a BF image. Calculated profiles across a bend contour in a crystal of thickness $3t'_g$ and with $t'_g/t''_g = 0.1$ are shown in Figure 4.13.

If we compare the curves of Figure 4.13 with the images of a bend contour shown in Figure 3.18, we can see that the dynamical theory with absorption predicts in considerable detail the contrast actually observed in

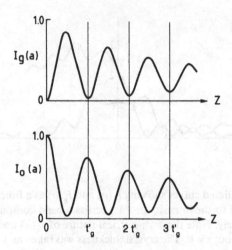

Figure 4.14. Calculated curves showing $I_0(a)$ and $I_g(a)$ as a function of thickness z for $s = 0$. These curves correspond to the intensity of thickness fringes in a wedge-shaped crystal in BF and DF, respectively. Note the decrease in the fringe contrast with increasing thickness. $t'_g/t''_g = 0.05$ and $t''_0 = t''_g$. (From Hirsch et al. 1965.) Compare with Figure 3.19.

both BF and DF images, in particular the symmetrical DF image and the asymmetrical BF image. For thicker crystals, the oscillations in both the BF and DF images tend to fade out; and when $z \approx 10t'_g$, they are no longer observed.

When $s = 0$, $I_0(a)$ and $I_g(a)$ become

$$I_0(a) = \exp\left(-\frac{2\pi z}{t''_0}\right)\left\{\cosh^2 t''_g z - \sin^2 \frac{\pi z}{t'_g}\right\} \tag{4.80a}$$

$$I_g(a) = \exp\left(-\frac{2\pi z}{t''_0}\right)\left\{\sinh^2 t''_g z + \sin \frac{\pi z}{t'_g}\right\} \tag{4.80b}$$

and theoretical profiles are plotted in Figure 4.14. Thus, for a wedge-shaped crystal set at $s = 0$, the thickness fringes tend to fade out in the thicker regions of the specimen (as observed in Figure 3.19); and this fading becomes more pronounced as the absorption increases, that is, for higher values of t'_g/t''_g.

Finally, we examine the physical reasons for the symmetry about $s = 0$ for bend contours in DF and for the asymmetry about $s = 0$ in BF, as illustrated in Figure 4.13. Consider the four waves of amplitudes $A_{0\alpha}$, $A_{0\beta}$, $A_{g\alpha}$, and $A_{g\beta}$ excited in the crystal by an incident beam. The values of these amplitudes are

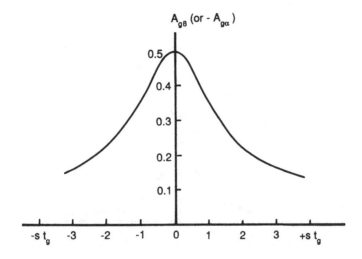

Figure 4.15. Amplitudes $A_{g\beta}$ and $(-A_{g\alpha})$ plotted as functions of st_g from Eq. (4.50a,b).

$$A_{0\alpha} = \frac{1}{2}\left(1 + \frac{s}{\sigma}\right) \qquad (4.49a)$$

$$A_{0\beta} = \frac{1}{2}\left(1 - \frac{s}{\sigma}\right) \qquad (4.49b)$$

$$A_{g\alpha} = -\tfrac{1}{2}(t_g\sigma)^{-1} \qquad (4.50a)$$

$$A_{g\beta} = \tfrac{1}{2}(t_g\sigma)^{-1} \qquad (4.50b)$$

where t_g is the extinction distance and

$$\sigma = \frac{1}{t_g}(1 + s^2 t_g^2)^{1/2} \qquad (4.41)$$

It is clear from Eq. (4.50) and Figure 4.15 that the amplitudes $A_{g\alpha}$ and $A_{g\beta}$ are both symmetrical about $s = 0$. Recall that in Section 4.8 it was shown that the waves with wavevectors originating on the β branch of the dispersion surface are preferentially absorbed, whereas the waves whose wavevectors originate on the σ branch suffer little or no absorption. However, this effect does not give rise to any asymmetry about $s = 0$ of the intensity of the diffracted beam, which is shown in Figure 4.13.

The situation is different for the forward diffracted (or transmitted) beam. In Figure 4.16, the amplitudes $A_{0\alpha}$ and $A_{0\beta}$ are plotted against st_g. Note that the wave of amplitude $A_{0\alpha}$ is mainly excited when $s > 0$, whereas the wave of amplitude $A_{0\beta}$ is mainly excited when $s < 0$. Therefore,

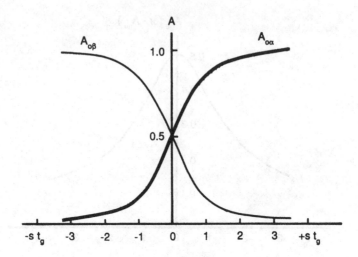

Figure 4.16. Amplitudes $A_{0\alpha}$ and $A_{0\beta}$ plotted as a function of st_g from Eq. (4.49a, b).

because the wave whose wavevector originates on the β branch is preferentially absorbed, the intensity of the transmitted beam (which is calculated by adding the two waves from the α and β branches) is asymmetric about $s = 0$, as shown in Figure 4.13.

4.10 The Darwin–Howie–Whelan equations

An alternative treatment of the dynamical diffraction of electrons is based on the earlier dynamical theory of x-ray diffraction given by Darwin. This alternative method, often described as the wave-optical approach, does not involve solving the Schrödinger equation and is not completely rigorous. However, because it is based on purely physical arguments, it is useful in giving a physical insight into dynamical diffraction. For the two-beam case, this approach leads to a pair of coupled linear first-order differential equations for the amplitudes of the transmitted and strongly diffracted waves. These equations, which were first derived by Howie and Whelan (see Hirsch et al. 1965), are particularly useful for discussing the diffraction effects due to crystal imperfections; they are justified by showing that they are consistent with the wave mechanical approach. The alternative treatment that follows is based on that given by Amelinckx (1964).

Consider a *thin* slab of crystal dz on a crystal of thickness z. A beam is incident at the Bragg angle θ, as shown in Figure 4.17. The incident

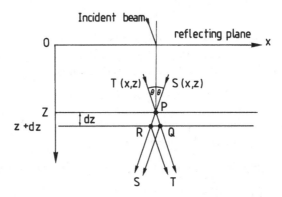

Figure 4.17. Diagram showing the waves to be considered in the
wave-optical dynamical theory. (After Amelinckx 1964.)

wavevector **k** and the normal to the diffracting planes **g** are in the plane
of the diagram. We are thus considering the symmetrical Laue case. We
assume that the incident wave has the same amplitude over the part of
the entrance face considered. We want to calculate the amplitudes of the
waves passing out of the slab dz in the transmitted beam direction and in
the scattered beam direction. These amplitudes are designated T and S,
respectively.

At P, the amplitude of the wave passing *into* the slab dz in the direction of the transmitted beam is

$$T(x, z)$$

At Q, the amplitude of the wave passing *out of* the slab dz in the direction of the transmitted beam is, therefore,

$$T(x, z) \times T(dz)$$

At P, the amplitude of the wave passing *into* the slab dz in the direction
of the scattered (or diffracted) beam is

$$S(x, z)$$

At Q, the amplitude of the wave passing *out of* the slab dz in the direction of the scattered beam is

$$T(x, z) \times S(dz)$$

At R, the amplitude of the wave passing *out of* the slab in the direction
of the transmitted beam is

$$S(x,z) \times S^{(-)}(dz)$$

The (−) sign indicates that this wave is scattered in the reverse sense to the scattered wave $S(x,z)$. At R the amplitude of the wave passing *out of* the slab dz in the direction of the scattered beam is

$$S(x,z) \times T(dz)$$

The amplitude of the wave passing *out of* the slab dz in the direction of the transmitted beam can be written as

$$T(x+dx, z+dz) \quad \text{or} \quad T(x+dz\tan\theta, z+dz) \quad \text{or} \quad T(x+\theta dz, z+dz)$$

since $\tan\theta = \theta$ for small θ.

$T(x+\theta dz, z+dz)$ is equal to the sum of the amplitudes of the waves passing out of the crystal at Q and R in the transmitted beam direction. Therefore,

$$T(x+\theta dz, z+dz) = T(x,z) \times T(dz) + S(x,z) \times S^{(-)}(dz) \quad (4.81)$$

If we assume that the slab dz has a thickness of atomic dimensions, we can set $T(dz) = 1$. Similarly, we can set $S(dz)$ proportional to dz. The constant of proportionality must have dimensions of $(\text{length})^{-1}$, so we put it equal to π/t_g. The π cannot be justified at this stage.

Now the scattered wave will be $\pi/2$ out of phase with the incident wave (see Section 1.3). We can incorporate this phase shift into the expression for $S(dz)$ by the factor i since $\exp(i\pi/2) = \cos(\pi/2) + i\sin(\pi/2) = i$. Therefore, we can write

$$S(dz) = \frac{i\pi}{t_g}(\exp -2\pi isz)\, dz \qquad (4.82a)$$

and

$$S^{(-)}dz = \frac{i\pi}{t_g}(\exp 2\pi isz)\, dz \qquad (4.82b)$$

The origin of the phase factors $\exp(-2\pi isz)$ and $\exp(2\pi isz)$ can be seen in Eq. (3.48). Thus, Eq. (4.81) becomes

$$T(x+\theta dz, z-dz) = T(x,z) + S(x,z)\frac{i\pi}{t_g}(\exp 2\pi isz)\, dz \qquad (4.83)$$

The left-hand side can be expanded by a Taylor expansion:

$$f(x+h) = f(x) + hf'(x) + \cdots$$

If we include only first differential terms, then Eq. (4.83) becomes

$$T(x,z)+\theta\,dz\,\frac{\partial T}{\partial x}+\frac{\partial T}{\partial z}\,dz = T(x,z)+\frac{i\pi}{t_g}S(x,z)(\exp 2\pi isz)\,dz$$

which reduces to

$$\frac{\partial T}{\partial z}+\frac{\theta}{}\frac{\partial T}{\partial x}=\frac{i\pi}{t_g}S(\exp 2\pi isz) \tag{4.84}$$

Since the Bragg angles are small, we can assume that both the transmitted and the scattered waves are propagating in a narrow column of crystal. Thus, we can consider T and S to be functions of z only; and for the *column approximation*,

$$\frac{dT}{dz}=\frac{i\pi}{t_g}S(\exp 2\pi isz) \tag{4.85}$$

Similarly, the amplitude of the wave passing *out of* the slab dz in the direction of the scattered beam can be written as

$$S(x-dx,z+dz)\quad\text{or}\quad S(x-dz\tan\theta,z+dz)\quad\text{or}\quad S(x-\theta\,dz,z+dz)$$

for small θ. This amplitude is equal to the sum of the amplitudes of the waves passing out of the crystal at Q and R in the scattered beam direction. Therefore,

$$S(x-\theta\,dz,z+dz)=T(x,z)\times S(dz)+S(x,z)\times T(dz) \tag{4.86}$$

Proceeding in the same manner as before, we find for the column approximation that

$$\frac{dS}{dz}=\frac{i\pi}{t_g}T\exp(-2\pi isz) \tag{4.87}$$

There are a variety of coupled differential equations equivalent to Eqs. (4.85) and (4.87) in which the amplitudes T' and S' are related to T and S by phase factors that are functions of z. As an example, we let

$$T=T'\exp(\pi isz) \tag{4.88a}$$

$$S=S'\exp(-\pi isz) \tag{4.88b}$$

Differentiating Eq. (4.88a), we obtain

$$\frac{dT}{dz}=T'(\pi is)\exp(\pi isz)+\frac{dT'}{dz}\exp(\pi isz)$$

and, with Eq. (4.88b), Eq. (4.85) now becomes

$$\frac{dT'}{dz}+\pi sT'=\frac{i\pi}{t_g}S' \tag{4.89}$$

Similarly,

$$\frac{dS'}{dz} - \pi i s S' = \frac{i\pi}{t_g} T' \qquad (4.90)$$

We can eliminate T' and S' in turn from Eqs. (4.89) and (4.90) in the following way. To eliminate S', first differentiate Eq. (4.89):

$$\frac{d^2T'}{dz^2} + i\pi s \frac{dT'}{dz} = \frac{i\pi}{t_g} \frac{dS'}{dz}$$

Then we substitute for dT'/dz from Eq. (4.89) and for dS'/dz from Eq. (4.90) and obtain

$$\frac{d^2T'}{dz^2} + i\pi s \left(\frac{i\pi S'}{t_g} - i\pi s T' \right) = \frac{i\pi}{t_g} \left(\frac{i\pi T'}{t_g} + i\pi s S' \right)$$

which becomes

$$\frac{d^2T'}{dz^2} + (\sigma\pi)^2 T' = 0 \qquad (4.91)$$

Similarly, by eliminating T' we obtain

$$\frac{d^2S'}{dz^2} + (\sigma\pi)^2 S' = 0 \qquad (4.92)$$

The wave-optical treatment can be justified by showing that the expressions ψ_0 and ψ_g obtained as solutions of the Schrödinger equation are also solutions of the two equations just derived.

Recall that

$$\psi_0 = \exp 2\pi i \mathbf{K} \cdot \mathbf{r} \exp \pi i s z \left\{ \cos \pi \sigma z - \left(\frac{is}{\sigma} \right) \sin \pi \sigma z \right\} \qquad (4.54)$$

$$\psi_g = \exp 2\pi i (\mathbf{K} + \mathbf{g}) \cdot \mathbf{r} \exp \pi i s z \{ i(t_g \sigma)^{-1} \sin \pi \sigma z \} \qquad (4.57)$$

Because the exponential terms in front of the curly brackets are simply phase factors, and we are ultimately interested in intensities, these equations become

$$\psi_0 = \cos \pi \sigma z - \left(\frac{is}{\sigma} \right) \sin \pi \sigma z$$

$$\psi_g = i(t_g \sigma)^{-1} \sin \pi \sigma z$$

It is easily shown by direct substitution that these expressions for ψ_0 and ψ_g are solutions of Eqs. (4.91) and (4.92), respectively. Hence, the various pairs of coupled differential equations obtained from the wave optical-approach are justified. Absorption can be included, as before, by making $1/t_0$ and $1/t_g$ complex.

5

The observation of crystal defects

In this chapter we discuss the origin and nature of the contrast arising from the main types of crystal defect. We also show how, in principle, the parameters characterizing these defects (such as the Burgers vector of a dislocation) can be determined from the images observed under different diffracting conditions. Some types of distortion of a perfect crystal that are frequently encountered in important rock-forming minerals (such as spinodal decomposition and exsolution) will not be discussed specifically here because we can easily understand their contrast in terms of the ideas to be developed for the main types of defect. This chapter is concerned with the principles of the observation of crystal defects; specific examples will be given in Chapters 8 and 9.

5.1 General ideas

We saw in Chapters 3 and 4 that contrast is observed in a perfect crystal if it is bent, giving rise to *bend contours,* or if it varies in thickness, giving rise to *thickness fringes.* The ways in which contrast from crystal defects arise are closely related to the origins of bend contours and thickness fringes. If the crystal planes in the immediate neighborhood of a crystal defect are bent, then the diffracted intensity from this distorted region will be different from that of the surrounding perfect crystal, and contrast will arise analogously to that from a bend contour. In addition, because the diffracted intensity varies with depth in the crystal, the contrast from the defect depends on the position of the defect below the top surface of the specimen. Further, the contrast from a defect for which there is no local bending of the crystal planes is due to the way in which the defect modifies the normal variation of diffracted intensity with depth in a perfect crystal.

This section considers in simple terms the essential ideas of the origin of the contrast from the basic types of crystal defect in order to provide a foundation for a later, more detailed discussion.

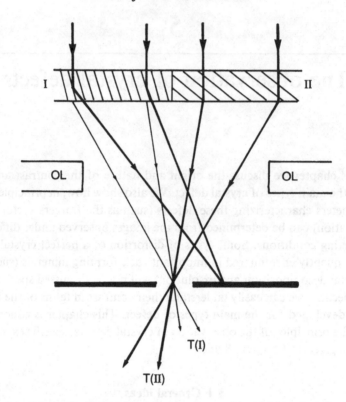

Figure 5.1. Schematic diagram showing the manner in which the aperture in the back focal plane of the objective lens gives rise to contrast in a bright field (BF) image of a specimen consisting of two differently oriented grains. Grain I is diffracting strongly (and the diffracted beam does not pass through the aperture), whereas Grain II is not. Thus, the intensity of the direct beam from I is much lower than the intensity of the direct beam from II; hence, in the final BF image, II is bright and I is dark.

Let us begin by considering the thin specimen shown schematically in Figure 5.1. It consists of two parts, both perfect crystals of the same material and of uniform thickness but different orientation, joined along a planar boundary that is normal to the plane of the specimen and parallel to the incident electron beam. Also shown are the objective lens and the aperture in the back focal plane. As shown, this aperture is positioned so that it allows only the transmitted beam to pass through. Thus, a bright field (BF) image is formed. Now suppose that the specimen is oriented with respect to the incident beam so that part I is diffracting strongly, but part II is not. Thus, the transmitted beam from part II is strong compared with the transmitted beam from part I; and in the BF image, part II is bright and part I relatively dark. If we tilt the incident beam so that the strongly diffracted beam from part I passes through the objective lens aperture, then in the DF image, part I will be bright and part II dark. The contrast observed in both the BF and DF images arises because the two parts of the specimen are diffracting differently *and* because there is an aperture in the back focal plane of the objective lens.

If parts I and II are twin-related and the operating diffraction vector **g** is normal to the twin plane (which is common to both parts), then the two parts will diffract with equal intensity and no contrast between I and II will be observed. For this reflection, the twin is said to be *out-of-contrast*.

If the specimen is thin enough to produce negligible absorbtion of the incident beam and if part II is not diffracting, then we can neglect part II; it will be uniformly bright (completely transparent) in BF and uniformly black in DF since no diffracted intensity is derived from it. Thus, we need to consider only part I of the specimen. Suppose the boundary between parts I and II is inclined to the plane of the specimen, so that in the region of overlap each part is wedge-shaped, as shown in Figure 5.2. Because only part I is diffracting, the region of overlap gives rise to normal thickness fringes whose spacing depends on the operating reflection **g** and the deviation **s** from the exact Bragg condition. From Figure 5.2, it is clear that the image of the boundary consists of a series of alternating bright and dark fringes running parallel to the intersection of the boundary plane and the specimen surface. If both parts I and II are diffracting (either with the same **g** and different values of **s**, or with different **g** and perhaps with different **s**), then the image of the boundary will again be a fringe pattern; but it will not be a simple superposition of the thickness fringes from the two overlapping parts. We discuss these fringe patterns later in more detail, including those produced by a stacking fault in which the two parts are simply displaced with respect to each other by a nonlattice vector in the plane of the fault. The main point to be appreciated at this stage is that an inclined boundary (such as a grain boundary, a twin boundary, an antiphase boundary, a stacking fault, or an interphase boundary between two different minerals) gives rise to a type of thickness fringe pattern, the details of which depend on the nature of the boundary and the diffracting conditions. Overlapping crystals can also give rise to moiré fringes, which will be discussed in Chapter 6.

In general, the crystal planes in the neighborhood of a dislocation are bent, as can be seen from the conventional drawing of an edge dislocation shown in Figure 5.3. This distortion can, for present purposes, be

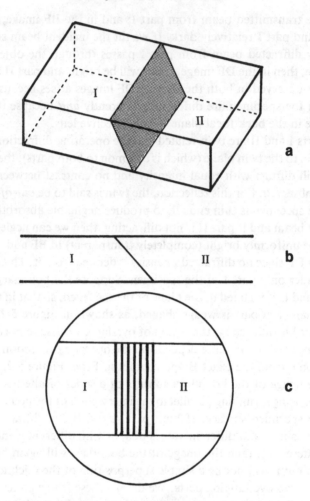

Figure 5.2. Schematic diagrams illustrating the manner in which fringes are produced by an inclined boundary between two differently oriented parts I and II of a specimen. (a) Perspective diagram of the specimen and boundary. (b) The specimen viewed edge-on. (c) Nature of the final image.

described approximately as a rotation. Consider the planes in Figure 5.3 that are perpendicular to the page. On the right-hand side of the dislocation these planes are rotated clockwise, whereas on the left-hand side they are rotated in the opposite sense. Now suppose that far away from the dislocation in the perfect crystal, the incident electron beam is at an angle that is slightly larger than the exact Bragg angle (i.e., $\theta_1 = \theta_B + \Delta\theta$, as shown). From the diagram we can see that the planes near the dislocation

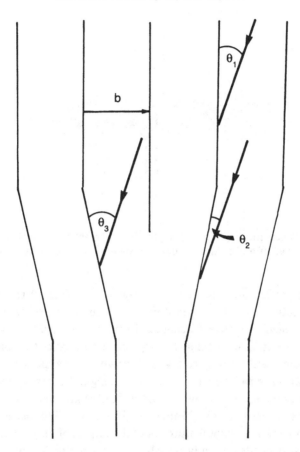

Figure 5.3. Schematic diagram of an edge dislocation of Burgers vector **b**. Away from the dislocation core, the beam is incident on the planes at a glancing angle $\theta_1 = \theta_B + \Delta\theta$. Thus, due to the bending of the planes near the core, the angle of incidence on the right-hand side is $\theta_2 < \theta_1$, and on the left-hand side it is $\theta_3 < \theta_1$.

on the right-hand side are rotated clockwise; in this region of the crystal the angle θ_2 between the incident beam and the reflecting planes is less than θ_1 and therefore is closer to the exact Bragg angle θ_B. However, the planes near the dislocation on the left-hand side are rotated even further away from the exact Bragg angle ($\theta_3 > \theta_1$). Thus, the diffracted intensity from the region close to the dislocation on the right-hand side is greater than the background intensity well away from the dislocation, and in a BF image the dislocation gives rise to a dark line on the right-hand side of the dislocation. It is clear that a dislocation shows up in good contrast

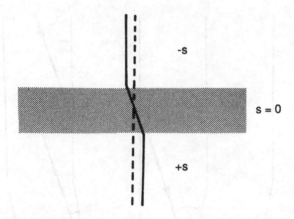

Figure 5.4. Schematic diagram showing the displacement across the bend contour of the image of a dislocation from the position of the dislocation itself.

when the crystal is oriented near the Bragg angle. Thus, if the crystal is bent, dislocations can be seen best when they are close to the bend contour. As the deviation from θ_B changes from $+s$ to $-s$ across a bend contour, the dislocation image may be displaced as it crosses the contour, as shown schematically in Figure 5.4. If the two-beam approximation (see Section 4.2) is not satisfied and two (say) strong diffracted beams are excited, then two images of a dislocation will be produced. Double images can also arise under strict two-beam conditions; this will be discussed later.

A particularly important feature about the images of dislocations is that the contrast depends critically on which planes are diffracting strongly. Not all planes in the neighborhood of a dislocation are equally distorted. For the edge dislocation shown in Figure 5.3, the Burgers vector **b** is normal to the reflecting planes (i.e., **b** is parallel to **g**), and the contrast is strong. However, the planes normal to the dislocation (i.e., the planes parallel to the page) are undistorted. Thus, if an image of the crystal is formed with these planes diffracting, the dislocation will be out-of-contrast. This condition for invisibility of a dislocation can be expressed mathematically by $\mathbf{g} \cdot \mathbf{b} \times \mathbf{u} = 0$, where the unit vector **u** is parallel to the direction of the dislocation. For a pure screw dislocation (**b** parallel to **u**), all planes parallel to **b** are undistorted; such a dislocation is out-of-contrast for all reflections that satisfy the condition $\mathbf{g} \cdot \mathbf{b} = 0$. Note, however, that the invisibility criteria apply only to elastically isotropic crystals and so are unlikely to be strictly applicable to most minerals. Nevertheless, they are often extremely useful, as will be seen in Section 5.6.

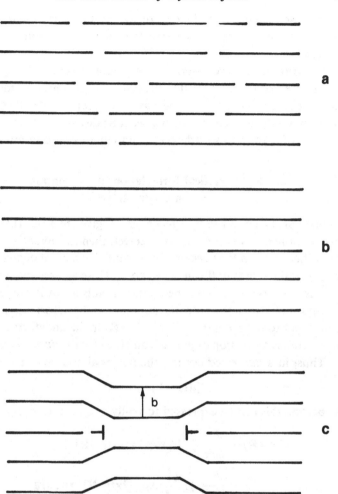

Figure 5.5. Schematic diagrams showing (a) a random distribution of vacancies, (b) condensation of a cluster onto a single plane, and (c) the collapse of the planes to form a prismatic dislocation loop.

Although individual point defects, such as vacancies and interstitial atoms, cannot be seen by these contrast mechanisms, we can observe certain clusters of defects. For example, if a number of vacancies cluster together (*condense*) on a single atomic plane, then the planes above and below will collapse, forming an edge dislocation loop (as shown in Figure 5.5). Sometimes small voids (or gas bubbles) that have a negligible strain field are present in a crystal. These features can be observed because

they decrease the effective thickness of the crystal. Small inclusions that
strain the surrounding crystal give rise to contrast in a similar way to that
of a dislocation.

In this section, we have described the general nature and origin of the
contrast observed from planar defects, dislocations, and small inclusions.
In the next section, we describe the way to calculate the detailed contrast
from a crystal defect. The following sections discuss the results of such
calculations for the types of defect commonly observed in minerals.

5.2 Mathematical formulation of the contrast
from a crystal defect

The potential at a point $\mathbf{r}$ in a *perfect* crystal is given by $V(\mathbf{r})$. If the crystal
is now deformed by the presence of a defect, then the potential at point $\mathbf{r}$
will be different from $V(\mathbf{r})$ because the potential at a point depends on the
positions of the atoms in the neighborhood. If we assume that the deformation is not too severe (i.e., the deformation is a slowly varying function of position), then the potential at point $\mathbf{r}$ in the deformed crystal will
be equivalent to the potential at point $(\mathbf{r}-\mathbf{R})$ in the undeformed crystal.
$\mathbf{R}$, in general, is a function of position and is called the *displacement function*. Thus, in a *deformed* crystal, the potential at a point $\mathbf{r}$ is given by

$$V_d(\mathbf{r}) = V(\mathbf{r}-\mathbf{R}(\mathbf{r})) \tag{5.1}$$

As before, this can be expressed formally as a Fourier series:

$$V(\mathbf{r}-\mathbf{R}) = \frac{h^2}{2me} \sum_g U_g \exp 2\pi i g \cdot (\mathbf{r}-\mathbf{R})$$

$$= \frac{h^2}{2me} \sum_g [U_g \exp -2\pi i g \cdot \mathbf{R}] \exp 2\pi i g \cdot \mathbf{r}$$

$$= \frac{h^2}{2me} \sum_g [U_g \exp(-\alpha_g)] \exp 2\pi i g \cdot \mathbf{r} \tag{5.2}$$

where

$$\alpha_g = 2\pi g \cdot \mathbf{R} \tag{5.3}$$

The Fourier coefficients are now functions of $\mathbf{r}$ instead of being constants.

For the two-beam approximation there is only one operating reflection
$\mathbf{g}$, and for the deformed crystal the corresponding Fourier coefficient is

$$U_g \exp(-i\alpha_g)$$

From Eq. (5.3), it is clear that if $-\mathbf{g}$ is the operating reflection, then $\alpha_{-g} = -2\pi g \cdot \mathbf{r} = -\alpha_g$. Therefore, we put

$$\alpha_{-g} = -\alpha_g = \alpha \qquad (5.4)$$

and the Fourier coefficient corresponding to the operating reflection g in the deformed crystal becomes

$$U_g \exp i\alpha$$

This means effectively that $1/t_g$ becomes $(1/t_g)\exp i\alpha$ in the deformed crystal; see Eq. (4.27). Therefore, Eqs. (4.85) and (4.87) become

$$\frac{dT}{dz} = \frac{i\pi S}{t_g}\exp 2\pi i(sz + \alpha') \qquad (5.5a)$$

$$\frac{dS}{dz} = \frac{i\pi T}{t_g}\exp -2\pi i(sz + \alpha') \qquad (5.5b)$$

where

$$\alpha' = \frac{\alpha}{2\pi} = \mathbf{g}\cdot\mathbf{R} \qquad (5.6)$$

If we put

$$T = T'\exp \pi i(sz + \alpha') \qquad (5.7a)$$

$$S = S'\exp -\pi i(sz + \alpha') \qquad (5.7b)$$

then

$$\frac{dT}{dz} = T'\pi i\left(s + \frac{d\alpha'}{dz}\right)\exp \pi i(sz + \alpha') + \exp \pi i(sz + \alpha')\cdot\frac{dT'}{dz} \qquad (5.7c)$$

By substituting Eqs. (5.7b) and (5.7c) into Eq. (5.5a), we obtain

$$\frac{dT'}{dz} + \pi i\left(sz + \frac{d\alpha'}{dz}\right)T' = \frac{i\pi S'}{t_g} \qquad (5.8a)$$

A similar expression is obtained from Eq. (5.5b), using T from Eq. (5.7a) and dS/dz from Eq. (5.7b):

$$\frac{dS'}{dz} - \pi i\left(s + \frac{d\alpha'}{dz}\right)S' = \frac{i\pi T'}{t_g} \qquad (5.8b)$$

Equations (5.8) differ from those calculated for an undeformed crystal, Eqs. (4.89) and (4.90), in that s is now replaced by $(s + d\alpha'/dz)$. Physically this means that the deformation of the crystal in the neighborhood of the defect causes a local change of s, the deviation from the exact Bragg angle, as has already been discussed in Section 5.1; see Figure 5.3, for example. It is clear that if $\alpha' = 0$, then Eqs. (5.5) and (5.8) reduce to the equations for an undeformed crystal, and the defect will be out-of-contrast.

In the kinematical limit, $T = 1$ and Eq. (5.5b) becomes

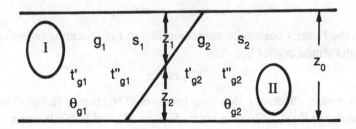

Figure 5.6. Diagram illustrating the notation used in the calculation of the fringe patterns due to an inclined boundary in a thin crystal.

$$\frac{dS}{dz} = \frac{i\pi}{t_g} \exp -2\pi i(sz + \alpha')$$

Integrating over the thickness of the crystal ($0 < z < t$), we obtain

$$S = \frac{i\pi}{t_g} \int_0^t \exp -2\pi i(sz + \alpha')\, dz$$

$$= \frac{i\pi}{t_g} \int_0^t \exp(-2\pi isz) \exp(-2\pi i\alpha')\, dz \qquad (5.9)$$

If we compare Eq. (5.9) with Eq. (3.48), which gives the amplitude of the diffracted wave for a perfect crystal, we can see that the deformation due to the crystal defect is represented by the factor $\exp(-2\pi i\alpha')$. If $\alpha' = 0$, this factor is unity, and the defect will be out-of-contrast.

Now the form of **R** depends on the nature of the defect. To determine the contrast due to a particular kind of defect in BF and DF images, we must find the intensities $I_0 \,(=TT^*)$ and $I_g \,(=SS^*)$ of the transmitted and diffracted beams as a function of position around the defect, using Eqs. (5.5) or Eqs. (5.8). In the following sections we discuss in detail the solutions of these equations for the various types of crystal defect already considered in general terms in Section 5.1.

5.3 Planar defects

Consider a thin specimen consisting of two parts, I and II, joined along an inclined boundary, as in Figure 5.6. The indices $j = 1, 2$ refer to parts I and II. $\mathbf{g}_j$ are the operating diffraction vectors; s_j the extinction errors (i.e., the deviations from the exact Bragg angles); t'_{gj} the extinction distances; t''_{gj} the absorption lengths; θ_{gj} the phase angles of the structure factors

$$F_{gj} = \sum_n f_{nj} \exp(-i\theta_{gj})$$

and z_j the thickness. The total thickness $z_0 = z_1 + z_2$. Further, we define

$$\Delta g = g_1 - g_2$$
$$\Delta s = s_1 - s_2$$
$$\alpha = \theta_{g1} - \theta_{g2}$$
$$\delta = s_1 t'_{g1} - s_2 t'_{g2} \tag{5.10}$$

For a general boundary, Δg, Δs, α, and δ are all nonzero.

Gevers, Landuyt, and Amelinckx (1965) have used the two-beam dynamical theory with absorption, Eqs. (5.5), to derive explicit expressions for the transmitted intensity TT^* and the diffracted intensity SS^* for a crystal containing a general boundary, from which the detailed nature of the intensity profiles of the fringe patterns due to specific types of planar interfaces can be determined.

5.3.1 α-boundaries. These boundaries are defined as those for which

$$\alpha = (\theta_{g1} - \theta_{g2}) \neq 0$$
$$\Delta g = 0$$
$$\Delta s = 0$$
$$\delta = 0 \tag{5.11}$$

Note that $\Delta s = 0$ and $\delta = 0$ indicate that $t'_{g1} = t'_{g2}$, and hence $|F_{g1}| = |F_{g2}|$.

A nonzero value of α can arise in two ways: (i) if, for structural reasons, $\theta_{g1} \neq \theta_{g2}$ and $|F_{g1}| = |F_{g2}|$ and (ii) if parts I and II are displaced by a vector **R** in the boundary. It can be seen from Eq. (5.9) that such a displacement introduces a phase shift $\alpha = 2\pi g \cdot R$ into the diffracted amplitude. Thus, in general,

$$\alpha = 2\pi g \cdot R + (\theta_{g1} - \theta_{g2}) \tag{5.12}$$

Clearly, if $\alpha = 0$ no fringes will be produced and the boundary will be out-of-contrast. This invisibility criterion is the basis for the experimental determination of the magnitude and direction of **R**.

We now discuss the nature of the fringes produced by the most common types of α-boundary.

5.3.2 α-boundaries with $\alpha = \pm 2\pi/3$. The most extensively studied α-boundaries are stacking faults in fcc metals for which $R = \pm \frac{1}{6}[2\bar{1}\bar{1}]$ in the (111) plane. Therefore,

$$\alpha = 2\pi g \cdot \mathbf{R} = \pm 2\pi (hkl) \cdot \frac{1}{6}[2\bar{1}\bar{1}]$$

$$= \pm \frac{2\pi}{6}(2h - k - l)$$

and because h, k, l are either all even or all odd (see Section 3.5),

$$\alpha = 0, \quad \pm \frac{2\pi}{3}, \quad \text{or} \quad \pm \frac{4\pi}{3}$$

These stacking faults are often described by a fault vector $\mathbf{R}_F$ normal to the plane of the fault obtained by adding a lattice vector to $\mathbf{R}$:

$$\mathbf{R}_F = \frac{1}{6}[2\bar{1}\bar{1}] + \frac{1}{2}[011] = \frac{1}{3}[111]$$

The lattice vector simply adds 2π to α and therefore has no effect on the nature or visibility of the fringes.

The characteristics of the fringe patterns due to α-boundaries with $\alpha = \pm 2\pi/3$ can be determined from the general theory of Gevers et al. (1965) or from the treatments given by Hashimoto, Howie, and Whelan (1962) and Gevers, Art, and Amelinckx (1963). In addition, reviews have been given by Hirsch et al. (1965) and by Amelinckx (1970). Computed profiles for BF and DF images using the same reflection are shown in Figure 5.7. Note the following characteristics:

1 In BF, the fringe pattern is symmetrical about the center of the specimen, and for positive α the outer fringes are both bright. For negative α, the outer fringes are dark.
2 In DF, the fringe pattern is asymmetrical and, if the same reflection is used as for the BF image, the outer fringe at the top of the specimen is the same in BF and DF, but the other fringe at the bottom is of opposite contrast. Thus, from a pair of BF and DF images, the top and bottom surfaces of the specimen can be identified and the sense of inclination of the boundary plane determined.
3 There is reduced contrast between bright and dark fringes near the center of the fault.
4 Near $s = 0$, the separation between dark (or bright) fringes corresponds roughly to the depth periodicity of the extinction distance t_g.
5 For an inclined boundary in a wedge-shaped crystal, with increasing thickness new fringes are added at the center of the fringe pattern and the outer fringes are continuous and unaffected.

McLaren and Phakey (1966) found that the fringe patterns due to boundaries with $\alpha = \pm \pi/3$ have similar characteristics.

a

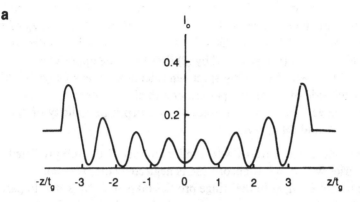

b

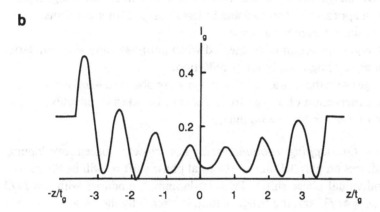

Figure 5.7. Computed image profiles for an α-fringe pattern with $\alpha = 2\pi/3$. (a) Bright field and (b) dark field. $z = 7.25t_g$; $t_0'' = t_g'' = 13.3t_g$; $st_g = -0.2$. (After Hashimoto et al. 1962.)

The minimum value of α for which fringe contrast is observed is about $0.04\pi = 7.2°$, according to Humphreys, Howie, and Booker (1967).

5.3.3 α-boundaries with $\alpha = \pi$. The fringe contrast due to boundaries of this type has been treated by van Landuyt, Gevers, and Amelinckx (1964) and is quite distinct from that of the α-boundaries just considered. For $s = 0$, π-fringes have the following characteristics:

1 The center of the thin specimen corresponds to a bright fringe in the BF image and a dark fringe in the DF image.
2 Over the whole depth of the specimen and for any specimen thickness, the BF and DF images are complementary.

3   The fringes are parallel to the center line of the specimen, and the nature of the outer fringes is determined by the thickness z_0 of the specimen. In BF they are dark for $z_0 = kt'_g$ (where k is an integer) and bright for $z_0 = (k + \frac{1}{2})t'_g$; and by (2) they have the opposite contrast in DF. Thus, with increasing specimen thickness, a new fringe is added to each side of the fringe pattern at a thickness contour.

4   The fringe separation corresponds to a depth periodicity of $t'_g/2$ in both BF and DF images.

Note that fringe patterns with similar characteristics are predicted for boundaries with $\alpha \neq \pi$ if absorption is neglected (i.e., $t''_g \rightarrow \infty$).

Significant changes in the fringe profiles take place for $s \neq 0$. In particular, DF images are no longer symmetrical, and in BF the fringe separation can approach t'_g for $s > 0$ and be less than $t'_g/2$ for $s < 0$. Thus, observations should be made at $s = 0$.

π-fringes are commonly observed when antiphase domain boundaries are imaged using a superlattice reflection.

Fringes with the characteristics of $\alpha = \pi$ are observed when $\alpha = \pi \pm \pi/6$. Thus, observation of π-type fringes cannot be taken as unequivocal evidence that $\alpha = \pi$ for the boundary.

5.3.4 Overlapping α-boundaries.   For closely spaced overlapping boundaries on parallel planes, the total phase shift α_n will be the sum of the individual phase shifts. Two overlapping boundaries with $\alpha = 2\pi/3$ have $\alpha_n = 4\pi/3$, and the fringe pattern is essentially the same as for a single boundary. However, for three overlapping boundaries $\alpha_n = 2\pi$, and the boundary is out-of-contrast.

For overlapping π-boundaries, the intensity is uniform over the area of overlap at $s = 0$, being either bright or dark depending on the depth separation of the boundaries.

The contrast at the overlap of intersecting boundaries can be quite complex; the reader should consult the original papers of Gevers, Art, and Amelinckx (1964) and van Landuyt et al. (1964) for details.

5.3.5 δ-boundaries.   These boundaries are defined as those for which

$$\delta = s_1 t'_{g1} - s_2 t'_{g2} \neq 0$$
$$\alpha = 0$$

(5.13)

The general case for which $s_1 \neq s_2$ and $g_1 \neq g_2$ has been treated by Gevers et al. (1965). However, we start by considering the so-called *symmetrical* case for which $s_1 = -s_2$ and $t'_{g1} = t'_{g2} = t'_g$, so that $\delta = \pm \Delta s \, t'_g$.

The symmetrical case applies to the 90°-boundaries in ferroelectric barium titanate which is tetragonal, but the distortion from cubic is very slight, c/a being 1.0098. Thus, $\mathbf{g}_1$ and $\mathbf{g}_2$ are essentially equal in magnitude and direction. δ-fringes due to domain boundaries of this type have been studied in detail, both theoretically and experimentally, by van Landuyt et al. (1965), and a review has been given by Amelinckx (1970). These fringes have the following characteristics:

1. The BF image is asymmetric, the outer fringes being of opposite contrast. However, the DF image is symmetric.
2. In BF, the fringe at the top surface of the specimen is bright for $\Delta s > 0$ and dark for $\Delta s < 0$.
3. In DF, the outer fringes are bright for $\Delta s > 0$, and dark for $\Delta s < 0$.
4. The fringes reverse contrast on changing from $+\mathbf{g}$ to $-\mathbf{g}$ because this also changes the sign of Δs.
5. The contrast is a function of Δs and is independent of the magnitude of s.
6. For an inclined boundary in a wedge-shaped specimen, with increasing thickness new fringes are added at the center of the pattern in the center of the bright thickness contours, in both BF and DF images.

For an *asymmetrical* orientation, $s_1 \neq -s_2$ and σ, given by Eq. (4.41), is different in parts I and II of the specimen. Thus, the thickness fringes in parts I and II have a different spacing, given by $1/\sigma_1$ and $1/\sigma_2$, and in the region of the boundary these fringes join up and the spacing is the average of $1/\sigma_1$ and $1/\sigma_2$. Again, because $1/\sigma_1$ and $1/\sigma_2$ are different, the spacing of the δ-fringes in the boundary is different at the top and bottom of the specimen.

If $s_1 = s_2 = s$ and $t'_{g1} \neq t'_{g2}$ then $\delta = s(t'_{g1} - t'_{g2})$, which will be nonzero provided $s \neq 0$.

Overlapping δ-boundaries give complex images which have been discussed in detail by Remaut et al. (1965, 1966).

5.3.6 (α-δ)-boundaries. These boundaries are defined as those for which both α and δ are nonzero. An interphase boundary (such as at an exsolution lamella) is likely to be of this kind. The α-component arises because of a phase difference between the structure factors F_{g1} and F_{g2} of the precipitate and matrix, and the δ-component because of a small misorientation of the reflecting planes ($\Delta s \neq 0$) and/or a difference of extinction distance t'_{g1} and t'_{g2}.

5.3.7 Twin boundaries and microtwins. Twins are very common in many important rock-forming minerals, so it is appropriate to discuss the contrast from twin boundaries and narrow twin lamellae (microtwins) as special cases of planar defects. The orientation relationship between the two elements of a twin is either (1) reflection in the twin plane or (2) rotation of 180° about an axis normal to a plane. The twin boundary (or composition plane) across which the change in orientation actually occurs in the crystal does not, in general, coincide with the twin plane. For example, the boundary for pericline twins in the triclinic feldspars is parallel to the twin axis [010] and its orientation about that axis is a function of the lattice parameters and, hence, of chemical composition. On the other hand, the boundary of the albite twin in these minerals is parallel to the twin plane (010). In some minerals, the two parts of a twin are also displaced with respect to each other by a nonlattice vector $\mathbf{R}$. Thus, in general, a twin boundary gives rise to $(\alpha\text{-}\delta)$-fringes.

If an image is formed using a diffraction vector $\mathbf{g}$ normal to the twin plane, then the crystal on either side of the boundary will diffract with equal intensity and the twin will be out-of-contrast, as in Section 5.1. For this reflection, the δ-component is zero, and the twin boundary is also out-of-contrast (a) if $\mathbf{R} = 0$ or (b) if $\mathbf{R} \neq 0$, but it lies in the twin plane, as is usually the case.

An inclined microtwin gives rise to an α-fringe pattern under two-beam conditions at or near $s = 0$, provided only the surrounding matrix is diffracting, the microtwin lamella being far from a diffracting orientation. The reason for this result can be understood from Figure 5.8(a), which shows a microtwin in a matrix, together with the reflecting planes. It is clear that this microtwin introduces a relative displacement of parts I and II of the matrix. Because the microtwin lamella is not diffracting, it acts just like a stacking fault and gives rise to α-fringes. The displacement vector $\mathbf{R}$ depends on the crystal structure and on the width of the lamella. For example, the microtwin shown in Figure 5.8(b) is wider than that shown in (a), and $\mathbf{R} = 0$.

α-fringes due to microtwins have been observed in a number of minerals. Similar contrast effects can occur at a thin lamella of one phase in a matrix of another phase if only the matrix is diffracting.

5.3.8 Grain boundaries. It is usual to restrict the term *grain boundary* (GB) to the interface between two crystals of the same material and crystal structure but of different orientation. The change of orientation across the GB usually gives rise to a change of contrast (i) because of a

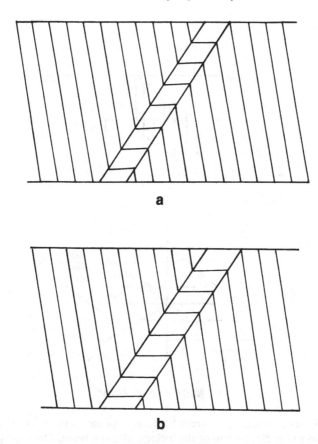

Figure 5.8. Schematic diagrams of an inclined microtwin in a thin crystal. (a) The thickness of the microtwin is such that it produces a displacement of half a planar spacing across it. (b) A wider microtwin produces no relative displacement.

difference in the extinction error s if the same diffraction vector **g** is operating in both grains or (ii) if different diffraction vectors are operating in the two grains. When the misorientation between the two crystals is only of the order of a few degrees, the misorientation can be accommodated by an array or network of lattice dislocations, and the boundary is called *low-angle* GB (Hull and Bacon 1984). For a tilt boundary consisting of a simple periodic array of edge dislocations, the misorientation θ is given by

$$\theta = \frac{b}{D} \qquad (5.14)$$

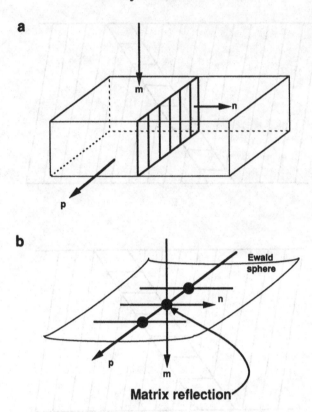

Figure 5.9. (a) Geometry of a grain boundary normal to the plane of the foil and parallel to the direction **m** of the incident electron beam. The boundary has a periodicity along the direction **p**, and the boundary normal is specified by **n**. (b) The reciprocal lattice and Ewald sphere in the neighborhood of a matrix reflection, showing the extra reflections due to the periodic boundary and the streaking of the reflections in the direction **n**. (After Carter and Sass 1981.)

where b is the magnitude of the Burgers vector of the dislocations and D the spacing of the dislocations. Thus, with $\theta = 1°$ and $b = 0.5$ nm, $D \approx$ 30 nm. These dislocations can be imaged by any of the imaging modes to be considered in detail in Sections 5.4 and 5.5. The periodic array of dislocations forming the boundary can act like a diffraction grating, giving rise to extra reflections in the SAD pattern *provided* the spacing D of the dislocations is less than the diameter of the coherence area of the electron beam (see Section 1.6). This is illustrated in Figure 5.9 for the case of a boundary that is normal to the plane of the specimen foil and parallel to the direction of the incident electron beam. This technique is particularly useful for determining the periodicity of a GB if the individual defects are not easily resolvable in a normal image. The extra reflections are usually streaked in a direction **n** normal to the plane of the boundary (see Section 3.7). Budai, Gaudig, and Sass (1979) pointed out that the length of this streaking was inversely proportional to the boundary "thickness" that Bristowe and Sass (1980) suggested was a measure of the region of large strain at the grain boundary. The slightly more complex geometry of an inclined GB is discussed by Carter and Sass (1981). However, Vitek and Rühle (1986) evaluated the diffraction profiles of the extra reflections in the direction normal to a planar interface. They calculated the diffraction intensities for the entire diffracting volume, consisting of a distorted interface zone bordered by two perfect crystals, so that interference effects were handled rigorously. The model is applicable to any type of interface. They specifically treated the case of a twist GB, for which the lattice expansion decreased exponentially away from the interface and found that "no simple quantitative relationship exists between the width of the distorted interface region and the resultant streak length in diffraction." Thus, the interpretation of Bristowe and Sass (1980) may not be valid.

With increasing misorientation, the dislocation spacing decreases. When θ is about 20° or greater, the dislocation cores overlap and the dislocations lose their individuality. The image of a high-angle GB depends on the diffracting conditions in the two grains. If two-beam dynamical conditions are operating in one grain ($s \approx 0$ for the diffraction vector g_1) and kinematical conditions are operating in the other grain (s large for all g), then an inclined boundary between the grains will give rise to thickness fringes, as discussed in Section 5.1. This type of image is commonly observed in polycrystalline specimens. If dynamical two-beam conditions are operating in both grains ($s \approx 0$ for g_1 and g_2), then the image of the inclined boundary will be a very complex fringe pattern and moiré fringes (discussed in Chapter 6) may also arise.

During the last decade, numerous reports have appeared of linear defects in GBs in metals observed by TEM. The observed variation in geometry as well as the variation of contrast for various diffracting conditions indicate that a single type of defect cannot account for all the observations. Among the numerous interpretations that have been suggested for the linear defects are (i) absorbed lattice dislocations; (ii) steps, at least a few unit cells high, in the boundary plane; (iii) structural dislocations in the boundary that accommodate small deviations from the special orientations predicted by the coincidence site lattice (CSL) theory, the structural unit model (SUM), or the plane matching theory (PMT); and (iv) special GB dislocations associated with grain boundary sliding in deformed specimens. These four possibilities are discussed in detail by Loberg and Norden (1976). Since the interpretation of the linear defects is still uncertain, considerable care needs to be taken in drawing any conclusions from them about the detailed structure of the GB.

In spite of the differences between metallic and ionic materials, high-angle GBs with periodic structures have been observed in a number of ceramics (see, e.g., Sun and Balluffi 1982); and McLaren (1986) has suggested that the CSL description may apply to high-angle GBs in quartz. However, amorphous second-phase material has been observed in a large number of hot-pressed ceramics. Most of the second-phase resides at three- and four-grain junctions; the small amount found along the GBs is usually very thin, <2 nm (Clarke 1987). Three techniques have been proposed for detecting intergranular phases on the nanometer scale:

1 Lattice imaging (Chapter 6) of the crystals on both sides of the boundary (Clarke 1979a, b)
2 Out-of-focus phase contrast imaging (see Section 5.7.1 and Clarke 1980)
3 DF imaging using the inelastic, diffusely scattered electrons (near the central beam), which, it is assumed, arise predominantly from noncrystalline regions of the specimen (Clarke 1979a; McLaren and Etheridge 1980).

All these techniques require the GB to be oriented accurately parallel to the direction of the incident electron beam. In practice, this is impossible to achieve if the boundary orientation changes with depth in the foil. Kouh Simpson et al. (1986) have identified other, more serious difficulties. In alumina, for example, grooves develop along many GBs during specimen preparation by ion-milling (Section 2.7). The grooving depends on variables such as the GB orientation relative to the specimen surface, GB character, and configuration. Also, surface damage can occur during milling and carbon coating. In the presence of such defects, experimental evidence suggests that the use of techniques (2) and (3) to identify glassy phases at GBs can give misleading results. These techniques by themselves cannot be used to identify conclusively the presence of glassy phases at GBs; so, whenever possible, high-resolution lattice imaging must be used in conjunction with them.

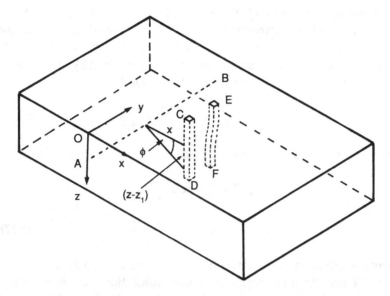

Figure 5.10. A screw dislocation *AB* parallel to the plane of a thin crystal. The column *CD* of unit cells in the perfect crystal is deformed to the shape *EF* by the screw dislocation. This diagram also shows the parameters used in the calculation of the contrast from a dislocation. (After Hirsch, Howie, and Whelan 1960.)

5.4 Dislocations

5.4.1 Screw dislocation. The simplest case to start with is that of a straight screw dislocation of Burgers vector **b** parallel to the surface of a thin parallel-sided crystal foil, as shown in Figure 5.10. Using the coordinate system defined there, the dislocation *AB* is parallel to y and at a depth z_1 below the top surface. The dislocation causes a column *CD* of unit cells parallel to z in the perfect crystal to be deformed. If we assume that the atomic displacements around the dislocation are the same in the thin specimen as in an infinitely large, elastically isotropic crystal, then the components u, v, w of the deformation of the column along the $x, y,$ and z directions will be

$$u = 0$$

$$v = \left(\frac{\mathbf{b}}{2\pi}\right) \tan^{-1}\left(\frac{z - z_1}{x}\right) \qquad (5.15)$$

$$w = 0$$

which are derived in a number of standard textbooks on dislocation theory, such as Hull and Bacon (1984). Thus, the deformed column has the shape *EF*. The displacement in the y direction of the unit cells in the column at different depths below the top surface is specified by the displacement vector $\mathbf{R}(z)$, which from Eq. (5.15) is given by

$$\mathbf{R}(z) = \frac{\mathbf{b}}{2\pi} \tan^{-1} \frac{z - z_1}{x} = \frac{\mathbf{b}\varphi}{2\pi} \qquad (5.16)$$

Thus, the phase angle α' introduced by the dislocation is given by

$$\alpha' = \mathbf{g} \cdot \mathbf{R} = \frac{\mathbf{g} \cdot \mathbf{b}}{2\pi} \tan^{-1} \frac{z - z_1}{x}$$

$$= \frac{n}{2\pi} \tan^{-1} \frac{z - z_1}{x} \qquad (5.17)$$

where $n = \mathbf{g} \cdot \mathbf{b}$ is an integer that may take values $0, \pm 1, \pm 2, \dots$.

The BF and DF image profiles across the dislocation (i.e., the variation of the transmitted intensity I_0 and diffracted intensity I_g as a function of x/t_g) are obtained by numerical integration of Eqs. (5.5) or (5.8) using α' given by Eq. (5.17). If the Burgers vector $\mathbf{b}$ lies in the reflecting plane, $n = \mathbf{g} \cdot \mathbf{b} = 0$ and the dislocation will be invisible. Image profiles calculated by Howie and Whelan (1962) for $n = 1$ and $n = 2$ are shown in Figure 5.11. The following facts should be noted:

(a)   The BF and DF images are similar, but not identical, and are darker than background.

(b)   For $n = 1$, the image is a sharp peak, the width at half-height being about $0.2t_g$.

(c)   For $n = 2$, the image consists of two peaks whose separation is about $0.2t_g$.

(d)   For $n = 1$, the image is displaced from $x = 0$ (the position of the dislocation) for all conditions except DF at $s = 0$. Thus, the image changes from one side to the other of the dislocation on crossing a bend contour (see also Figure 5.4).

(e)   For $n = 2$ the two peaks are symmetrically placed about $x = 0$ for $s = 0$, but not for $s \neq 0$.

5.4.2 Edge dislocation.   The displacement vector $\mathbf{R}$ for an edge dislocation of Burgers vector $\mathbf{b}$ parallel to the surface of a thin foil, as shown in Figure 5.12, is given by

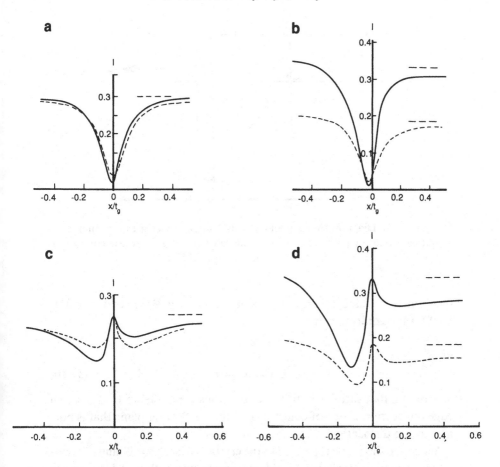

Figure 5.11. Computed image profiles of screw dislocations. Bright field and dark field images are shown as solid and broken lines, respectively. $z = 8t_g$; $z_1 = 4t_g$; $t_0'' = t_g'' = 10t_g$. (a) $n = 1$; $st_g = 0$. (b) $n = 1$; $st_g = 0.3$. (c) $n = 2$; $st_g = 0$. (d) $n = 2$; $st_g = 0.3$. (After Howie and Whelan 1962.)

$$R = \frac{1}{2\pi} \left| b(\varphi - \gamma) + b\, \frac{\sin 2(\varphi - \gamma)}{4(1 - \nu)} + b \times u \left\{ \frac{1 - 2\nu}{2(1 - \nu)} \ln r + \frac{\cos 2(\varphi - \gamma)}{4(1 - \nu)} \right\} \right|$$

(5.18)

where u is a unit vector in the positive sense of the dislocation, γ the angle between the slip plane and the crystal surface, $r^2 = x^2 + (z - z_1)^2$, ν is the Poisson ratio, and φ is as defined previously. It is, of course, assumed that the thin crystal-plate behaves as an infinite, elastically isotropic

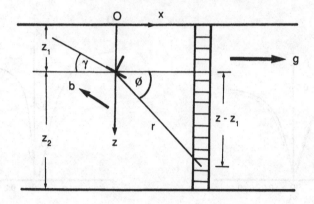

Figure 5.12. Diagram showing an edge dislocation adjacent to a column of unit cells and the parameters used in the calculation of the image contrast. (After Howie and Whelan 1962.)

medium. The dislocation is out-of-contrast if $\alpha' = g \cdot R = 0$; and from Eq. (5.18), this requires both

$$g \cdot b = 0 \qquad (5.19a)$$

and

$$g \cdot (b \times u) = 0 \qquad (5.19b)$$

The only g that satisfies both these equations is parallel to u; that is, an edge dislocation is out-of-contrast only for a reflecting plane that is normal to the dislocation.

When the slip plane is parallel to the crystal surface ($\gamma = 0$) and the crystal is oriented normal to the incident electron beam, then u, b, and g are coplanar and $g \cdot b \times u = 0$. Thus, from Eq. (5.18) with $\nu = \frac{1}{3}$, α' is given by

$$\alpha' = g \cdot R = \frac{1}{2\pi} g \cdot b \left[\varphi + \frac{3}{8} \sin 2\varphi \right]$$

$$\approx \frac{1}{2\pi} g \cdot b \tan^{-1} \left[\frac{2(z - z_1)}{x} \right] \qquad (5.20)$$

where $\tan \varphi = (z - z_1)/x$.

BF and DF image-profiles computed using this α' with $g \cdot b = 1$ are shown in Figure 5.13. By comparing these profiles with those in Figure 5.11(a), we see that the image of an edge dislocation is about twice as wide as that of a screw dislocation.

For an edge dislocation parallel to the plane of the crystal foil, a special case arises when $\gamma = \pi/2$. The Burgers vector b is now normal to the

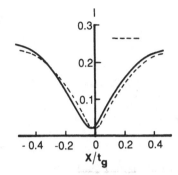

Figure 5.13. Computed image profiles for an edge dislocation. Bright field and dark field images are shown as solid and broken lines, respectively. $\mathbf{g} \cdot \mathbf{b} = 1$; $z = 8t_g$; $z_1 = 4t_g$; $t_0'' = t_g'' = 10t_g$; $st_g = 0$. (After Hirsch et al. 1965.)

foil and $\mathbf{g} \cdot \mathbf{b} = 0$ when g lies in the plane of the foil. However, $\mathbf{g} \cdot \mathbf{b} \times \mathbf{u} \neq 0$. Using Eq. (5.18) with $\nu = \frac{1}{3}$, we find that α' is given by

$$\alpha' = \mathbf{g} \cdot \mathbf{R} = \frac{1}{2\pi} \mathbf{g} \cdot \mathbf{b} \times \mathbf{u} \left\{ \frac{1}{4} \ln r + \frac{3}{8} \cos 2 \left(\varphi - \frac{\pi}{2} \right) \right\}$$

$$= \frac{1}{2\pi} \left(\mathbf{g} \cdot \mathbf{b} \times \frac{\mathbf{u}}{8} \right) \left\{ \ln r^2 + 3 \cos 2 \left(\varphi - \frac{\pi}{2} \right) \right\}$$

$$= \frac{1}{2\pi} m \left\{ \ln(x^2 + (z - z_1)^2) + 3 \frac{(z - z_1)^2 - x^2}{(z - z_1)^2 + x^2} \right\} \qquad (5.21)$$

where $m = \mathbf{g} \cdot \mathbf{b} \times \mathbf{u}/8$ is usually in the range $-0.5 < m < 0.5$.

Figure 5.14 shows the image profile computed by Howie and Whelan (1962) using α' from Eq. (5.21) for a dislocation in the middle of the crystal for $s = 0$ and $m = 0.25$ (it is independent of the sign of m). The effect of stress-relaxation at both surfaces of the foil was included by adding an image dislocation in each surface.* Because α' is a function of x^2, the profiles are symmetrical about $x = 0$ for all values of m, so a double image is produced. The contrast increases with increasing magnitude of m. Slightly more complex images are formed for dislocations near the surfaces of the foil. They are still symmetrical about $x = 0$ but depend on the sign of m.

The vector $\mathbf{R}$ in Eq. (5.21) describes the displacements normal to the slip plane of the dislocation. These displacements cause a buckling of the reflecting plane, which is the origin of the contrast. A plane that is parallel to the slip plane is the most severely buckled, and thus the contrast is greatest if this plane is the reflecting plane. For these conditions, g is

---

* Stress-relaxation also occurs when a dislocation intersects the foil surface; it is this effect that makes dislocations viewed end-on visible. (See Tunstall et al. 1964.)

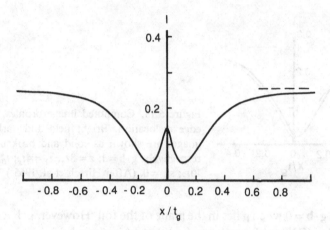

Figure 5.14. Computed bright field image profile for an edge dislocation.
$\mathbf{g} \cdot \mathbf{b} = 0$; $\gamma = \pi/2$; $z = 8t_g$; $z_1 = 4t_g$; $t_0'' = t_g'' = 10t_g$; $m = 0.25$; $st_g = 0$.
(After Howie and Whelan 1962.)

parallel to $\mathbf{b} \times \mathbf{u}$, and hence m $(= \mathbf{g} \cdot \mathbf{b} \times \mathbf{u}/8)$ is a maximum; m is less for other g lying in the plane of the foil, and the dislocation image is of lower contrast. Similarly, the contrast is lower for values of γ less than $\pi/2$.

Appreciable contrast is expected only for very small values of s. This can be uderstood as follows. From Eq. (5.8) we see that both the transmitted intensity I_0 and the diffracted intensity I_g from the region in the neighborhood of a dislocation depend on $(s + d\alpha'/dz) = (s + \mathbf{g} \cdot d\mathbf{R}/dz)$. Because the logarithmic term in Eq. (5.21) changes rapidly with z only very close to the dislocation, and because the cosine term in the same equation changes sign twice as z increases from 0 to $(z_1 + z_2)$, $\mathbf{g} \cdot d\mathbf{R}/dz$ has little influence unless s is very small. Thus, the contrast associated with the $(\mathbf{g} \cdot \mathbf{b} \times \mathbf{u})$ term in Eq. (5.18) is expected to be vanishingly small for large s. This has important implications when attempting to determine the Burgers vector of a dislocation that has an appreciable edge component and will be discussed more fully in Section 5.6.

Howie and Whelan (1962) also consider the important special case of a pure edge dislocation loop lying in the plane of the foil, as shown in Figure 5.15. Clearly, $\mathbf{b}$ is normal to the foil; and with g in the plane of the foil, $\mathbf{g} \cdot \mathbf{b} = 0$. For those diametrically opposite segments A of the loop that are essentially parallel to g, $\mathbf{g} \cdot \mathbf{b} \times \mathbf{u} = 0$ and they are out-of-contrast. However, for the other parts B of the loop, $\mathbf{g} \cdot \mathbf{b} \times \mathbf{u} \neq 0$; they usually give rise to a double image that is symmetrical about the dislocation. If the loop is close to the surface of the foil, a more complex image may be formed.

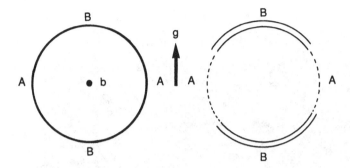

Figure 5.15. Schematic diagram showing (a) a prismatic dislocation loop lying in the plane of the foil and (b) its image when $\mathbf{g} \cdot \mathbf{b} = 0$.

5.4.3 Dislocations of mixed character.   Image profiles for dislocations of mixed character have been calculated by Howie and Whelan (1962). For $\mathbf{g} \cdot \mathbf{b} = 1$, the images are similar to those of screw and edge dislocations but of intermediate width.

5.4.4 Dislocations inclined to the foil surface.   The image of a dislocation that is inclined to the surface of the crystal is a little narrower than that of a dislocation parallel to the surface. However, the most striking features of the image of an inclined dislocation are the oscillations, which give rise to dotted, zigzag, or alternating black–white contrast. These effects, shown in Figure 5.16, are analogous to the fringe contrast of an inclined planar defect such as a stacking fault. Like the image of a stacking fault, the oscillations are more pronounced near the crystal surfaces and tend to die out at the center of the crystal (see Figure 5.7). The reader should consult the paper by Howie and Whelan (1962) for a detailed discussion of the contrast of inclined dislocations.

5.4.5 Partial dislocations.   If a dislocation with Burgers vector $\mathbf{b}$ splits into two partial dislocations $\mathbf{b}_1$ and $\mathbf{b}_2$ according to the reaction $\mathbf{b} = \mathbf{b}_1 + \mathbf{b}_2$, then a strip of stacking fault with a displacement vector $\mathbf{R}$ will be created between the two partial dislocations. If $\mathbf{g} \cdot \mathbf{R}$ is an integer, the fault will be out-of-contrast (see Section 5.3.1): $\mathbf{g} \cdot \mathbf{b}_1$ and $\mathbf{g} \cdot \mathbf{b}_2$ will also be integers but the partial dislocations may not be out-of-contrast. If $\mathbf{g} \cdot \mathbf{R}$ is not an integer, the fault will be in contrast but it may be difficult to decide whether the partial dislocations are visible or not in the presence of the stacking fault fringes.

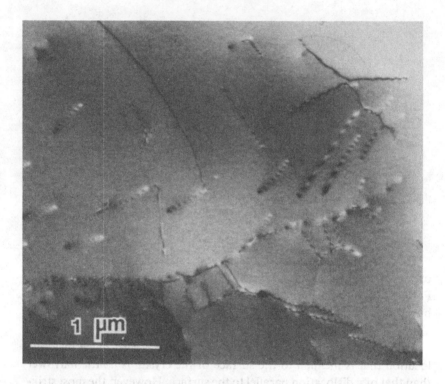

Figure 5.16. Bright field electron micrograph showing the dotted and zigzag contrast of inclined dislocations in a thin foil of chromium.

A partial dislocation (called a *stair rod dislocation*) is formed at the intersection of two stacking faults on different planes with different fault vectors $\mathbf{R}_1$ and $\mathbf{R}_2$ and its Burgers vector is $\mathbf{b}_p = \mathbf{R}_1 - \mathbf{R}_2$. An extended summary of the contrast from partial dislocations in fcc metals has been given by Edington (1975).

5.5 Kinematical and weak beam dark field (WBDF) images of dislocations

5.5.1 Introduction. In Sections 5.3 and 5.4, we considered the contrast of planar defects and dislocations in terms of the two-beam dynamical theory (including absorption) when the crystal is set at or near the exact Bragg angle (i.e., when s is zero or small). These diffracting conditions (which are usually described as *dynamical*) are, in general, satisfactory for determining the nature of the planar defects from the details of the fringe patterns they produce. However, this is often not so for dislocations, due to the width and complexity of their dynamical images. We see from Figure 5.13 that the width at half-height of the image of an edge dislocation is about $0.4t_g$ at $s = 0$. Thus, the narrowest image of an edge dislocation in quartz (observed using the strongest reflection $\mathbf{g} = 10\bar{1}1$ for which $t_g = 68.4$ nm) is 27 nm. It is clear, therefore, that two such dislocations would be resolved only if they were separated by more than about 50 nm. Dynamical images are also quite complex, especially if the dislocations are inclined to the plane of the foil. However, when s is large, the kinematical theory of diffraction is valid, and it is found that the images of dislocations are usually narrower and less complex. Thus, for kinematical images, it is not only possible to resolve more closely spaced dislocations but also it is often possible to use the simple invisibility criterion $\mathbf{g} \cdot \mathbf{b} = 0$ to determine the Burgers vector of a dislocation, as will be discussed in Section 5.6. For these reasons it is often preferable to image dislocations under kinematical conditions, particularly in dark field, the so-called weak beam dark field (WBDF) technique. However, bright field images are also useful and sometimes more convenient. In this section, we discuss the calculated kinematical intensity profiles of BF and DF images of dislocations, together with the procedures for establishing kinematical and weak-beam conditions in the electron microscope.

5.5.2 Calculated intensity profiles. The amplitude of the wave diffracted by a column of perfect crystal (CD in Figure 5.10) is, according to the kinematical theory (Chapter 3), given by Eq. (3.48):

$$A(s,z) = \frac{F_g}{V_c} \int_{\text{column}} \exp(-2\pi isz)\,dz \qquad (5.22)$$

and the diffracted intensity $I_g(s,z)$ is given by

$$I_g(s,z) = \left(\frac{F_g}{V_c}\right)^2 \frac{\sin^2 \pi ts}{(\pi s)^2} \qquad (5.23)$$

$I_g(s,z)$ is plotted as a function of s in Figure 3.11. We see that for a given thickness t, $I_g(s,z)$ oscillates with s, the periodicity being t^{-1}. Similarly, for a given s ($\neq 0$), $I_g(s,z)$ oscillates with thickness t, the periodicity being s^{-1}. It is also clear from Eq. (5.23) that the diffracted intensity *averaged* over the length of the column is proportional to s^{-2}.

For an imperfect crystal, the amplitude of the wave diffracted by the distorted column EF in Figure 5.10 is given by

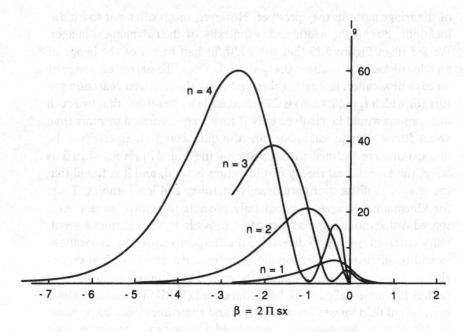

Figure 5.17. Image profiles for screw dislocations calculated using the kinematical theory for various values of n in a crystal of thickness $z = 20/s$. The dislocation is located at $\beta = 0$. (After Hirsch, Howie, and Whelan 1960:)

$$A(s, z) = \frac{F_g}{V_c} \int_{\text{column}} \exp(-2\pi i \alpha') \exp(-2\pi s z) \, dz \qquad (5.24)$$

as shown in Section 5.2, where $\alpha' = \mathbf{g} \cdot \mathbf{R}$. The intensity profiles of images of dislocations of various types can be calculated by integration of Eq. (5.24) using the appropriate form of $\mathbf{R}$, given in Section 5.4.

The intensity profiles of images of screw dislocations for values of $n = \mathbf{g} \cdot \mathbf{b} = 1, 2, 3, 4$ are shown in Figure 5.17. The intensity I_g of the diffracted beam (in arbitrary units) is plotted as a function of $\beta = 2\pi s x$, which, for constant s, is proportional to the distance from the dislocation. It is clear from this figure that in DF a dislocation appears as a bright image on a dark background. Because there is no absorption under kinematical conditions, the BF image is complementary, provided two-beam conditions apply.

Note the following important points in Figure 5.17:

1 The dislocation image lies to one side of the dislocation, in agreement with the intuitive explanation of the origin of dislocation contrast given in Section 5.1.

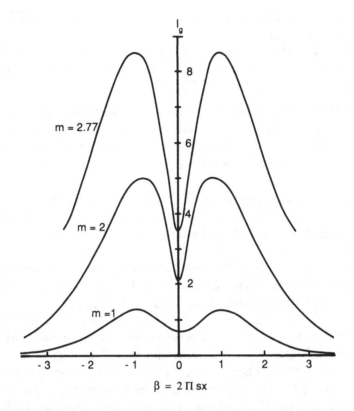

Figure 5.18. Image profile for edge dislocations calculated using the kinematical theory. $\mathbf{g} \cdot \mathbf{b} = 0$; $\gamma = \pi/2$; $m = 1$, 2, and 2.77. (From J. D. Fitz Gerald, unpublished.)

2   The displacement of the image peak from the position of the dislocation is of the same order as the width of the image.
3   The image displacement and image width increase with n.
4   The image is a single line for $n = 1$ and 2; but for $n = 3$ and 4, a double image is produced.

The intensity profiles for pure edge dislocations calculated using α' given by Eq. (5.20) for $\mathbf{g} \cdot \mathbf{b} \times \mathbf{u} = 0$ have characteristics similar to those for screw dislocations except that both the image width and the displacement are about twice that of a screw dislocation with the same n.

For an edge dislocation parallel to the crystal foil and $\gamma = \pi/2$, $\mathbf{g} \cdot \mathbf{b} = 0$, but $\mathbf{g} \cdot \mathbf{b} \times \mathbf{u} \neq 0$. α' is given by Eq. (5.21). The kinematical image profiles for edge dislocations of this type for several values of $\mathbf{g} \cdot \mathbf{b} \times \mathbf{u}$ are shown in Figure 5.18. We see that

(i) all images are double and symmetrical about $x = 0$, the position of the dislocation;

(ii) the peak height increases with increasing value of $\mathbf{g} \cdot \mathbf{b} \times \mathbf{u}$. When $\mathbf{g} \cdot \mathbf{b} \times \mathbf{u} = 2.77$, the peak height is similar to that of a screw dislocation with $n = 1$ (Figure 5.17).

Note also that for all these profiles, an increase in s causes a decrease in peak height because I_g is proportional to s^{-2}. Because $\beta = 2\pi sx$, an increase in s causes the image peak to move closer to the dislocation, coupled with a decrease in image width.

5.5.3 Establishing kinematical BF and weak beam DF conditions.

The procedures to be followed for establishing these diffracting conditions are best described in terms of the Ewald sphere construction (Sections 3.4 and 3.7).

Figure 5.19(a) shows the Ewald sphere construction with $s = 0$ for the first-order reflection $\mathbf{g}$. The BF (dynamical) image is obtained by allowing the straight-through beam 0 to pass down the optic axis and through the centered objective aperture, as shown schematically. If we tilt the incident beam so as to bring the diffracted beam $\mathbf{g}$ down the optic axis, the Ewald sphere will move to the position shown in Figure 5.19(b); and now $s = 0$ for the diffracted beam 3g. A WBDF image is formed by allowing the beam $\mathbf{g}$ to pass through the centered objective aperture, as shown. For this beam, s_g is positive (since the reciprocal lattice point lies inside the Ewald sphere), and its magnitude from Eq. (3.62) is $2g^2/2k = \lambda/d^2$.

Figure 5.19(c) shows the Ewald sphere construction corresponding to a BF image with $s = 0$ for the second-order reflection 2g. If we tilt the incident beam so as to form a WBDF image with the first-order reflection $\mathbf{g}$, then $s = 0$ for the beam 4g, as shown in Figure 5.19(d); and the magnitude of s_g increases to $3\lambda/2d^2$. WBDF images using the first-order reflection $\mathbf{g}$ with larger values of s_g can be obtained by having $s = 0$ for higher-order reflections. An increase in s_g causes the thickness fringes in a wedge-shaped crystal to move closer together and decreases the image width and image displacement of a dislocation; it also decreases the intensity. However, the mechanical stability of modern specimen holders is such that exposures of as much as 60 seconds are practical at magnifications up to about $\times 20,000$.

Experience has shown that BF dislocation images formed under diffraction conditions approximating those in Figure 5.19(c) - particularly when the structure factor of the first-order reflection $\mathbf{g}$ is much larger than that of the second-order reflection 2g (e.g., $\mathbf{g} = 10\bar{1}1$ in quartz) - have

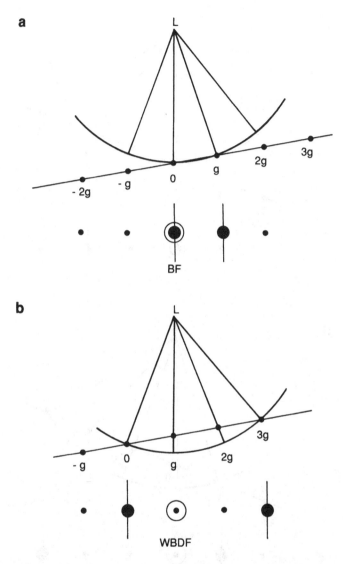

Figure 5.19. Ewald sphere diagrams and corresponding diffraction patterns illustrating the procedures for setting up the conditions for weak beam dark field imaging using the first-order diffracted beam **g**. (*Continued, p. 160*)

kinematical-like characteristics. They are more or less complementary to DF images obtained under conditions like those of Figure 5.19(b), as well as being significantly narrower and simpler than the images obtained under conditions corresponding to those in Figure 5.19(a). Sometimes it is

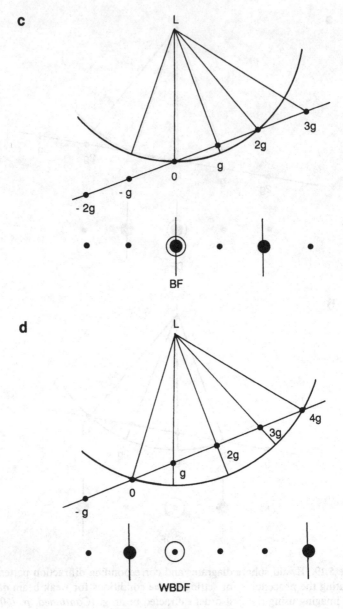

Figure 5.19. *Continued.*

difficult to achieve two-beam conditions for a weaker reflection; and in BF, double images of dislocations are frequently observed when two sets of reflecting planes that produce contrast on opposite sides of the dislocation are operating simultaneously. Therefore, for weaker reflections, it is usually advantageous to form a DF image under conditions such as those in Figures 5.19(b) or (d). The observed images agree closely with those predicted by the kinematical theory (Figures 5.17 and 5.18), even for very weak reflections such as $g = 0003$ in quartz with dislocations for which $g \cdot b = 3$.

Cockayne (1973) has given a detailed account of the principles and practice of the WBDF technique, including weak beam image profiles of dislocations calculated using a six-beam dynamical theory. The diffracting conditions that optimize the capabilities of the microscope are the following:

(a) $s_g \geq 0.2 \text{ nm}^{-1} = 0.02 \text{ Å}^{-1}$

(b) $st_g \geq 5$

(c) No other reflections (systematic or nonsystematic) strongly excited.

Under these conditions, the image peak of a dislocation is close (≤ 2 nm) to the dislocation core and is sufficiently narrow (≈ 15 nm at half-height) to enable the position of the dislocation to be accurately defined experimentally. However, we can use less stringent diffracting conditions if a lower accuracy is sufficient for the particular dislocations being studied, or if the diffracted intensity is too low under the optimum conditions to record the image with a reasonable exposure time.

5.6 Determination of Burgers vectors

It is not usual to attempt to determine the Burgers vector b of a dislocation entirely from the nature of the images observed under different diffracting conditions. For any crystalline mineral, the most likely Burgers vectors are the shortest lattice vectors, and some indications of the possible Burgers vectors may have been obtained from macroscopically observed slip planes and slip directions. Thus, the task of determining the Burgers vector of a dislocation observed in the electron microscope is essentially a matter of matching the images observed under various diffracting conditions with the images predicted by the appropriate theory for the possible Burgers vectors.

In its simplest form, this matching of observation and theoretical prediction involves finding diffraction vectors g for which the dislocation is invisible (or very weak) and determining the *direction* of b by assuming that the condition $g \cdot b = 0$ is satisfied. However, $g \cdot b = 0$ is a sufficient condition for invisibility only for a pure screw dislocation, and the invisibility of a pure edge dislocation also requires $g \cdot b \times u = 0$. Thus, invisibility of a dislocation of mixed character is not expected for any reflection, even though the condition $g \cdot b = 0$ is satisfied, because $g \cdot b \times u \neq 0$. The contrast increases with increasing value of $\mathbf{g} \cdot \mathbf{b} \times \mathbf{u}$, as can be seen in Figure 5.18, and for the special case of an edge dislocation loop lying in the plane of the foil shown in Figure 5.15. However, the contrast arising from $\mathbf{g} \cdot \mathbf{b} \times \mathbf{u}$ being nonzero is expected to be appreciable only when the crystal is near the reflecting position ($s \approx 0$), as discussed in Section 5.4.2. Therefore, the $\mathbf{g} \cdot \mathbf{b} = 0$ condition for invisibility may be applicable for WBDF imaging conditions, even when the value of $\mathbf{g} \cdot \mathbf{b} \times \mathbf{u}$ is large. The $\mathbf{g} \cdot \mathbf{b} = 0$ criterion for invisibility can be used only for determining the direction of the Burgers vector. However, in many cases, the magnitude of $\mathbf{b}$ is obvious from examination of the crystal structure if the direction is known.

Another method of determining the Burgers vector has been given by Ishida et al. (1980). This method is based on the fact that when a dislocation intersects the surface of a foil of varying thickness, the number n of thickness fringes that terminate at the intersection in a WBDF image is equal to the magnitude of $\mathbf{g} \cdot \mathbf{b}$. However, n can be positive or negative depending on the signs of $\mathbf{g}$ and $\mathbf{b}$. Thus, for a given dislocation, the sign of n is determined by the sign of $\mathbf{g}$. Consequently, a change in the sign of $\mathbf{g}$ will cause the terminating fringes to shift from one side of the image to the other, with respect to the intersection of the dislocation with the surface. This technique can be used to determine the direction, magnitude, and sign of the Burgers vector of the dislocation, provided three images of this type can be obtained using three noncoplanar diffraction vectors $\mathbf{g}$.

Because the image of a dislocation usually lies to one side of the dislocation, as in Figures 5.11(b) and 5.17, the sign of $\mathbf{b}$ can also be determined from this effect, provided the signs of $\mathbf{g}$ and $\mathbf{s}$, and the side on which the image lies, are known. These ideas can also be used to determine the nature of prismatic dislocation loops (i.e., whether they are of vacancy or interstitial type); the method is discussed in detail by Edington (1975). Examples of experimental determinations of Burgers vectors will be given in later chapters in sections dealing with dislocations in specific minerals.

In the preceding discussions of the contrast of dislocations, we have assumed that the materials are elastically isotropic in spite of the fact that most crystals (including cubic crystals, with the exception of tungsten) are elastically anisotropic. Thus, a failure of the $\mathbf{g} \cdot \mathbf{b} = 0$ invisibility condition might not be surprising for dislocations in most of the important rock-forming minerals. However, as stated earlier, this simple condition sometimes works remarkably well for WBDF images. Ordered β-brass is an example of a material that is extremely anisotropic; under dynamical conditions, screw dislocations for which $\mathbf{g} \cdot \mathbf{b} = 0$ are visible as a pair of closely spaced images with no contrast at the position of the core of the dislocation (Head 1967a). When this occurs, the only satisfactory method of determining the Burgers vector is the matching of dynamical images of the dislocation observed for a number of different reflections with the images calculated using anisotropic elasticity theory. This matching can be facilitated by using the technique developed by Head (1967a, b) whereby the calculated intensities are used to form a complete two-dimensional image: a theoretical or computed micrograph. The advantage of this technique is that the theoretical calculations are presented in the same form as the experimental image, so that it is much easier for an observer to compare the theoretical and experimental situations. The technique is also applicable to partial dislocations and planar defects. A detailed account of the principles of the technique and the procedures involved in producing the simulated images, together with examples, are given by Head et al. (1973).

5.7 Small inclusions and precipitates

5.7.1 Inclusions with negligible strain fields.   Small inclusions which have a negligible strain field can be made visible by several mechanisms. Some contrast may arise simply because these features influence the normal absorption. However, they are usually revealed either by structure factor contrast (Ashby and Brown 1963b) or by phase contrast (Charai and Boulesteix 1983) mechanisms. These will now be discussed.

The origin of structure factor contrast can be understood in the following way. When a wedge-shaped crystal is set at or near the exact Bragg angle, we observe a set of thickness fringes. At the exact Bragg angle ($s = 0$), the spacing between the fringes is a maximum and the intensity I_0 of the transmitted beam (BF images) as a function of the thickness z (without absorption) is given by Eq. (4.59a),

$$I_0 = \cos^2 \frac{\pi z}{t_g} \tag{5.25}$$

and is shown graphically in Figure 4.9. The *change* in intensity with thickness is obtained by differentiating I_0 with respect to z. Thus,

$$\Delta I_0 = -\frac{2\pi}{t_g} \Delta z \sin \frac{\pi z}{t_g} \cos \frac{\pi z}{t_g} \tag{5.26}$$

A small inclusion of thickness Δt ($\ll t_g$) and extinction distance t_{gi} increases the effective thickness of the foil by an amount

$$\Delta z = \Delta t \left[\frac{1}{t_{gi}} - \frac{1}{t_g} \right] \tag{5.27}$$

which, with Eq. (5.26), leads to

$$\Delta I_0 = -2\pi \Delta t \left[\frac{1}{t_{gi}} - \frac{1}{t_g} \right] \sin \frac{\pi z}{t_g} \cos \frac{\pi z}{t_g} \qquad (5.28)$$

From this expression we see that the inclusions will be invisible if $\Delta I_0 = 0$, which will be true for thicknesses $z = t_g/2, t_g, 3t_g/2, \ldots$. However, the inclusions will be visible with maximum contrast when $z = t_g/4, 3t_g/4, 5t_g/4, \ldots$.

For a small void or bubble that makes no contribution to the intensity of the Bragg reflection, $t_{gi} = \infty$, and from Eq. (5.27), $\Delta z = -\Delta t/t_g$. Therefore, such an inclusion decreases the effective thickness of the foil; it is clear that the inclusion will appear either darker or brighter than background depending on whether it is on the thick side or the thin side of a dark thickness fringe in a BF image. If a void is located either in the middle of a dark thickness fringe or midway between an adjacent pair, it will be invisible. These ideas are illustrated diagrammatically in Figure 5.20. Of course, any small spherical precipitate for which $t_{gi} > t_g$ gives rise to this type of contrast. If $t_{gi} < t_g$ then the reverse contrast is observed. Examples of small glassy spheres in quartz that are imaged in this way are shown in Figure 8.55.

Under the conditions for which the inclusions are most visible (i.e., for thickness $z = t_g/4$, and so on), the contrast (or relative intensity change) is given by

$$\left| \frac{\Delta I_0}{I_0} \right| = 2\pi \Delta t \left| \frac{1}{t_{gi}} - \frac{1}{t_g} \right| \qquad (5.29)$$

This equation can be used to estimate the minimum size of inclusion that can be observed, provided this is not limited by the resolution of the microscope. As an example, consider a spherical void of diameter Δt in quartz imaged using the strongest reflection $10\bar{1}1$ for which $t_g = 68.5$ nm at 100 kV. If the minimum contrast $|\Delta I_0/I_0|$ that can be detected is 10%, then $\Delta t (\min) = 0.1 t_g/2\pi = 1.1$ nm. In practice, this figure is probably optimistically small.

Inclusions of the order of 2 nm diameter or larger can also be made visible by a phase contrast mechanism. When a crystal is oriented so that no strong, low-order diffracted beams are operating, all the Fourier coefficients V_g of the potential are negligibly small except the mean inner potential V_0, which is effectively the refractive index of the crystal for electrons (as explained in Sections 4.1 and 4.2). If the mean inner potential V_{0i} of an inclusion is different from the mean inner potential V_0 of the matrix, then the inclusion can be considered as a phase object. In the light microscope, a phase object is usually barely visible at exact focus; but if the objective lens is slightly defocused, it will be seen with high contrast.

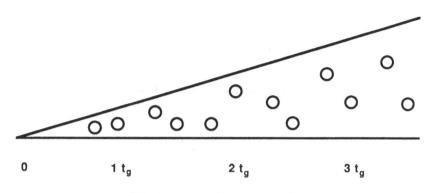

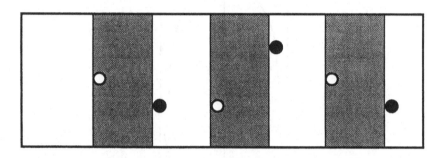

Figure 5.20. Schematic diagram illustrating the imaging of small inclusions (voids or bubbles) in a wedge-shaped crystal using structure factor contrast.

This behavior is also observed in the electron microscope and has been analyzed by Charai and Boulesteix (1983). They have shown that the intensity $I(x, \Delta f)$ of a phase object (whose thickness t' is a function of position x), relative to a background intensity of unity in a BF image, is given by

$$I(x, \Delta f) = 1 + \frac{\Delta f(V_{0i} - V_0)}{2E} \cdot \frac{d^2 t'}{dx^2} \qquad (5.30)$$

where E is the accelerating voltage.* It is clear that when the defocus Δf is zero, $I(x, \Delta f)$ is unity and the inclusion is invisible. For a spherical inclusion of diamter Δt, in Figure 5.21(a),

---

* In Charai and Boulesteix (1983) there is an error of sign in the derivation of Eq. (5.30). The authors also use the sign convention that $\Delta f < 0$ for underfocus. Equation (5.30), as written here, is correct for the opposite sign convention that was adopted in Section 2.6.

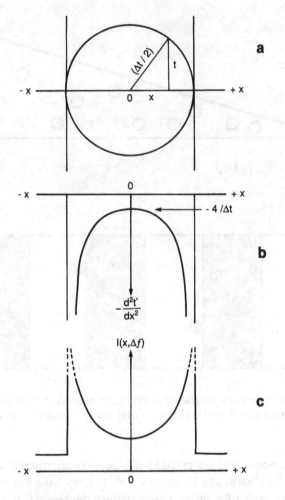

Figure 5.21. Imaging a spherical inclusion by phase contrast. (a) Coordinate system used to describe the inclusion. (b) Plot of Eq. (5.32). (c) Calculated intensity profile across the inclusion.

$$t' = 2t = 2\left[\left(\frac{\Delta t}{2}\right)^2 - x^2\right]^{1/2} \tag{5.31}$$

for $-(\Delta t/2) \le x \le +(\Delta t/2)$, and

$$\frac{d^2t'}{dx^2} = -\frac{2(\Delta t/2)^2}{[(\Delta t/2)^2 - x^2]^{3/2}} \tag{5.32}$$

which is plotted in Figure 5.21(b).

Therefore, if $V_{0i} < V_0$ (as is the case for a void, bubble, or amorphous inclusion), then $I(x, \Delta f)$ is greater than unity at underfocus (i.e., for $\Delta f > 0$). The intensity profile through the center of such an inclusion is shown in Figure 5.21(c), from which it is clear that the BF image is a disk that is brighter than background with a bright (Fresnel-like) fringe around the edge. For overfocus ($\Delta f < 0$), the image is of opposite contrast. However, if $V_{0i} > V_0$, the image will be darker than background for $\Delta f > 0$ and brighter than background for $\Delta f < 0$. Note that Eq. (5.30) predicts that the intensity $I(x, \Delta f)$ tends to infinity at the circumference of the image of a spherical inclusion. Nevertheless, the theory is in good qualitative agreement with observations if the amount of defocus is small ($|\Delta f| <$ 100 nm). Bursill (1983), for example, has used this technique to image voidites (voidlike defects) in diamond with diameters ranging from 0.75 to 7.5 nm. For large defocuses ($|\Delta f| \geq 1 \ \mu$m) the contrast is high, but it may not be possible to infer the size of the inclusion directly from the size of the image. Large defocuses have been used to image bubbles in deformed synthetic quartz when the bubbles are difficult to identify in the presence of a high density of dislocations (see Section 9.4.3).

5.7.2 Inclusions with a strain field.   Small inclusions that strain the surrounding crystal give rise to contrast in a way similar to a dislocation. As an example, consider a small spherical particle that produces an elastic compressive strain in the surrounding crystal. If we assume that the surrounding crystal is elastically isotropic, then the strain field will be purely radial, as in Figure 5.22(a). The planes parallel to Oy, for example, are bent around the inclusion. The bending is obviously a maximum in the Ox direction and zero along Oy. Thus, if an image is formed with these planes reflecting strongly (i.e., with g normal to Oy, as shown), then there will be a *line of no contrast* through the center of the inclusion and parallel to Oy. Because the distortion is purely radial, a line of no contrast is observed normal to g for any reflecting plane. On each side of the line of no contrast there is a series of bright and dark lobes, as shown schematically in Figure 5.22(b).

Ashby and Brown (1963a) have shown that the image width w (defined as the width corresponding to a change of transmitted intensity of x percent from background intensity) depends principally on the dimensionless parameter $P_s = \epsilon g r_0^3 / t_g^2$, where ϵ is the misfit strain between the inclusion and the matrix, and r_0 the radius of the constrained precipitate particle. They constructed curves of w/t_g as a function of $\log_{10} P_s$ for values of x of 2, 10, 20, and 50 percent, from which it is possible to determine ϵ by

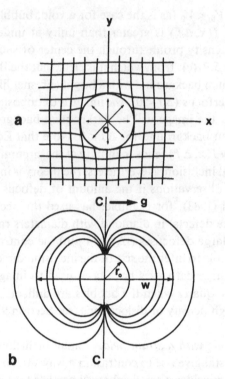

Figure 5.22. Imaging a spherical inclusion by its strain field in the surrounding isotropic crystal matrix. (a) Diagram illustrating the compressive strain around the inclusion. (b) Schematic diagram illustrating the nature of the image and, in particular, the line of no contrast CC normal to the diffraction vector **g**.

measuring w and r_0 from an electron micrograph taken with a reflection **g** whose extinction distance is t_g. The 20-percent curve is shown in Figure 5.23. Curves were also constructed from which the smallest inclusion that will be visible can be calculated if ϵ, g, and t_g are known. If, for example, $\epsilon = 0.01$, $g = 2$ nm^{-1}, and $t_g = 100$ nm, the smallest visible particle will have a diameter of about 20 nm. The sign of the strain can also be determined from the asymmetry of the DF images of particles near either surface of the thin crystal.

Ashby and Brown (1963b) extended the analysis to cover the type of strain field found around platelike precipitates whose mismatch with the surrounding crystal is appreciable only in a direction normal to the plate. The analysis is also applicable to the strain normal to the plane of a prismatic dislocation loop of Burgers vector **b**. When the plane of the loop or precipitate is more or less normal to the foil, the images are similar to that

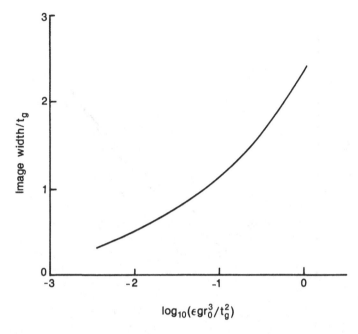

Figure 5.23. Variation of the 20-percent image width with ϵ, g, r_0, and t_g for a spherical inclusion in an elastically isotropic crystal matrix at $s = 0$. (After Ashby and Brown 1963a.)

of a spherical inclusion except that the line of no contrast is always parallel to the intersection of the plane of the dislocation loop or precipitate with the plane of the foil, independently of the direction of the diffraction vector **g**.

Curves of the 20-percent image width (w/t_g) as a function of

$$P_l = \log_{10}\left[\frac{r_l^2 g b \cos\theta}{t_g^2}\right]$$

for several values of (r_l/t_g) are shown in Figure 5.24. r_l is the radius of the dislocation loop or precipitate and θ the angle between **g** and the normal to the plane of the loop (i.e., the angle between **g** and **b**). From these curves, we can determine the magnitude of **b** by measuring w and r_l from an electron micrograph taken with a known reflection **g**. For a precipitate plate of thickness a the "equivalent $|\mathbf{b}|$" is given by $a\epsilon$, where ϵ is the constrained strain for the precipitate, provided $a/r_l \ll 1$. The contrast is most easily understood when the plane of the loop or inclusion is normal to the plane of the foil, and $\theta = 0$.

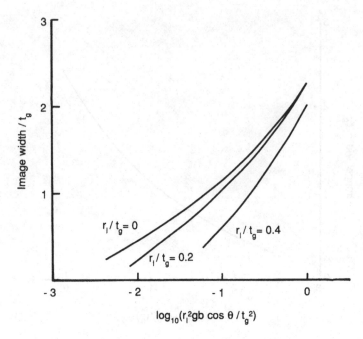

Figure 5.24. Variation of the 20-percent image width with g, b, and θ for a prismatic dislocation loop (or a platelike inclusion) in an isotropic crystal matrix for several values of r_l. Thickness of foil $= 5t_g$; $s = 0$; $t_g'' = 10t_g$. (After Ashby and Brown 1963b.)

In these strong-beam images, the inclusions are revealed by the strain they induce in the surrounding crystal. The image is much larger than the inclusion itself, which may not actually be seen. Thus, it may be extremely difficult to estimate from such images the thickness a and the radius r_l of the inclusion. This difficulty can be overcome in two ways. For example, if an image is formed using a reflection for which $\mathbf{g}$ is perpendicular to $\mathbf{b}$, no strain contrast will be produced (ideally), and the inclusion itself may be seen due to other contrast mechanisms. Again, since weak-beam images of strain fields only show contrast where the strain field is high, the image peak lies close to the inclusion itself, and so its size and shape can be estimated more accurately in a WBDF image than in a BF image formed under strong dynamical conditions.

5.8 Density of dislocations and other defects

To study defects by TEM, the defects must be present in a high enough concentration that there is a good chance of actually finding a defect in the field of view at the magnification being used. To get a feel for the sort of concentrations that are necessary, consider a specific example. The density ρ of grown-in dislocations in synthetic quartz crystals is characteristically of the order of 10^3 cm of dislocation line per cm^3; that is, $\rho = 10^3$ cm/cm$^3 \equiv 10^3$ cm^{-2} (McLaren et al. 1971). This density corresponds to an average spacing between dislocations of a little greater than 300 μm. Thus, the probability of finding a dislocation within a randomly selected field of view of area 10 μm × 10 μm (corresponding approximately to the full area of a standard photographic film at a relatively low magnification of ×10,000) is only about 1 in 1,000. Clearly a dislocation density of at least 10^6 cm^{-2} is required in order to be reasonably sure of observing at least one dislocation intersecting a randomly selected area of this size. It is not surprising, therefore, that the grown-in dislocations in synthetic quartz have not been investigated by TEM.

Ham (1961) and Ham and Sharpe (1961) have discussed the two main methods for determining dislocation densities from electron micrographs of thin foils. In the first method, a set of random lines with a total length L is marked on a given area A of the micrograph, and the number of intersections N that dislocations make with these lines is measured. The dislocation density is then given by

$$\rho = \frac{2N}{Lt} \qquad (5.33)$$

where t is the thickness of the foil. This method assumes that the dislocations are randomly oriented with respect to the plane of the foil, which may not always be so. However, a more serious disadvantage is that the method requires a knowledge of the foil thickness, which is often not easy to determine. The simplest method for estimating t at any point is to count the number of *bright* thickness fringes from the edge of the foil in a BF micrograph at $s = 0$ and calculate t from $t/t_g = 0, 1, 2, \ldots$ (see Figures 4.9 and 4.14). In the second method the number of intersections N' that the dislocations make with both foil surfaces in an area A is measured, and ρ is calculated from

$$\rho = \frac{2N'}{A} \qquad (5.34)$$

This method has the advantage of not requiring a knowledge of the foil thickness t, but it becomes very difficult to count surface intersections for dislocation densities higher than about 10^9 cm^{-2}. Clearly, measurements of the number-density of small dislocation loops or small inclusions (such as bubbles or voids) requires a knowledge of thickness t.

6

High-resolution transmission
electron microscopy

6.1 Introduction

The term *high-resolution transmission electron microscopy* (HRTEM) is now generally taken to mean an imaging mode in which at least two beams pass through the objective lens aperture. The image produced in this way shows the periodicity of the crystal and, for this reason, is often referred to as a *lattice image*. A typical example is shown in Figure 6.1. The basic optical principles involved were discussed in detail in Chapter 1 and illustrated in Figure 1.11, but several important aspects of HRTEM have not yet been considered. It is clear that we require a microscope with a resolution of 0.4 nm or better to produce lattice images.

The optical treatment in Chapter 1 is concerned with the nature of the image of a simple two-dimensional grid formed by a perfect lens, in focus, using monochromatic light. However, to understand the nature of the lattice image formed in the electron microscope, we must take into account the thickness and the orientation of the three-dimensional crystal, the defect of focus and the aberrations of the objective lens (see Chapter 2), and the beam convergence, because all these factors influence the relative phases of the diffracted beams that are permitted to pass through the objective aperture.

In this chapter, a brief introductory discussion is given of (i) the experimental variables that influence the nature of HRTEM images, (ii) the experimental techniques involved, and (iii) the basis of image interpretation. More detailed accounts have been given by Spence (1981) and Veblen (1985a).

6.2 Experimental variables

6.2.1 Crystal orientation and thickness. The way in which changes of orientation and crystal thickness influence the nature of the image can

Figure 6.1. A high-resolution TEM lattice image of a crystal of Y_2BaCuO_5. (Courtesy J. D. Fitz Gerald.)

be understood easily by considering the simple case of a one-dimensional lattice image formed using the straight-through beam and a single diffracted beam $\mathbf{g}$. From Eqs. (4.54) and (4.57), the total wavefunction $\Psi = \psi_0 + \psi_g$ is given by

$$\Psi = \exp(2\pi i \mathbf{K} \cdot \mathbf{r}) \exp(\pi i s z) [T + S \exp 2\pi i \mathbf{g} \cdot \mathbf{r}] \qquad (6.1)$$

where

$$T = \cos \pi \sigma z - \frac{is}{\sigma} \sin \pi \sigma z$$

$$S = \frac{i}{\sigma t_g} \sin \pi \sigma z$$

$$\sigma = \frac{1}{t_g} \sqrt{1 + (s t_g)^2}$$

and s specifies the deviation from the exact Bragg angle. From Eq. (6.1), the resulting intensity I is

$$I = \Psi \Psi^* = TT^* + SS^* + TS^* \exp(-2\pi i \mathbf{g} \cdot \mathbf{r}) + ST^* \exp(2\pi i \mathbf{g} \cdot \mathbf{r}) \qquad (6.2)$$

This can be written as

$$I = I_0 + I_g + 2\sqrt{I_0 I_g} \sin(2\pi \mathbf{g} \cdot \mathbf{r} + \varphi) \qquad (6.3)$$

where

$$2\sqrt{I_0 I_g} \sin \varphi = TS^* + T^*S = -\frac{2s}{\sigma^2 t_g} \sin^2 \pi\sigma z$$

$$2\sqrt{I_0 I_g} \cos \varphi = i(T^*S - S^*T) = -\frac{2}{\sigma t_g} \sin \pi\sigma z \cos \pi\sigma z$$

Because we have neglected absorption,

$$I_0 + I_g = 1$$

and Eq. (6.3) becomes

$$I = 1 + 2\sqrt{I_g(1 - I_g)} \sin(2\pi \mathbf{g} \cdot \mathbf{r} + \varphi) \qquad (6.4)$$

If we select a coordinate system in which the x-axis is parallel to $\mathbf{g}$, then $\mathbf{g} \cdot \mathbf{r} = |\mathbf{g}|x = x/d(hkl)$, and Eq. (6.4) becomes

$$I = 1 + 2\sqrt{I_g(1 - I_g)} \sin\left(\frac{2\pi x}{d} + \varphi\right) \qquad (6.5)$$

Thus, the image consists of a uniform background modulated by a sinusoidal ripple of period equal to $Md(hkl)$, where M is the magnification – that is, an image of alternate bright and dark fringes with a spacing between adjacent dark (or bright) fringes of $Md(hkl)$. Compare with Figure 1.11(8a, b).

The visibility of the fringes depends on crystal orientation and thickness through the influence of s and z on $I_g = SS^*$. Clearly, the contrast is a maximum when $I_g = \frac{1}{2}$. When $s = 0$, maximum contrast is achieved only for certain thickness, as can be seen from Figure 4.8. Changes of s and z, through their influence on φ, also cause the fringe pattern to shift; consequently, the position of the fringes bears no simple relation to the position of the atomic planes. Thus, changes in the visibility, orientation, and spacing of the fringes can occur due to local variations in crystal thickness and/or deviation from the exact Bragg condition. The modifications to lattice fringes that we expect from changes in thickness and orientation are shown diagrammatically in Figure 6.2.

6.2.2 Defect of focus. In the Gaussian image plane, defined by Eq. (1.1), the image has very low contrast. The nature of the image changes above and below this plane because the waves have different relative phases. When the objective lens is defocused by an amount Δf relative

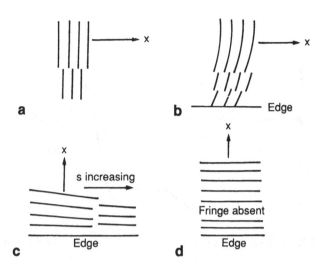

Figure 6.2. Schematic diagrams showing the modifications of lattice fringes that result from changes in thickness and orientation. (a) Shift of fringes due to a step in thickness; (b) bending and shift of fringes due to a continuous change in thickness near the specimen edge; (c) fringe shift and change of spacing due to change in orientation; and (d) change of fringe spacing due to a change in thickness near an edge. (After Hirsch et al. 1965.)

to the Gaussian focus, there is a phase shift χ_f given by Eq. (2.11). Since, for small angles, $\beta = 2\theta = \lambda/d = \lambda|\mathbf{g}|$ by Bragg's law, χ_f becomes

$$\chi_f = \Delta f \pi \lambda |\mathbf{g}|^2 \qquad (6.6)$$

6.2.3 Lens aberrations. The stability of the accelerating voltage and lens currents in modern electron microscopes is such that chromatic aberration is not a serious problem in most applications of HRTEM. However, electron lenses with negligible spherical aberration have not yet been developed. Spherical aberration introduces phase shifts because the beams passing through the outer zones of the lens are brought to a focus in a different plane from the Gaussian image plane which, by definition, applies to beams close to the axis of the lens (i.e., paraxial rays). The phase shift introduced by spherical aberration is $-(\frac{1}{2})\pi C_s \lambda^3 |\mathbf{g}|^4$.

6.2.4 Phase contrast transfer function. The phase shifts due to the combination of spherical aberration and defect of focus can be combined into a single phase factor χ given by

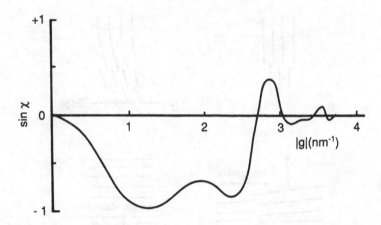

Figure 6.3. A phase contrast transfer function for a modern objective lens.

$$\chi = -(\tfrac{1}{2}\pi C_{s}\lambda^{3}|\mathbf{g}|^{4} - \Delta f\,\pi\lambda|\mathbf{g}|^{2}) \qquad (6.7)$$

Sin χ is called the *phase contrast transfer function* (PCTF) of the objective lens.

It is clear from Eq. (6.7) that the phase shifts associated with spherical aberration can be partially offset by having a positive value of the defocus Δf, corresponding to a slight underfocus of the objective lens. For a given degree of underfocus, sin χ varies with $|\mathbf{g}|$, as shown schematically in Figure 6.3. We can see that the phase is negative until a certain value of $|\mathbf{g}|$ is reached; this point is called the first-zero crossing (FZC) of the PCTF. The value of $|\mathbf{g}|$ at the FZC varies as the focus of the objective lens is changed. The focus at which the value of $|\mathbf{g}|$ at the FZC is a maximum ($\approx\sqrt{2}C_{s}^{-1/4}\lambda^{-3/4}$) is called the Scherzer focus and, from Eq. (6.6), corresponds to a defocus of $\Delta f = (C_{s}\lambda)^{1/2}$. It has been shown (see Spence 1981) that if the objective lens aperture (placed symmetrically about the central beam) has an angular width such that it passes only those diffracted beams for which $|\mathbf{g}|$ is less than or equal to the value of $|\mathbf{g}|$ that corresponds to the Scherzer focus, then, for a very thin crystal (≈ 10 nm), the image will be interpretable; that is, the contrast will be such that tunnels in a structure will be white and regions of high atomic density will be black on a positive print. From the dependence of the value of $|\mathbf{g}|$ at the FZC on C_{s} and λ, it is clear that a lens will produce a faithful image for larger values of $|\mathbf{g}|$ (i.e., for smaller d-spacings) if the values of C_{s} and λ are reduced.

6.2.5 Beam convergence. In HRTEM, it is essential to illuminate the region of the specimen being studied with a highly coherent electron beam. The degree of spatial coherence depends on the convergence of the beam: the more convergent the beam, the lower the degree of coherence, as discussed in Section 1.6.

6.3 Experimental techniques

The previous section showed that interpretable HRTEM images are not obtained unless quite stringent experimental conditions are fulfilled. The important questions – "What instrumentation does one need, and what does one actually do, to obtain an interpretable HRTEM image?" – have recently been considered by Veblen (1985a) who has described in some detail the experimental techniques that he has found essential for successful high-resolution microscopy. Because descriptions of tricks-of-the-trade are relatively rare in the literature, his main points are summarized in the following subsections.

6.3.1 Instrumentation. For many mineralogical studies, the minimum requirement is a 100-kV microscope with a point-to-point resolution of 0.4 nm, fitted with a double-tilt (or tilt-rotate) specimen holder.

6.3.2 Specimen preparation. Most HRTEM investigations have used specimens in the form of crushed fracture fragments supported on a holey carbon film attached to a standard copper grid. Specimens thinned by ion (or atom) bombardment are also used, but the amorphous film which tends to form on the surfaces of foils prepared in this way is sometimes too thick for successful high-resolution imaging. See also Section 2.7.

6.3.3 Alignment of the electron microscope. It is absolutely essential that the microscope be accurately aligned and that the astigmatism of the objective lens be fully corrected with the same objective aperture as used for the high-resolution imaging.

6.3.4 Specimen orientation. The specimen must be oriented so that the incident electron beam is parallel to the zone axis of interest. This is done by tilting the specimen while observing the selected area (or the convergent beam) diffraction pattern.

Figure 6.4. The granularity in a carbon film in a slightly underfocused image.
Note also the bright Fresnel fringe at the edge of the hole in the film.

6.3.5 Specimen illumination.   To obtain a relatively coherent electron
beam a small source size is required. Thus, a small condenser aperture is
used, and LaB_6 and pointed filaments are preferable to the standard hairpin filament, although not essential. The illumination is focused onto the
specimen using the second condenser lens, and exposure times should be
only a few seconds.

6.3.6 Objective aperture.   In general, the aperture chosen should be
just large enough to pass all those diffracted beams with $|\mathbf{g}|$ just below the
FZC of the PCTF at Scherzer focus, as explained in Section 6.2. It is critical that the aperture be accurately centered.

6.3.7 Focusing.   To obtain the optimum defocus setting for HRTEM
imaging, it is necessary to recognize some reference focus and to know
what magnitude of defocus each step of the focus knob corresponds to.
The best reference is the focus at which there is a minimum contrast in the
granular carbon support film which behaves as a phase object (see Figure 6.4). For crystals with large unit cells, the image at optimum defocus

often has higher contrast than at other focus settings and, therefore, can be recognized directly.

6.3.8 Magnification. Most work can be performed at magnifications of about ×500,000, but magnifications of ×800,000 or higher may be necessary in particular cases.

6.3.9 Specimen damage in the beam. Some specimens that appear to be quite stable at low magnifications (< ×50,000) are often observed to degrade quite rapidly when viewed at the magnifications necessary for HRTEM. In general, the rate of damage appears to be decreased by increasing the accelerating voltage and by cooling the specimen. With practice, patience, and experience, it may be possible to record acceptable high-resolution images of beam-sensitive materials by adjusting the experimental conditions in one area of the specimen and then moving rapidly to an adjacent area and photographing immediately, with as short an exposure time as possible.

Any researcher (with or without experience of conventional two-beam TEM) should read Veblen's words of wisdom carefully before beginning to use HRTEM in a mineralogical investigation.

6.4 Interpretation of HRTEM images

The image obtained from a very thin crystal under optimum operating conditions (of defocus, etc.) is very similar to a drawing of the crystal structure (determined from x-ray diffraction) projected onto the plane normal to the incident electron beam. The image can be described as a map of the charge density (or potential or atomic density) throughout the crystal: the darker areas correspond to regions of higher charge density. However, the intuitive interpretation of HRTEM images, though often possible, can be very risky, especially if the image detail is on a scale that is finer than the theoretical point-to-point resolution of the electron microscope (≈ 0.3 nm). In view of this, it is now usual to compare the images observed under known conditions with images that have been computer simulated for model structures. Computer simulation can be used to study the effects of different experimental parameters (such as crystal thickness, defocus, and lens aberrations) and different structural models. The matching of these calculated images with those actually observed can be used to decide upon the best model for the structure. If an experimental through-focus series of images is found to match well with a calculated series, then it can be safely assumed that the correct structural model has been chosen.

The image calculation has three main parts. First, the nature of the electron wave function at the bottom face of the crystal of a given thickness is calculated using the n-beam dynamical theory of electron diffraction as formulated by Cowley and Moodie (1957). In this calculation, the crystal is imagined to be divided into slices perpendicular to the incident electron beam, and the intensities of the diffracted beams are calculated as the electrons propagate through successive slices. Second, the way in which the experimental parameters of the electron microscope alter the electron wave function to form the image is calculated. Last, the calculated image is presented as a halftone picture which can be compared visually with the experimental image.

The simulation of HRTEM images has been reviewed recently by Self and O'Keefe (1988), and an excellent summary has been given by Veblen (1985a).

6.5 Discussion of HRTEM images

The characteristic of HRTEM that makes it of outstanding value in the study of minerals and other crystals is that the image formed under optimum conditions can be directly related to the crystal structure of the specimen, because the image is a projection of the charge density of the crystal on a plane normal to the incident electron beam. For a relatively simple structure in which the atoms lie in columns separated by empty tunnels parallel to the beam, the image resembles a drawing of the projected structure. Thus, domains of a second phase (even if they are only a few unit cells in extent) are revealed in HRTEM images, and their structure and orientational relationship to the matrix crystal can be determined. HRTEM images have also provided important information about twin and grain boundaries, and individual point defects have been identified in the lattice images of complex oxides. However, complications can arise in the images of boundaries between two structures if there is also a change in orientation. At the boundaries across which there is a misorientation, "dislocations" are sometimes observed in two-beam lattice images. However, these features do not necessarily correspond to real dislocations at the boundary; the dislocation in the image may be due to shifts and spacing changes of the fringes arising from changes in orientation, as illustrated in Figure 6.2. Conversely, a dislocation may give rise to an image that is not obviously interpretable (Cockayne, Parsons, and Hoelke 1971).

Although it is now possible to simulate the high-resolution image of a crystal defect of known structure, the calculations are usually very long. Thus, the methods of studying crystal defects (such as dislocations and the more common types of planar defects) discussed in Chapter 5, are clearly simpler and more generally appropriate. Nevertheless, the methods of HRTEM have provided information about the structure and defects of certain minerals that could not have been obtained in any other way; examples are given later.

6.6 Moiré patterns

Periodic images, known as moiré patterns, are produced from overlapping crystals that have slightly different lattice spacings or are slightly rotated with respect to each other. Due to double diffraction (see Section 3.5), additional diffracted beams are produced at small angles to the central beam and pass through the objective aperture. The periodic image is formed by the interference of these beams in the same way as a high-resolution lattice image. Because the periodicity of moiré patterns is related to the periodicities of the overlapping crystals, these images are a special case of lattice resolution and are discussed briefly here. More detailed accounts of moiré patterns have been given by Hirsch et al. (1965), Amelinckx (1964), and Edington (1975).

First, we consider two overlapping crystals of slightly different lattice spacings in the same orientation (parallel moiré). For simplicity, assume that the interface between the crystals is normal to the incident electron beam and that the specimen is oriented so that a row of systematic reflections P is excited in the top crystal, and a parallel row of systematic reflections Q is excited in the bottom crystal, as shown in Figure 6.5(a). Now suppose the beam P_1, which was diffracted by a set of planes (of spacing d_1) in the top crystal, is diffracted by a parallel set of planes (of spacing d_2) in the bottom crystal – that is, P_1 becomes the incident beam for the bottom crystal. This doubly diffracted beam is Q', which is equivalent to the beam $-Q_1$ when the origin O is shifted to P_1. The beam $-Q_1$ now lies close enough to the central beam to pass through the objective aperture, and so the image consists of a set of fringes (normal to OQ') whose spacing D is inversely proportional to

$$OQ' = |\Delta \mathbf{g}| = (g_2 - g_1) = \frac{1}{d_2} - \frac{1}{d_1}$$

Hence,

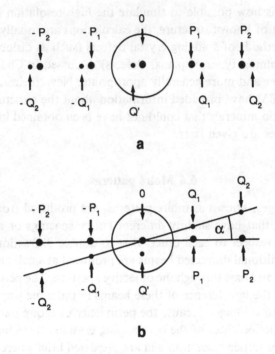

Figure 6.5. Diffraction patterns for (a) parallel and (b) rotational one-dimensional moiré fringes. $OP_1 = g_1$; $OQ_1 = g_2$; $OQ' = \Delta g = g_2 - g_1$.

$$D = \frac{d_1 d_2}{d_1 - d_2} \tag{6.8}$$

Second, we consider two crystals that have the same lattice spacings but the bottom crystal is rotated through a small angle α (rotational moiré). The diffraction pattern will appear as shown in Figure 6.5(b). Again, suppose the beam P_1 is diffracted by the top crystal and then becomes the incident beam for the bottom crystal. This doubly diffracted beam is Q' and is equivalent to $-Q_1$ when the origin is shifted from O to P_1. The moiré fringes will be normal to $OQ' = \Delta g$ (i.e., approximately parallel to the systematic rows) and the spacing D will be inversely proportional to $|\Delta g|$. Since $|g_1| = |g_2| = |g| = (1/d)$, we have

Figure 6.6. *Facing page.* Optical analogues showing the formation of a moiré pattern. (a) Superposition of two gratings of unequal spacing; (b) superposition of two identical gratings with a small relative rotation. The effects of an edge dislocation in one of the gratings are shown in (c) and (d).

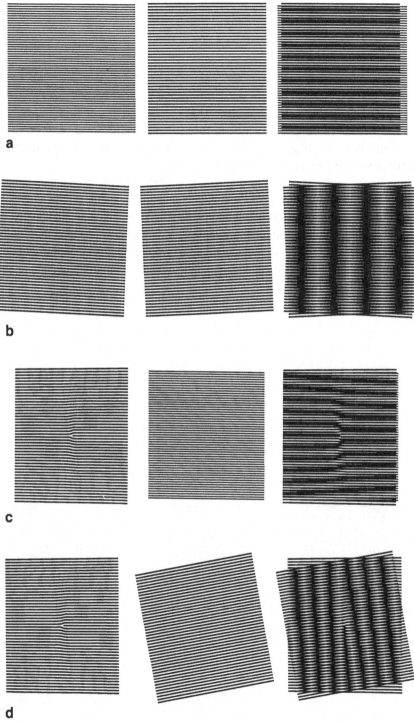

a

b

c

d

$$D = \frac{d}{\alpha} \qquad (6.9)$$

Thus, one-dimensional parallel and rotational moiré fringes can be clearly distinguished because they are always normal and parallel to **g**, respectively. Whether a one-dimensional or two-dimensional moiré pattern is produced depends on the number of doubly diffracted beams that pass through the objective aperture. It is clear from Figure 6.5 that moiré fringes will also be produced in dark-field images.

Optical analogues of one-dimensional parallel and rotational moiré patterns in two perfect "crystals" are shown in Figures 6.6(a, b).

Because the moiré patterns observed by TEM are, in effect, magnified images of the reflecting planes, crystal defects in either of the overlapping crystals become visible; see Figure 6.6(c, d). However, this technique has not been widely used for imaging crystal defects in minerals.

7

Chemical analysis in the transmission electron microscope

7.1 Introduction

Quantitative chemical analysis using the characteristic x-rays emitted by the specimen in the transmission electron microscope (TEM) is a natural extension of the standard electron microprobe, which is probably familiar to most mineralogists and geologists. However, because the TEM specimens are in the form of a thin foil, quantitative chemical information can be obtained from much smaller volumes of specimen and with relatively minor corrections compared with those necessary with the standard electron microprobe. The TEM technique has already made a significant contribution to mineralogical research, and its use is increasing rapidly. In this chapter, the essential features of x-ray chemical analysis in the TEM are summarized, together with some brief comments on the complementary technique of electron energy loss spectroscopy (EELS). A detailed account of the theory and practice of x-ray microanalysis and EELS has been given by Williams (1984) and is essential reading for anyone proposing to use these techniques.

When the high-energy electron beam in an electron microscope interacts with the specimen, the electrons may be deflected by the potential field of the atoms without loss of energy (i.e., the electrons are elastically scattered), or the electrons may be inelastically scattered (i.e., with loss of energy). The inelastically scattered electrons provide a means of detecting and determining the concentration of the atomic species present in a specimen. There are several mechanisms by which inelastically scattered electrons lose their energy. In the present context, perhaps the most important mechanism of energy loss involves the emission of an x-ray photon. The spectrum (intensity as a function of wavelength or photon energy) of the emitted x-rays consists of a continuous background (bremsstrahlung) on which is superimposed a number of sharp peaks whose positions are characteristic of the atoms present in the specimen.

The idea of focusing a small electron probe onto a specimen and deducing the chemical nature of the specimen from the wavelengths of the characteristic peaks of the emitted x-ray spectrum was suggested by Guinier, and the first electron microprobe x-ray analyzer was built by Castaing in the late 1940s. The first commercial microprobe, based on Castaing's principles, became available in 1956. Castaing's results stimulated research into x-ray microanalysis in many countries, and Duncumb (1962) described an instrument in which a wavelength-dispersive x-ray microanalyzer was added to a transmission electron microscope. A similar instrument was developed and produced commercially by Associated Electrical Industries in Britain, and Siemens also produced an analytical attachment for their electron microscopes.

The wavelength dispersive x-ray (WDX) spectrometer is based on the Bragg law and employs curved crystals that can be rotated so as to reflect the x-rays emitted by the specimen and focus them, one wavelength at a time, onto a detector. Although this technique allows high resolution of one wavelength from another, it has a number of practical disadvantages:

1 It takes a long time to acquire a spectrum, (a) because the intensity at each wavelength has to be determined sequentially as the spectrometer crystal is turned through different Bragg angles and (b) because long counting times are necessary at each wavelength due to low diffraction efficiency.
2 It is usually necessary to use a higher beam current than is normal for microscopy, which leads to increased contamination of the specimen and/or to radiation damage of beam-sensitive materials.
3 It is necessary to have a number of different crystals (usually computer controlled) in order to cover the required spectral range.

For all these reasons, the number of transmission electron microscopes fitted with a WDX spectrometer was small, and they have been superseded by the solid-state energy-dispersive x-ray (EDX) detector, which was developed in the late 1960s. Transmission electron microscopes fitted with an EDX spectrometer are now relatively common.

EDX spectrometers are much more efficient than WDX spectrometers. This increase in efficiency permits the use of smaller probe sizes with less current, resulting in higher spatial resolution and less contamination and beam damage of the specimen. A major advantage of EDX spectrometers is that they accept all the incoming x-rays simultaneously and then sort them electronically according to their energies so that the complete spectrum (over the energy range of interest) can be displayed on a video

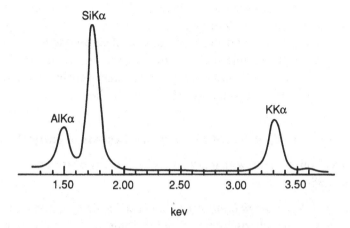

Figure 7.1. Typical EDX spectrum for a K-feldspar.

monitor. When a sufficient number of "counts" have been recorded, the spectrum is stored in the associated computer, after which subtraction of background and calculation of the chemical composition of the sample can be carried out. A typical EDX spectrum is shown in Figure 7.1.

There are, however, three main limitations of EDX spectrometers. The resolution of characteristic peaks is poorer than with WDX spectrometers, and the background is higher. Furthermore, the efficiency of the normal Li-doped Si detector with an 8 μm thick Be window falls off dramatically for elements of low atomic number Z; thus, only x-rays from Na ($Z = 11$) and heavier atoms can be detected. Windowless detectors are available, which can detect light elements such as B ($Z = 5$), but their use is not free of complications for quantitative analysis.

The intensities of the characteristic x-ray peaks depend on the crystallographic orientation of the specimen with respect to the incident electron beam. Significant changes may be observed if the crystal is set at or near the exact Bragg angle. For this reason it is usual to orient the crystal well away from any strong Bragg reflection when performing a quantitative analysis (Cherns, Howie, and Jacobs 1973). However, this orientation effect can be put to good use: it is the basis of a relatively new technique that has already been used successfully in a number of mineralogical investigations and will be discussed in some detail in Section 7.4.

The need for both qualitative and quantitative analyses of light elements in the transmission electron microscope has stimulated the development of electron energy loss spectrometry (EELS). In this technique, a magnetic prism mounted at the base of the electron microscope column separates electrons of various energies after the electrons have passed through the specimen, and the characteristic electron energy losses are used to identify the atomic species present. Quantitative analysis is still difficult, and so far EELS has not been used extensively in mineralogical research. Therefore, it will not be considered further.

7.2 Inelastically scattered electrons: mechanisms of energy loss

When electrons interact with matter, they lose energy by four mechanisms (Loretto 1984):

1   *Thermal diffuse or phonon scattering.* The incident electrons interact with atoms that are oscillating about their mean positions and lose or gain energy of the order of kT (≈ 0.025 eV at room temperature). This amount of energy is too small to be detected by any available electron spectrometer, and there is no evidence that these phonon-scattered electrons contain any microanalytical information.
2   *Plasmon scattering.* The incident electrons lose energy by exciting collective oscillations (called plasmons) of the valence electrons. The energy loss is of the order of 15 eV, and plasmon loss peaks are prominent in the low-loss region of electron energy-loss spectra.
3   *Single electron scattering.* Energy is transferred to single electrons (instead of to a large number as in plasmon scattering) by the fast incident electrons. Lightly bound valence electrons can be ejected with energies up to 50 eV; these are the electrons that are used to form secondary electron images in a conventional scanning electron microscope (Lloyd 1985). If an inner-shell electron is ejected, then the appropriate ionization energy must have been lost by the incident electron. This is the characteristic energy loss used in EELS to identify the atomic species present in the specimen.
4   *Direct radiation loss.* If the fast incident electrons are decelerated when they interact with the specimen, they emit x-ray photons directly with a continuous range of energies up to the energy E_0 of the incident electrons. This is the bremsstrahlung or continuous background radiation. The intensity I of the bremsstrahlung as a function of energy E is shown schematically in Figure 7.2. Note that I is zero when $E = E_0$ and rises rapidly as E decreases. For any given energy E, the intensity I increases with atomic number and beam current.

If the energy of the incident electron is high enough to ionize an atom by ejecting an inner-shell electron (the characteristic energy-loss electron),

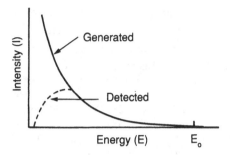

Figure 7.2. Schematic diagram of the energy spectrum of the bremsstrahlung, showing the difference between the radiation generated and the radiation detected by the EDX detector.

an x-ray photon will be produced when an outer-shell electron in the ionized atom falls into the hole in the inner shell. The x-rays emitted by an atom are characteristic of that atom, and these characteristic x-rays are labeled according to the electronic transition that occurred in the ionized atom. For example, the K series of characteristic x-ray lines arise when an electron is ejected from the innermost (K) shell. The lines $K\alpha_1$, $K\alpha_2$, and K_β correspond to electrons falling from different outer shells. Similarly, the L series arises when a hole in the second (L) shell is filled, and the M series when a hole in the third (M) shell is filled. Tables of the energies and wavelengths of the characteristic x-rays for all the atoms are readily available. Some representative values for the $K\alpha_1$ lines are given in Table 7.1. The L and M lines have lower energies (and longer wavelengths). For example, the $L\alpha_1$ line for Ag has an energy of 2.984 keV and a wavelength of 0.4154 nm.

7.3 X-ray microanalysis with EDX spectrometers

7.3.1 Construction and characteristics of EDX spectrometers. A schematic diagram of a Si(Li) detector and the major electronic components of an EDX spectrometer are shown in Figure 7.3. X-rays emitted by the specimen enter the detector through the Be window and produce electron-hole pairs in the crystal. The crystal is maintained at the temperature of liquid nitrogen, and an energy of 3.8 eV is required to produce each electron–hole pair. Thus, the number of pairs produced by a photon of energy $E(\rho)$ is $E(\rho)/3.8$. The crystal is biased as shown, and the negative charge (about 10^{-16} coul per photon) produced by the incident photons is amplified by a field effect transistor (FET). The output pulses are further

Table 7.1. *Values of the energies and wavelengths of the*
$K\alpha_1$ line for some elements

Element	Z	Energy (keV)	Wavelength (nm)
Na	11	1.041	1.1909
Mg	12	1.254	0.9889
Al	13	1.487	0.8337
Si	14	1.740	0.7125
K	19	3.313	0.3741
Ca	20	3.691	0.3358
Cr	24	5.414	0.2290
Cu	29	8.047	0.1541
Sr	38	14.164	0.0875
Ag	47	22.162	0.0559
Ba	56	32.191	0.0385
Ta	73	57.524	0.0215
Au	79	68.794	0.0180
U	92	98.428	0.0126

amplified and shaped. The height of each pulse is proportional to the energy of the incident photon. These pulses are stored and sorted according to height by a multichannel analyser (MCA) and the spectrum of x-ray intensity versus energy is displayed on a visible display unit (VDU). The amplification and shaping of each pulse takes about 50 μs. If a pulse arrives before the preceding pulse is processed, both pulses are rejected. Thus, there is a dead time which increases with the x-ray intensity and becomes significant for count rates greater than about 4,000 s^{-1}. Typically, spectra are obtained with count rates of 1,000 to 2,000 s^{-1} for a counting time of about 100 live seconds. Normal statistical fluctuations and electron noise limit the energy resolution of a Si(Li) detector to about 160 eV and a ratio of peak-height to background of about 50:1. Because the sensitivity of the detector falls off rapidly for energies less than about 1 keV, the spectrum of the background is as shown in Figure 7.2.

The active area of the detector is usually between 10 and 30 mm^2. The collection angle is maximized by placing the detector window as close to the specimen as possible in the limited space available in the gap of the objective lens pole piece. Even so, the collection angle is a small fraction of the solid angle of 4π steradian over which the x-rays are generated.

The detection and elimination of spurious effects and the optimization of the electron microscope necessary before quantitative x-ray microanal-

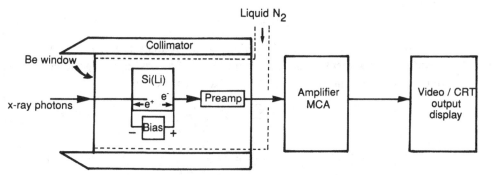

Figure 7.3. Schematic diagram of an EDX spectrometer.

yses can be undertaken are discussed in detail by Williams (1984) and need not be considered here.

7.3.2 Interpretation and quantification of x-ray data from thin specimens. After having acquired the x-ray spectrum, it is necessary to subtract the background bremsstrahlung to obtain the relative intensities of the characteristic x-ray peaks, from which the concentrations of the different kinds of atom in the specimen are determined. The steps involved in doing this will be considered briefly in this section.

Because the average energy lost by an electron in passing through a TEM thin foil specimen is only about 2 percent, it can be assumed that the ionization cross section is effectively constant and that the absorption and x-ray fluorescence are small enough to be neglected. Thus, for the simple case of a specimen containing only two kinds of atom, A and B, the weight percent concentrations C_A and C_B are proportional to the intensities I_A and I_B above background of the characteristic x-ray peaks. Hence,

$$\left.\begin{aligned} C_A &= k_A I_A \\ C_B &= k_B I_B \end{aligned}\right\} \tag{7.1}$$

k_A and k_B are not equal because of the different atomic numbers of the atoms, but they can be combined into a single constant k_{AB} by writing these equations as

$$\frac{C_A}{C_B} = k_{AB}\left(\frac{I_A}{I_B}\right) \tag{7.2}$$

k_{AB}, known as the Cliff–Lorimer factor (Cliff and Lorimer 1975), contains all the factors needed to correct for the atomic number difference.

We also have

$$C_A + C_B = 100 \qquad (7.3)$$

These are the basic equations of quantitative microanalysis and are easily extended to specimens containing more kinds of atoms. For a specimen containing three kinds of atoms A, B, and C, for example,

$$\frac{C_C}{C_A} = k_{CA}\left(\frac{I_C}{I_A}\right) \qquad (7.4)$$

$$\frac{C_C}{C_B} = k_{CB}\left(\frac{I_C}{I_A}\right) \qquad (7.5)$$

$$C_A + C_B + C_C = 100 \qquad (7.6)$$

and hence,

$$k_{AB}k_{CA} = k_{CB} \qquad (7.7)$$

Modern EDX spectrometers are equipped with a minicomputer which contains a *background subtract* program for the routine removal of the background intensity. Two processes are commonly used: (i) mathematical modeling of the background, which is based on the physics of bremsstrahlung production, or (ii) digital filtering, which is a purely mathematical method based on the fact that the background intensity distribution is a smooth and slowly varying function of energy.

Similarly, the computer contains a program for determining the intensities of the characteristic peaks. Again, two processes are used: (i) the peak intensities are obtained by fitting a Gaussian profile, to which the peaks closely approximate, or (ii) the peak is compared with a *library* standard peak, which has been acquired previously and stored. There is little to pick between the two processes (as with those for background subtraction), and the operator usually has no choice because the processes available have been selected by the manufacturer of the EDX spectrometer system.

The Cliff–Lorimer k-factors must be determined for any given microscope and accelerating voltage. Several methods have been used, the most accurate (but time-consuming) is the direct experimental method using standard thin foils with known concentrations of two or more elements whose characteristic peaks do not overlap. Cliff and Lorimer (1975) used mineral standards and determined the k-factors k_{XSi} for a large number of elements X, from which various k_{AB} could be determined from Eq. (7.7):

$$k_{AB} = \frac{k_{ASi}}{k_{BSi}} \qquad (7.8)$$

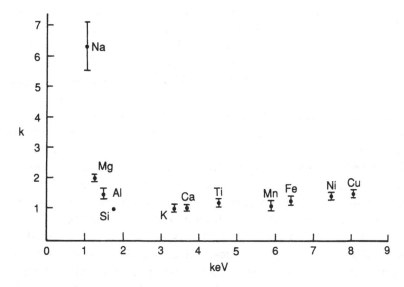

Figure 7.4. Calibration curve of the Cliff–Lorimer factor $k = (C_X/C_{Si})(I_{Si}/I_X)$ as a function of the Kα x-ray energy for an EDX detector for 100-kV electrons. (From Cliff and Lorimer 1975.)

Experimental values of k_{ASi} at 100 kV for the K characteristic peaks are shown graphically in Figure 7.4.

7.4 ALCHEMI

An interesting development of analytical electron microscopy, and a potentially very useful one for mineralogical research, has been made by Spence and Tafto (1982, 1983). The technique, known as ALCHEMI (atom location by channeling enhanced microanalysis), is the electron analogue of an x-ray technique originally used by Batterman (1969). The theoretical basis of the technique was discussed in Chapter 4, but it is appropriate to summarize that discussion before considering the ALCHEMI technique in detail.

In Section 4.9 it was shown that when a crystal is oriented near the Bragg angle for an operating reflection **g**, two standing waves of electron intensity (associated with α and β branches of the dispersion surface) are set up in the crystal normal to the reflecting planes and with a period equal to the spacing d ($= 1/|\mathbf{g}|$) between the planes. The combination of these two standing waves (which will not, in general, be as simple as that shown in Figure 4.12) will produce a third standing wave of the same period. This resultant standing wave produces a modulation of the electron intensity normal to the reflecting planes. A change in s causes this standing wave to shift in directions normal to the reflecting planes; the change in the intensity $I_0(a)$ of the transmitted electron beam with s, as illustrated in Figure 4.13, is a manifestation of this shift. If we assume that the absorption of an electron is due to the production of x-rays, then the intensity of x-rays emitted from a crystal in the neighborhood of a bend contour will be asymmetric about $s = 0$, the emission being higher on that side of the bend contour for which the absorption is greater – that is, on the side corresponding to $s < 0$, for the simple case discussed here. Because of the very strong interaction of kilovolt electrons with matter and because of the strong interaction between the many beams excited in polyatomic crystals, the positions of the intensity maxima of the resultant standing wave also depend on the depth in the crystal, the absorption length t_g'', and the number of excited beams (Tafto 1982). Nevertheless, the two-beam diffraction theory does predict correctly the qualitative variation of the emitted x-ray intensity with orientation near the Bragg angle for first-order reflections.

The technique of ALCHEMI takes advantage of this fact and ingeniously avoids the need for any a priori knowledge of the standing wave or of the variation of the position of the intensity maxima with depth in the crystal. This simplification is achieved by making use of the fact that the intensity of the x-rays from the known (or *reference*) atoms in the crystal is proportional to the magnitude of the *thickness-averaged* electron intensity at a particular crystallographic plane. The principle of the technique is best explained by taking a simple example.

Consider a set of alternating parallel planes (A and B) in a crystal containing atoms A and B, respectively, and impurity atoms X that are either on plane A or plane B. We now measure the x-ray counts corresponding to atoms A, B, and X for two orientations of the crystal near the exact Bragg angle, one orientation with $s < 0$ and the other with $s > 0$.

Let $N_A^{1,2}$, $N_B^{1,2}$, and $N_X^{1,2}$ be the x-ray counts from atoms A, B, and X for each of the two orientations, denoted by 1 and 2. Then we have the six equations

$$N_A^{1,2} = P_A M_A I_A^{1,2} \tag{7.9}$$

$$N_B^{1,2} = P_B M_B I_B^{1,2} \tag{7.10}$$

$$N_X^{1,2} = P_X M_X C_X I_B^{1,2} + P_X M_X (1 - C_X) I_A^{1,2} \tag{7.11}$$

where $I_A^{1,2}$ and $I_B^{1,2}$ are the thickness-averaged electron intensities on the A and B sites for each of the two orientations. M_A, M_B, and M_X are the number of sites per unit cell for the A, B, and X atoms; and P_A, P_B, and P_X are factors that take account of differences in x-ray yield and any other scaling factors. C_X is the fraction of X atoms on the possible sites on the B planes, so that $(1 - C_X)$ is the fraction on the A planes.

Let us suppose, as an example, that all the X atoms are on the B plane, that is $C_X = 1$. Then, from Eqs. (7.10) and (7.11),

$$\frac{N_X^1}{N_X^2} = \frac{I_B^1}{I_B^2} = \frac{N_B^1}{N_B^2}$$

and the ratio of the x-ray counts from the X atoms for the two orientations 1 and 2 is equal to the ratio of the counts from the B atoms for the same two orientations, but not equal to the ratio of the counts from the A atoms. Thus, if we record the variation with incident beam direction of the x-ray counts from the A, B, and X atoms, we need only observe which of the A and B counts is followed by the X counts in order to determine which plane contains the X atoms. This is true for all thicknesses and for any two orientations, provided that some variation of x-ray counts with changing orientation is observed.

However, this very simple procedure is not applicable if C_X is not exactly unity or zero. To determine intermediate values of C_X, we must again record x-ray counts from A, B, and X atoms for only two orientations. One of these orientations must be near the exact Bragg angle for a first-order reflection, and the other must be an orientation that avoids the excitation of any strong low-order Bragg beams. For this second orientation, a standing wave is not established in the crystal; the electron intensity is uniform across the unit cell and no channeling takes place. Thus,

$$I_A = I_B = I \tag{7.12}$$

and therefore,

$$N_A = P_A M_A I \tag{7.13}$$

$$N_B = P_B M_B I \tag{7.14}$$

$$N_X = P_X M_X C_X I + P_X M_X (1 - C_X) I$$
$$= P_X M_X I \tag{7.15}$$

The absence of superscripts on N_A, N_B, and N_X indicates the orientation for which no Bragg beams are excited.

From Eqs. (7.13), (7.14), and (7.15),

$$\frac{P_X M_X}{P_A M_A} = \frac{N_X}{N_A} \tag{7.16}$$

and

$$\frac{P_X M_X}{P_B M_B} = \frac{N_X}{N_B} \tag{7.17}$$

For an orientation near the exact Bragg angle, Eq. (7.11) becomes

$$N_X^1 = C_X P_X M_X (I_B^1 - I_A^1) + P_X M_X I_A^1$$

from which

$$C_X = \frac{N_X^1 - P_X M_X I_A^1}{P_X M_X (I_B^1 - I_A^1)}$$

Using Eqs. (7.9), (7.10), (7.16), and (7.17), we obtain

$$C_X = \frac{N_X^1/N_X - N_A^1/N_A}{N_B^1/N_B - N_A^1/N_A}$$

Note that for $C_X = 0$, N_X^1 is proportional to N_A^1, and that for $C_X = 1$, N_X^1 is proportional to N_B^1, in agreement with the first example considered previously.

The ALCHEMI technique just described is called *planar* ALCHEMI because the electrons are channeled along crystallographic planes. However, if the crystal is oriented so that the incident electron beam is parallel to a low-index zone axis, the electrons are channeled along columns of atoms; this modification is called *axial* ALCHEMI. Whether planar or axial ALCHEMI is used depends on the material being studied.

A crucial assumption of this technique is that the x-ray emission process is highly localized; that is, the electron currents are the same for the unknown element X and for the standards A and B, as in Eq. (7.11). Early planar ALCHEMI determinations of site-occupancies of minor elements in spinels (Tafto 1982) and olivine (Tafto and Spence 1982), for example, suggested that the assumption is valid. However, more recent studies by Rossouw and Maslen (1987) and Rossouw, Turner, and White (1988) indicate that the x-ray generation is not perfectly localized and that Eq. (7.11) is, in general, not valid.

This so-called localization effect depends on the crystal structure of the specimen (which determines the magnitude of the channeling) and to a lesser extent on the chemical composition. Thus, determination of site-occupancy is not fully quantitative, as previously assumed, and correct site-occupancies can be obtained only by experimentally determining correction factors for the particular crystal structure being investigated.

For a detailed discussion of these correction factors and other experimental details, the reader should consult the practical guide to ALCHEMI recently prepared by Otten (1989).

8

Mineralogical applications of TEM
I. Defects and microstructures in undeformed specimens

8.1 Introduction

The previous chapters have given an account of the essential physics of the transmission electron microscope and of the basic concepts and theories of electron diffraction that are required for interpretation of the images of crystalline materials, especially when these materials are structurally imperfect due to the presence of various types of point, line, and planar defects. The specific aim of this chapter and the next is to discuss some characteristic examples of the images of the various types of defect and defect microstructures that have been observed in a range of important rock-forming minerals. Although emphasis is placed on the nature of the images and their interpretation, rather than on the relevance of the various types of defects and microstructures to mineralogical properties and processes, these latter aspects are not entirely neglected.

The examples in this chapter cover a wide range of defects and microstructures, from the simpler type of planar defect (such as the stacking fault) to the defects associated with radiation damage. The following chapter deals with dislocations and the complex microstructures associated with plastic deformation. In both chapters, the examples are presented in a sequence that attempts to link them as much as possible in order to bring out the relationships, similarities, and differences among the various types of crystal defects from the points of view of their image contrast, structure, and origin.

8.2 Brazil twin boundaries in quartz

The simplest type of boundary is the α-boundary (Section 5.3.1), which involves no change in crystallographic orientation. Furthermore, $t'_{g1} = t'_{g2}$ and hence $|F_{g1}| = |F_{g2}|$; so there is no change of contrast across the boundary. The fringes arise at the boundary because of a phase shift α

197

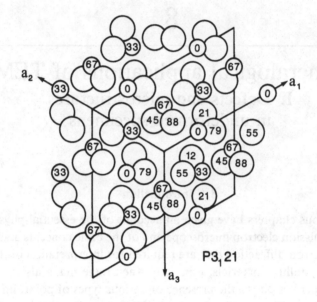

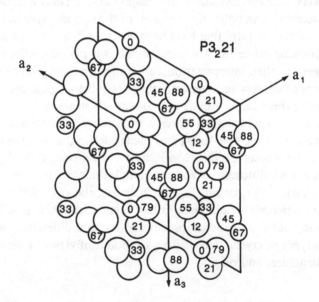

Figure 8.1. Projections onto (0001) of the atomic positions in right-handed (P3₁21) and left-handed (P3₂21) α-quartz. The small circles represent Si, and the large circles O. The numbers are the fractional *c*-axis coordinates.

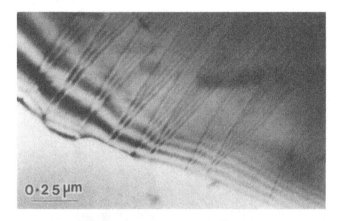

0·25 μm

Figure 8.2. α-fringe patterns due to Brazil twin boundaries in amethyst quartz. BF image: $\mathbf{g} = 10\bar{1}1$. Note the reversal of contrast at successive boundaries. (From McLaren and Phakey 1966.)

which, in general, is given by Eq. (5.12):

$$\alpha = (\theta_{g1} - \theta_{g2}) + 2\pi\mathbf{g}\cdot\mathbf{R}$$

The atoms in quartz are arranged according to either of the enantiomorphous space groups $P3_121$ or $P3_221$. The SiO$_4$ tetrahedra on the threefold axes form a right-handed (R) helix in $P3_121$ and a left-handed (L) helix in $P3_221$, as can be seen in the projections onto (0001) shown in Figure 8.1. The structures are related by reflection in (11$\bar{2}$0). Single crystals of quartz often consist of R and L domains oriented in this way so that the axes remain parallel. This twinning is usually referred to as Brazil twinning.

Polysynthetic Brazil twinning in which alternating R and L lamellae are arranged parallel to {10$\bar{1}$1} is characteristic of amethyst quartz (Frondel 1962). Therefore, it was assumed by McLaren and Phakey (1965a) that the fringe patterns due to planar defects on {10$\bar{1}$1}, which they observed in high concentrations in this mineral, were, in fact, Brazil twin boundaries. In a subsequent paper, McLaren and Phakey (1966) showed that the fringes were characteristic of α-boundaries in which the phase angle α was a function of both $(\theta_{g1} - \theta_{g2})$ and $2\pi\mathbf{g}\cdot\mathbf{R}$. The Brazil twin boundary is, therefore, the most general type of α-boundary and so is an appropriate defect with which to begin this discussion.

Figure 8.2 shows an array of Brazil twin boundaries parallel to one of the major rhombohedral planes, say ($\bar{1}$101), in amethyst quartz, imaged

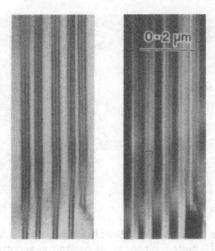

Figure 8.3. $\alpha = \pi$ fringes due to Brazil twin boundaries in amethyst quartz: (a) in BF and (b) in DF. $\mathbf{g} = 10\bar{1}0$. (From McLaren and Phakey 1966.)

with $\mathbf{g} = 10\bar{1}1$ (or $0\bar{1}11$). Note that across the boundaries the thickness extinction contours are continuous, and there is no change of contrast. It follows that there is no change of crystal orientation across the boundary, which was confirmed in the associated SAD pattern. The fringe profiles are symmetrical, but the contrast reverses at each successive boundary. Other features of the fringe patterns are (i) the intensity along the fringes varies with the background intensity and (ii) on passing a thickness extinction contour, the additional fringe is created by forking of the central fringe. It was also found that these fringe profiles became asymmetrical in DF images. Such features are characteristic of fringe patterns with $\alpha = \pm 2\pi/3$ for $s = 0$ (see Section 5.3.2). The reversal of contrast at successive boundaries indicates that α changes sign successively. When $\mathbf{g} = 10\bar{1}0$, the fringe profiles are again symmetrical; but now there is no reversal of contrast (and, therefore, no change in the sign of α) at successive boundaries. This is illustrated in Figure 8.3. These fringes must, therefore, have $\alpha = \pm \pi$ because only for this value of α is $\exp(i\alpha) = \exp(-i\alpha)$. These fringe patterns are complementary in BF and DF images and are symmetrical, which are characteristic features of $\alpha = \pi$ fringes (see Section 5.3.3).

Now the fringe contrast depends upon

$$\alpha = 2\pi \mathbf{g} \cdot \mathbf{R} + (\theta_{gL} - \theta_{gR}) \tag{8.1}$$

where θ_{gL} and θ_{gR} are the phase angles in the structure factors F_{gL} and F_{gR} of the left-handed and right-handed structures, respectively. Table 8.1

Table 8.1. *Phase angles for R and L quartz*

g	θ_L	θ_R	$\varphi = \theta_{gL} - \theta_{gR}$	$2\pi g \cdot R$ $R = \pm\frac{1}{2}[110]$	α
$10\bar{1}0$	π	π	0	π	π
$\bar{1}100$	π	π	0	0	0
$0\bar{1}10$	π	π	0	π	π
$10\bar{1}1$	2π	$4\pi/3$	$2\pi/3$	π	$-\pi/3$
$\bar{1}101$	$2\pi/3$	$2\pi/3$	0	0	0
$0\bar{1}11$	$4\pi/3$	2π	$-2\pi/3$	π	$\pi/3$

gives values for θ_{gL}, θ_{gR}, and $\varphi = (\theta_{gL} - \theta_{gR})$ calculated from the atom positions in the two structures referred to the same axes, for all reflections of the type $\{10\bar{1}0\}$ and $r\{10\bar{1}1\}$. It will be seen that $\varphi = 0$ for all $g = 10\bar{1}0$. Because the boundaries are observed when $g = 10\bar{1}0$, the two parts of the Brazil twin must be displaced relative to one another across the boundary by a vector R so as to introduce a nonzero value of α ($= 2\pi g \cdot R$). However, there is insufficient information to calculate R from the fringe patterns alone.

From the (0001) projections of the L and R structures shown in Figure 8.1, we see that if the structures are displaced relative to each other by

$$R_1 = \pm\frac{1}{2}[110] \equiv \pm\frac{1}{2}[11\bar{2}0] \equiv -a_3$$

or

$$R_2 = \frac{1}{6}[03\bar{2}] \quad \text{or} \quad \frac{1}{6}[03\bar{2}]$$

or

$$R_3 = \frac{1}{6}[\bar{3}02] \quad \text{or} \quad \frac{1}{6}[302]$$

then two of the three Si atoms on the *a*-axes will come into register, and the third Si atom will be taken into the *c*-axis channel. This matching of the Si atoms in the two structures is clearly seen in Figure 8.4, where the displacement vector is $R_1 = \frac{1}{2}[110]$. We can also see that the O atoms at $z = 0.55$ and 0.79 are brought into register. These O atoms are very close to being in a plane that is parallel to $r(\bar{1}101)$, which is the clue to the structure of the twin boundary: Figure 8.5 is a projection on a plane normal to a_3 ($= [\bar{1}\bar{1}0]$) of the SiO_4 tetrahedra in R and L quartz which shows the boundary parallel to $r(\bar{1}101)$ between the R structure (lower-right) and the L structure (upper-left). The R and L structures are displaced by $R_1 = \pm\frac{1}{2}[110]$. R_1 is normal to the plane of the diagram, which is parallel to the twin (mirror) plane. Note that the boundary is a thin slab of structure

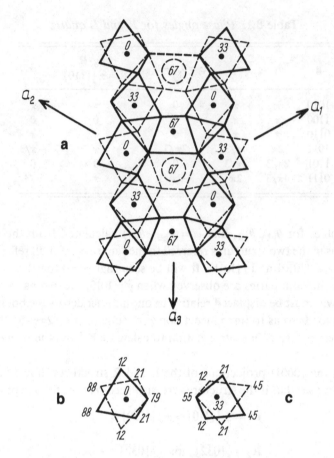

Figure 8.4. Projection onto (0001) of the SiO_4 tetrahedra of left-handed quartz (full lines) and of right-handed quartz (dotted lines) displaced by a vector $\mathbf{R} = \pm\frac{1}{2}[110]$. (From McLaren and Pitkethly 1982.)

that contains only O atoms (at $z = 0.55$ and 0.79), which are in their correct positions for both R and L quartz. This slab, therefore, provides an ideal interface between the two structures. No such simple slab of structure exists for a boundary parallel to $z(1\bar{1}01)$, which suggests that Brazil twin composition planes occur preferentially on major rhombohedral planes $r\{10\bar{1}1\}$. The same arguments apply to Brazil twin boundaries with the other possible fault vectors $\mathbf{R}_2$ and $\mathbf{R}_3$ (in which the nonzero z-component is a consequence of the threefold screw axis parallel to [0001]). In all cases, the fault vector lies in the plane of the boundary, as should be so for α-boundaries.

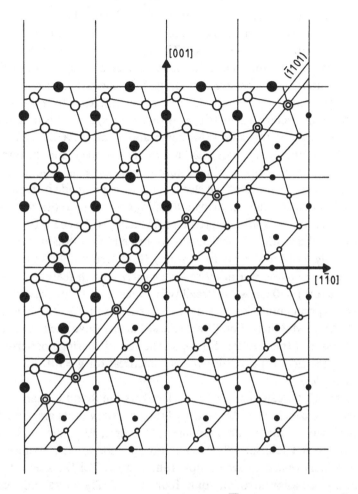

Figure 8.5. Projection onto a plane normal to $\mathbf{a}_3 = [\bar{1}10]$ of the SiO₄ tetrahedra, showing the structure of a Brazil twin boundary parallel to $r(\bar{1}101)$ in which the right-handed quartz (lower-right) is displaced with respect to the left-handed quartz (upper-left) by a fault vector $\mathbf{R} = \pm\frac{1}{2}[110]$.

This model of the Brazil twin boundary can be tested by seeing if the observed fringe patterns are consistent with the calculated values of $\alpha = 2\pi\mathbf{g}\cdot\mathbf{R}+\varphi$ for different operating reflections $\mathbf{g}$. The values of $2\pi\mathbf{g}\cdot\mathbf{R}$ and α for $\mathbf{R} = \frac{1}{2}[110]$, corresponding to a boundary on $(\bar{1}101)$, for $\mathbf{g} = 10\bar{1}0$ and $10\bar{1}1$ (and symmetrically equivalent $\mathbf{g}$) are included in Table 8.1. It will be seen that $\alpha = \pi$ for $\mathbf{g} = 10\bar{1}0$, which is consistent with the fringe patterns observed for this reflection (Figure 8.3). For $\mathbf{g} = 10\bar{1}1$ and $0\bar{1}11$, $\alpha = -\pi/3$

and $+\pi/3$, respectively. These values of α refer to a boundary for which the electron beam passes from L quartz into R quartz. For the reverse situation, φ ($\neq 0$) is of opposite sign. This leads to a reversal in the sign of α, as can be easily verified using the values of φ given in Table 8.1. Thus, the fringe patterns at successive boundaries are expected to reverse contrast, as is observed in Figure 8.2. Also note that fringes with $\alpha = \pm\pi/3$ have the same characteristics as fringes with $\alpha = \pm 2\pi/3$ (see Section 5.3.2).

At the time of this study (McLaren and Phakey 1966), the ion-bombardment thinning technique had not been developed, and the thin specimens were in the form of thin fracture-fragments. This, together with the limited tilting facilities of the electron microscope, made it impossible to image the defects under as many diffracting conditions as desirable. Nevertheless, the observations do provide convincing evidence for the proposed model of the boundary, which was confirmed in a subsequent investigation using TEM and x-ray topography (McLaren and Pitkethly 1982).

The Brazil twins just discussed are clearly growth features. However, very narrow (≈ 10 nm wide) Brazil twin lamellae on (0001) have been produced experimentally by very high shear stresses on (0001) and examined by TEM by McLaren et al. (1967). These lamellae give rise to overlapping α-boundaries (see Section 5.3.4), so the resulting fringe patterns in BF are symmetrical for $\mathbf{g} = 10\bar{1}0$ and asymmetrical for $\mathbf{g} = 10\bar{1}1$, as shown in Figure 8.6.

Although we can identify a boundary between R and L quartz, it is not possible to decide from the image of the boundary or its associated SAD pattern which side of the boundary is R-quartz, for example. However, Goodman and Johnson (1977) developed a method for identifying enantiomorphically related space groups from a single CBED pattern, together with a calculation involving only four dynamically interacting beams.

8.3 Stacking faults, twinning, and polytypism in wollastonite

Stacking faults are α-boundaries for which $\alpha = 2\pi\mathbf{g}\cdot\mathbf{R}$. $(\theta_{g1}-\theta_{g2})$ is zero for all $\mathbf{g}$. In some structures, stacking faults and twins are closely related, and different regular sequences of these defects produce various polytypes. Wollastonite is a relatively simple example of such a structure, for which the stacking faults have been studied in some detail by TEM, both by their α-fringe contrast and in two-dimensional high-resolution lattice images.

Wollastonite is common in thermally metamorphosed impure limestone and can occur in contact-altered calcareous sediments. In most occurrences, it has been formed as a result of the reaction

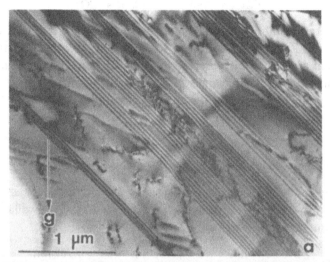

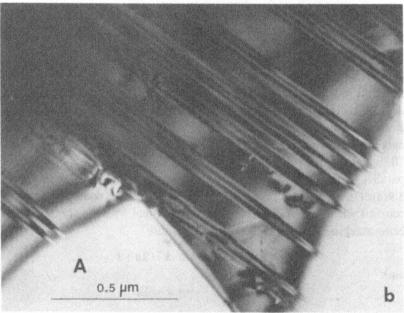

Figure 8.6. Overlapping α-fringes due to thin Brazil twin lamellae in experimentally deformed natural quartz. (a) BF; $\mathbf{g} = 10\bar{1}1$. (b) BF; $\mathbf{g} = 10\bar{1}0$. Note the asymmetrical fringe profile in (a) and the symmetrical fringe profile in (b). (From McLaren et al. 1967.)

$$CaCO_3 + SiO_2 \rightarrow CaSiO_3 + CO_2$$

at temperatures above $\approx 700°C$ and pressures from ≈ 100 MPa (Deer, Howie, and Zussman 1963). These conditions suggest that the stacking faults that are commonly observed in wollastonite are growth features.

The structure is triclinic and consists basically of chains of SiO_4 tetrahedra running parallel to [010] with a 3-repeat, as shown in Figure 8.7(a). The unit cell dimensions are

$$a = 0.794 \text{ nm} \qquad b = 0.732 \text{ nm} \qquad c = 0.707 \text{ nm}$$
$$\alpha = 90°03' \qquad \beta = 95°17' \qquad \gamma = 102°28'$$

with $Z = 6$ and the space group is $P\bar{1}$ (Deer, Howie, and Zussman 1963). There is also a closely related monoclinic modification (parawollastonite) whose unit-cell dimensions are

$$a = 1.542 \text{ nm} \qquad b = 0.732 \text{ nm} \qquad c = 0.707 \text{ nm}$$
$$\alpha = 90° \qquad \beta = 95°24' \qquad \gamma = 90°$$

with $Z = 12$, and the space group is $P2_1$ (Trojer 1968).

Specimens of wollastonite from a number of localities have been examined by TEM by Jefferson and Thomas (1975), Wenk et al. (1976), Hutchison and McLaren (1976, 1977), and others. A typical selected-area electron diffraction pattern from a foil oriented with [001] parallel to the electron beam is shown in Figure 8.8. It can be seen that the diffraction maxima $hk0$ with $k = 2n + 1$ ($n = 0, 1, 2, \ldots$) are streaked parallel to $\mathbf{a}^*$. DF images with $\mathbf{g} = (h, 2n+1, 0)$ reveal a high density of planar defects parallel to (100). These are usually randomly distributed, as in Figure 8.9(a). However, as in Figure 8.9(b), these defects are completely out-of-contrast when $k = 2n$. This behavior is consistent with the planar defects being stacking faults with $\mathbf{R} = \frac{1}{2}[010]$, because

$$\alpha = 2\pi\mathbf{g}\cdot\mathbf{R} = \pi \quad \text{for } k = 2n+1$$

and

$$\alpha = 0 \qquad\qquad \text{for } k = 2n$$

Figure 8.7. *Facing page.* (a) Structure of triclinic wollastonite, projected onto (001). Calcium atoms are represented by circles; SiO_4 by tetrahedra. (b) Structure of one unit cell of a monoclinic wollastonite generated by a stacking fault with $\mathbf{R} = \frac{1}{2}[010]$ on the plane marked. Compare with (a). (c) Structure of one unit cell of monoclinic parawollastonite generated by twinning with a twin axis [010]. Comparison with (b) shows that this structure is identical to that generated by a single stacking fault in triclinic wollastonite. (From Hutchison and McLaren 1976, 1977.)

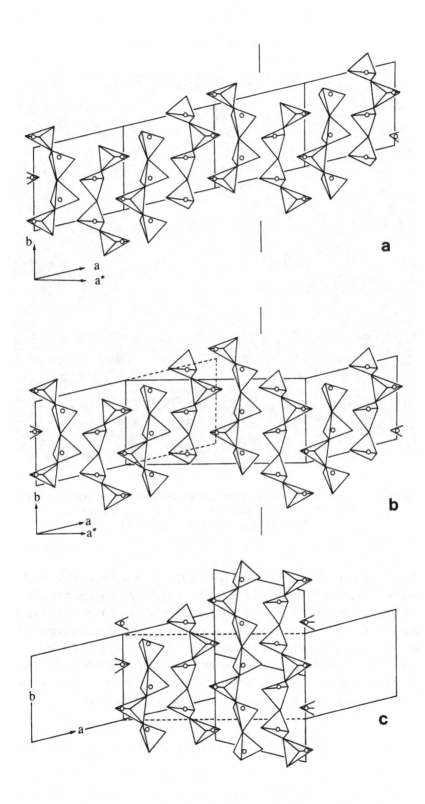

a

b

c

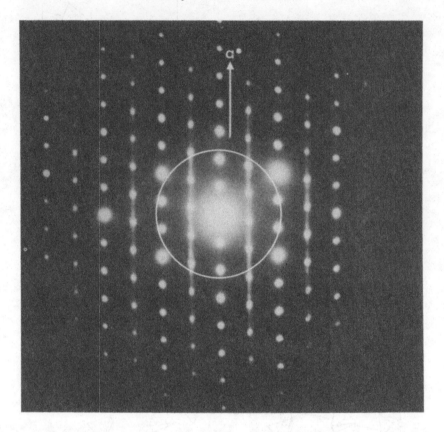

Figure 8.8. Electron diffraction pattern of wollastonite with the electron beam parallel to [001]. Note the streaking parallel to **a*** on odd layer lines. (From Hutchison and McLaren 1976.)

This fault vector is confirmed in Figure 8.10, which shows a single stacking fault in a two-dimensional $hk0$ lattice image formed using the diffraction maxima within the circle in Figure 8.8. Figure 8.7(b) shows that this stacking fault produces a lamella of monoclinic parawollastonite one unit cell wide ($a = 1.54$ nm), which is easily identified in Figure 8.10. It is clear that a wider lamella of parawollastonite in a triclinic matrix can be produced by a regular series of stacking faults, one in every *second* triclinic unit cell, giving a packing sequence along the **a*** direction of

$$
\overline{\underset{\ldots\text{AAAA}}{\text{T}}}\ \overline{\underset{\text{ABABABAB}}{\text{M}}}\ \overline{\underset{\text{AAAA}\ldots}{\text{T}}}
$$

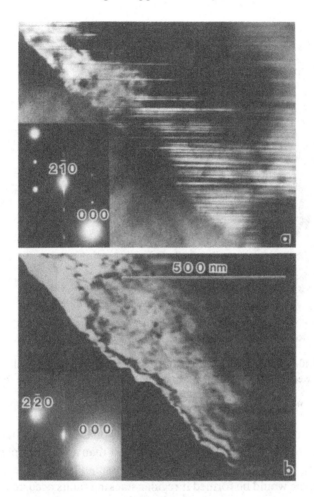

Figure 8.9. DF images of the same area of a wollastonite grain showing (100) faults: (a) in contrast for $\mathbf{g} = 2\bar{1}0$ and (b) out-of-contrast for $\mathbf{g} = 2\bar{2}0$. (From Wenk et al. 1976.)

where A and B are normal and faulted triclinic unit cells, respectively. Lamellae of this structure are easily identified as regions exhibiting regular one-dimensional lattice fringes with a spacing of 1.54 nm in DF images, formed using a group of reflections in a $k = (2n + 1)$ row. The region illustrated in Figure 8.11 clearly consists predominantly of parawollastonite whose 1.54-nm periodicity along $\mathbf{a}^*$ is interrupted by a fairly low density of stacking faults.

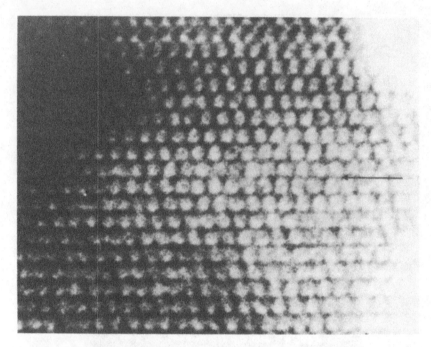

Figure 8.10. *hk*0 lattice image of wollastonite showing a single stacking fault (arrowed). Note the absence of any local contrast change at the fault. (From Hutchison and McLaren 1976.)

Because of their structural relationship, wollastonite and parawollastonite are better described as polytypes rather than as polymorphs. Other polytypes are possible (Deer, Howie, and Zussman 1963). For example, a new polytype would be formed if regular stacking faults occurred in every *third* triclinic unit cell, producing the packing sequence

$$...AABAABAABAA...,$$

which is equivalent to a sequence of alternating wollastonite (A) and parawollastonite (AB) unit cells. The periodicity along $\mathbf{a}^*$ is 2.3 nm, and such lamellae have been observed in one-dimensional lattice images by Wenk et al. (1976).

Twinning is common in wollastonite (Deer, Howie, and Zussman 1963; Müller and Wenk 1975). The twin axis is [010] and the composition plane (100). Figure 8.7(c) shows a twin lamella (one unit cell wide) in an untwinned matrix. To eliminate very severe misfit at the composition plane, a relative displacement of $\mathbf{R} = \frac{1}{2}[010]$ is necessary. Comparison of Figures

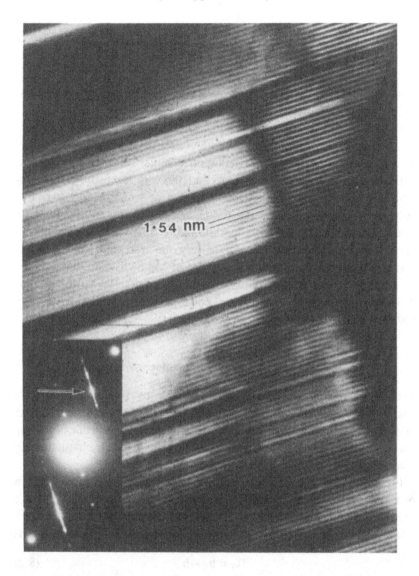

Figure 8.11. High-resolution DF image showing the 1.54-nm fringes and faults interrupting the periodic stacking in parawollastonite. (From Wenk et al. 1976.)

8.7(b) and (c) shows that the monoclinic unit cell produced by this twin operation is identical to that produced by a single stacking fault. A wide twin lamella can be produced by stacking faults in every adjacent triclinic unit cell:

Twin

...AAAABBBBBBAAAAAA...

M   M

Note also that the twin composition planes on each side of the twin lamella are in the center of parawollastonite cells.

The isotypic minerals pectolite and serandite (resp., $Ca_2NaHSi_3O_9$ and $Mn_2NaHSi_3O_9$) are chain silicates with crystal structures that are closely related to wollastonite. TEM observations (Müller 1976) have shown that both minerals contain stacking faults on (100) with $R = \frac{1}{2}[010]$. The polytypic behavior of pectolite is similar to that of wollastonite: there is a polytype analogous to parawollastonite, and submicroscopic twinning on (100) is frequent. However, no "paraserandite" was found and submicroscopic twinning on (100) appears to be very rare in serandite. It is possible that this is due to the partial replacement of Mn by Ca and that, in such an ordered structure, a stacking fault on (100) with $R = \frac{1}{2}[010]$ is energetically unfavorable. Accordingly, polytypism may be limited to pure end-members of the pectolite–serandite series.

8.4 Albite-law microtwins in plagioclase feldspars

The investigation by Fitz Gerald (1980) of microtwins in plagioclase feldspars is probably the most detailed yet undertaken of this type of defect in important rock-forming minerals and provides an excellent example of the theoretical and experimental procedures involved in characterizing these defects from their TEM contrast.

Figure 8.12(a) is a two-dimensional representation showing the origin of a fault vector R_T associated with the presence of a slab of twin one unit cell thick. In the albite twin law, the two parts of the twin are related by a rotation of 180° about b^*. From Figure 8.12(b), we see that R_T is defined as

$$R_T = b_T - b \qquad (8.2)$$

The composition plane is parallel to the twin plane (010).

Clearly, R_T depends on the structural state and composition of the feldspar concerned, through their influence on the lattice parameters. However, R_T is always in the (010) plane and has the form

$$R_T = [u0w] \qquad (8.3)$$

where

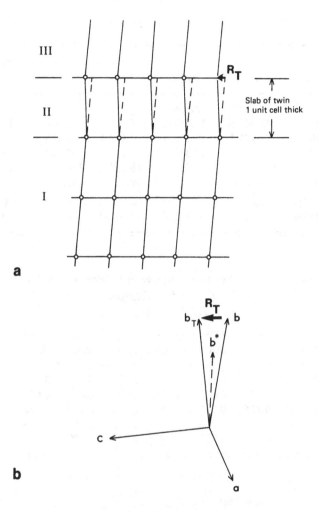

Figure 8.12. (a) Diagram showing the origin of a fault vector $\mathbf{R}_T$ due to a slab of twin one unit cell thick. (b) Diagram of the fault vector $\mathbf{R}_T$ for the unit cell axes of low albite. (From Fitz Gerald 1980.)

$$\left.\begin{array}{l} u = \dfrac{-2b(\cos\gamma - \cos\beta\cos\alpha)}{a\sin^2\beta} \\[3mm] w = \dfrac{-(2b\cos\alpha - ua\cos\beta)}{c} \end{array}\right\} \qquad (8.4)$$

If the matrix only is diffracting, an α-fringe pattern will be produced for which the phase angle α is given by

Table 8.2. *Cell parameters from Smith (1974) and values for u and w defined in Eq. (8.4) for plagioclase feldspars of several compositions*

	An_0	An_{10}	An_{20}	An_{30}	An_{40}	An_{50}
a	8.138 Å	8.148 Å	8.156 Å	8.161 Å	8.167 Å	8.173 Å
b	12.781 Å	12.809 Å	12.833 Å	12.848 Å	12.858 Å	12.865 Å
c	7.160 Å	7.145 Å	7.132 Å	7.123 Å	7.115 Å	7.110 Å
α	94.32°	94.07°	93.87°	93.71°	93.60°	93.55°
β	116.60°	116.52°	116.43°	116.35°	116.27°	116.20°
γ	87.67°	88.20°	88.80°	89.40°	89.65°	89.65°
u	0.02721	0.00110	0.03571	0.07155	0.08491	0.08302
w	0.25508	0.25504	0.26106	0.26981	0.27008	0.26621

Table 8.3. *Values of $\alpha' = \mathbf{g} \cdot \mathbf{R}_T$ for some low-index reflections for plagioclase feldspars of several compositions*

$\mathbf{g}$	An_0	An_{10}	An_{20}	An_{30}	An_{40}	An_{50}
$0\ k\ 0$	0	0	0	0	0	0
$1\ k\ 0$	0.027	0.001	0.036	0.072	0.085	0.083
$0\ k\ 1$	0.255	0.255	0.261	0.270	0.270	0.266
$1\ k\ 1$	0.228	0.256	0.297	0.341	0.355	0.349
$\bar{2}\ k\ 1$	0.310	0.253	0.190	0.127	0.100	0.100

$$\alpha = 2\pi \mathbf{g} \cdot \mathbf{R}_T$$
$$= 2\pi(h\mathbf{a}^* + k\mathbf{b}^* + l\mathbf{c}^*) \cdot (u\mathbf{a} + w\mathbf{c})$$
$$= 2\pi(hu + lw) \tag{8.5}$$

Correspondingly, for a microtwin that is n unit cells thick,

$$\mathbf{R}_T = n[u0w] \quad \text{and} \quad \alpha = n2\pi(hu + lw) \tag{8.6}$$

Values of u and w and of $2\pi\mathbf{g} \cdot \mathbf{R}_T$ for some low-index reflections $\mathbf{g}$ are listed for several plagioclase compositions in Tables 8.2 and 8.3, respectively.

Three microtwins (A, B, and C) in Cazadero albite (An_0) imaged in DF with six different diffraction vectors $\mathbf{g}$ are shown in Figure 8.13. From

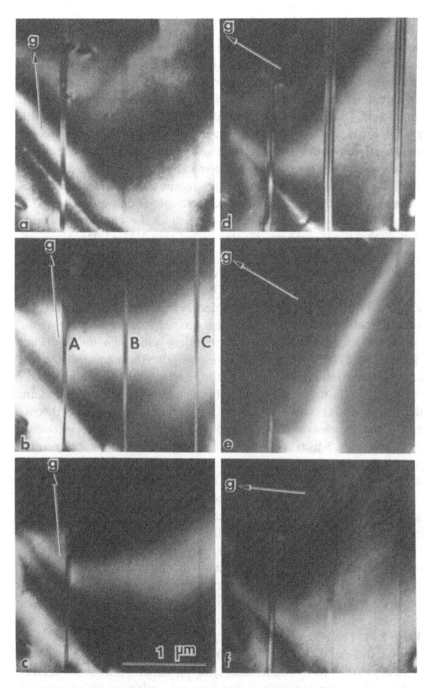

Figure 8.13. DF fringe patterns due to microtwins in Cazadero albite. The electron beam is near $[\overline{3}10]$. (a) $\mathbf{g} = 002$, (b) $\mathbf{g} = 003$, (c) $\mathbf{g} = 004$, (d) $\mathbf{g} = 1\overline{3}1$, (e) $\mathbf{g} = 2\overline{6}2$, and (f) $\mathbf{g} = 1\overline{3}0$. (From Fitz Gerald 1980.)

215

the values of $g \cdot R_T$ given in Table 8.3, it follows immediately that the microtwins B and C are two unit cells thick; that is, $n = 2$ in Eq. (8.6). Microtwin A is much thicker, and the fringe contrast cannot be used to determine its exact thickness because the twin boundaries do not overlap completely. The fringes due to microtwins B and C in Figure 8.13(d), with $g = 1\bar{3}1$, are clearly π-type (in agreement with $\alpha = 2\pi g \cdot R_T$); but better examples are shown in Figure 8.14(a), for which $g = 002$. These microtwins are out-of-contrast for $g = 0\bar{4}0$, as shown in Figure 8.14(c). It is clear from Table 8.3 that these observations are consistent with the defects being microtwins whose thicknesses are an odd multiple of unit cells ($n = 1, 3, 5,$ ...). The fringes of defect A are so close to ideal π-type fringes that it is likely that $n = 1$, corresponding to $\alpha = 2\pi(2 \times 0.255) = 1.02\pi$. However, the fringes of defect B are certainly less than ideal: they show a slight modulation with the background (thickness fringes) and a slight variation in the intensities of adjacent pairs. This suggests that $n = 3$ or greater. In Figure 8.14(b), the fringes due to defect A are π-type, while defect B is virtually out-of-contrast. These images are not consistent with $n = 1$ for A and $n = 3$ for B because the corresponding values of α for $g = 111$ are (from Table 8.3) 0.456π and 1.36π, respectively. The most probable reason for this contradiction between the observations and the microtwin theory is that in this micrograph both the microtwin and the matrix are diffracting; and there is evidence for this in the corresponding SAD pattern. Thus, the observed contrast probably consists of α-fringes modified by overlapping δ-fringes (see Section 5.3.5), which are not easily interpreted. Therefore, to determine the nature of microtwins it is essential to orient the specimen so that *only* the matrix is diffracting.

Figure 8.15 shows particularly good examples of microtwins in Wyangala plagioclase (composition near An_{30}) imaged in DF with $g = 20\bar{1}$ when only the matrix is diffracting. In Figure 8.15(a) the electron beam is near $[\bar{1}0\bar{2}]$ and the microtwins are viewed nearly edge-on. They are obviously several unit cells wide, but the exact width cannot be determined from such images. The same region after tilting so that the beam is near $[\bar{1}1\bar{2}]$ is shown in Figure 8.15(b). The microtwins A and B are now imaged as π-fringes, which is consistent with $n = 4$ for a composition of An_{30} (see Table 8.3). The effect on the fringe patterns of overlapping microtwins (such as those at C) is clearly seen. It is also seen that the fringe contrast is locally disturbed by dislocations lying on the microtwins. DF images using several different g indicate that most dislocations in this situation do not mark changes in thickness of the microtwin. Instead, these dislocations are probably lattice dislocations that were forced into the microtwin

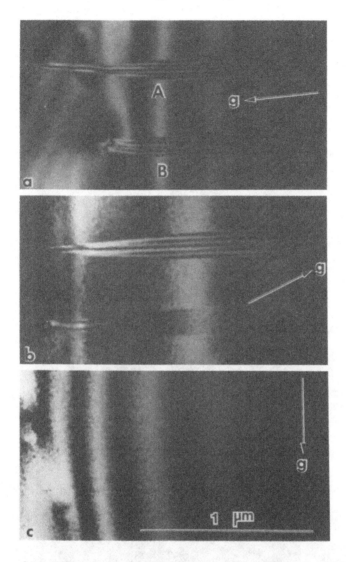

Figure 8.14. DF $\alpha = \pi$ type fringes due to microtwins in Cazadero albite. (a) $g = 002$, beam axis near [2$\bar{1}$0]; (b) $g = 111$, beam axis near [0$\bar{1}$1]; (c) $g = 0\bar{4}0$, beam axis near [100]. (From Fitz Gerald 1980.)

boundary by an external stress but were unable to glide across the microtwin. Mismatch between Burgers vectors of the matrix and twin lattices is the most probable barrier to dislocation glide (Marshall and McLaren 1977b).

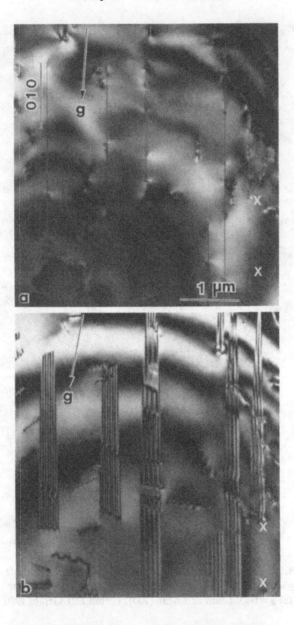

Figure 8.15. DF images of microtwins in Wyangala plagioclase (An$_{30}$). (a) $\mathbf{g} = 20\bar{1}$, beam axis near [$\bar{1}0\bar{2}$]. Microtwins are viewed edge-on. (b) Same region as (a), again with $\mathbf{g} = 20\bar{1}$, but the beam axis is near [$\bar{1}1\bar{2}$]. (From Fitz Gerald 1980.)

In regions with relatively high microtwin densities, microtwins with $n = 2$ are the most common. However, the reason for this has not been established. The fringe patterns of individual microtwins are continuous unless affected by overlapping contrast from adjacent microtwins or dislocations. Rarely, the fringe pattern of an individual microtwin changes sharply across a line of contrast similar to that of a dislocation. Such changes of fringe contrast are interpreted as due to a *ledge* (an integral number of unit cells high) in one of the boundaries of the microtwin. Attempts to determine microtwin thicknesses directly using HRTEM were frustrated by electron beam damage of the specimen.

8.5 Twins and twin boundaries

The term *microtwin* used in the previous two sections clearly refers to twin lamellae that are only one or a few unit cells wide. Relatively wide twin lamellae can be imaged in strong contrast (in either BF or DF) by orienting the specimen so that either the twin or the matrix is diffracting more strongly than the other; an example of a pericline twin in microcline is shown in Figure 8.16(a, b). Because the twin lamella is slightly inclined, fringes appear at the two (nonoverlapping) twin boundaries, as explained in Sections 5.1 and 5.3.7. If **g** is normal to the twin plane (or, as in this example, normal to the twin axis), the twin and matrix will diffract with equal intensity, and the twin will be out-of-contrast, as shown in Figure 8.16(c). The weak fringes at the twin boundaries in this micrograph indicate there is an additional phase shift across the boundaries. This phase shift could arise because of a relative displacement **R** across the boundary and/or because of a difference in the phase angles of the structure factors of the twin and matrix, as in the case of the Brazil twin boundary in quartz (Section 8.2). However, structural considerations and structure factor calculations indicate that the observed fringes do not arise in either of these two ways. It appears, therefore, that the weak fringes are probably due to some slight structural distortion (relaxation) in the twin boundary.

Marshall and McLaren (1974) showed that the boundaries of deformation pericline twin lamellae in experimentally deformed anorthite (An_{95}) involve a displacement. The specimens were cut normal to [010] so that the twin boundaries were viewed edge-on and the twins were out-of-contrast for all $g = h0l$, because these reflecting planes are unaffected by the twin. However, in DF images with *b*-reflections ($h + k = 2n + 1$, $l = 2n + 1$) and with *c*-reflections ($h + k = 2n$, $l = 2n + 1$), the *twin boundaries* are seen

Figure 8.16 Pericline twin lamella in microcline. (a) BF image, $\mathbf{g} = 040$. (b) DF image, $\mathbf{g} = 040$. (c) DF image, $\mathbf{g} = h0l$.

with the contrast characteristic of α-boundaries. On the other hand, the twin boundaries are out-of-contrast for both a-reflections ($h + k = 2n$, $l = 2n$) and d-reflections ($h + k = 2n + 1$, $l = 2n$). These observations indicate that $\mathbf{R} = \frac{1}{2}[001]$, which does not lie in the plane of the boundary.

Furthermore, direct resolution of the (001) planes showed that the lattice fringes on either side of the boundary are, in fact, displaced by half of the fringe spacing, in agreement with $\mathbf{R} = \frac{1}{2}[001]$.

Twin boundaries involving a relative displacement have also been observed by Hutchison, Irusteta, and Whittaker (1975) in HRTEM images of amosite, a fibrous form of grunerite amphibole whose composition is approximately $Fe_6MgSi_8O_{22}(OH)_2$. The (200) and (001) lattice fringes show that the twin boundaries on (100) involve a displacement $\mathbf{R} = \frac{1}{2}[001]$.

8.6 Dauphiné twin boundaries in quartz

The Dauphiné twin boundaries in α-quartz, which have been studied in detail by McLaren and Phakey (1969), are particularly interesting because the nature of their diffraction contrast fringes depends critically on the operating reflection $\mathbf{g}$. For certain reflections, these twin boundaries behave as α-boundaries, whereas for other reflections, the fringe patterns originate not from any phase difference α, or any misorientation (change of s), but from the fact that the extinction distances in the two parts are significantly different.

The two parts of a Dauphiné twin are related by a rotation of 180° about the c-axis. Figure 8.17 shows projections onto (0001) of the SiO_4 tetrahedra in left-handed quartz (P3$_2$21) and in its Dauphiné twin. In Figure 8.18(a), the two structures are shown superimposed, without any relative displacement. It can be seen that only a slight rearrangement is necessary to produce a structure in the boundary that is the same as that of high-temperature, hexagonal β-quartz, as shown in Figure 8.18(b). A precisely similar twin could, of course, be produced in right-handed quartz (P3$_1$21).

To determine the nature of the contrast of Dauphiné twin boundaries, it is necessary to calculate the amplitudes ($|F_L|$ and $|F_T|$) and phases (θ_L and θ_T) of the structure factors for L quartz and its Dauphiné twin (T) for various reflections $\mathbf{g}$, when the atom positions in the two structures are referred to the same axes $\mathbf{a}_1$, $\mathbf{a}_2$, $\mathbf{a}_3$, and $\mathbf{c}$. The calculated values of $|F_L|$, $|F_T|$, and $\alpha = (\theta_L - \theta_T)$ are given in Table 8.4, together with the extinction distances t_{gL} and t_{gT} for 100-kV electrons.

The data given in Table 8.4 predict the following behavior:

1 For $\mathbf{g} = 10\bar{1}0$,

$$|F_L| = |F_T| \quad \text{and} \quad \alpha = 0$$

Hence, the twin will be out-of-contrast and no fringes will be produced by the boundary. This is observed.

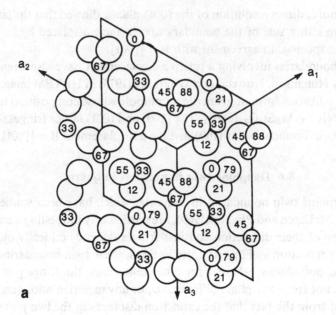

a

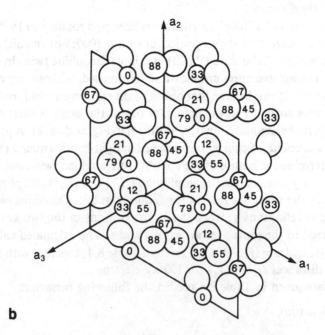

b

Figure 8.17. Projection onto (0001) of the SiO₄ tetrahedra in left-handed (P3₂21) quartz and its Dauphiné twin. The small circles represent Si and the large circles O. The numbers denote the fractional *c*-axis coordinates.

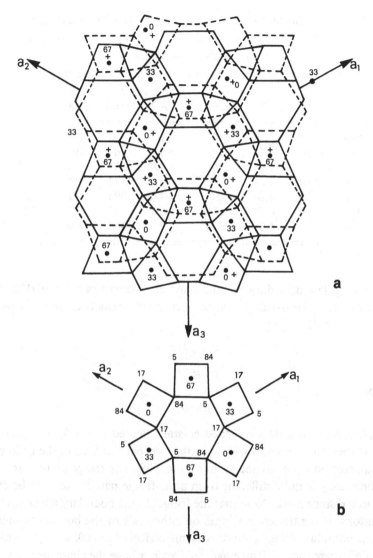

Figure 8.18. (a) Superposition of the two projections shown in Figure 8.17. Only a slight rearrangement is necessary to produce a structure at the Dauphiné boundary which is the same as that of β-quartz shown in (b). (From McLaren and Phakey 1969.)

2 For $\mathbf{g} = 11\bar{2}0$,

$$|F_L| = |F_T| \quad \text{and} \quad \alpha = 36°$$

Thus, the twin boundary will behave as a pure α-boundary, as is ob-

Table 8.4. *Structure factor amplitudes and phase differences, and extinction distances for L quartz and its Dauphiné twin for 100-kV electrons*

g	$\|F_L\|$	$\|F_T\|$	$\alpha = (\theta_L - \theta_T)$	$\dfrac{\|F_L\|}{\|F_T\|}$	t_{gL} (nm)	t_{gT} (nm)	$\dfrac{t_{gL}}{t_{gT}}$
$10\bar{1}0$	5.16	5.16	0	0	155	155	1
$10\bar{1}1$	11.6	7.5	0	$\approx 3/2$	69	107	$\approx 2/3$
$01\bar{1}1$	7.5	11.6	0	$\approx 2/3$	107	69	$\approx 3/2$
$10\bar{1}2$	4.64	2.23	180°	≈ 2	172	395	$\approx 1/2$
$01\bar{1}2$	2.23	4.64	180°	$\approx 1/2$	395	172	≈ 2
$11\bar{2}0$	4.67	4.67	±36°	1	150	150	1
$11\bar{2}2$	5.3	5.3	±20°	1	150	150	1

served. The boundary has the same characteristics for $g = 11\bar{2}2$. In both cases, the α-fringe contrast is relatively weak because α is small.

3 For $g = 10\bar{1}1$,

$$|F_L| \approx \tfrac{3}{2}|F_T|$$

$$t_{gL} = \tfrac{2}{3}t_{gT}$$

and

$$\alpha = 0$$

Therefore, in a wedge-shaped crystal oriented at $s = 0$, the spacing between the thickness fringes in the two parts will be in the ratio of approximately $\tfrac{3}{2}$, as shown in Figure 8.19. The fringe pattern at the boundary is quite different from an α-fringe pattern, as is to be expected since $\alpha = 0$. Note that the fringes in the boundary are continuations of the thickness fringes on either side of the boundary; thus, the boundary fringe pattern is asymmetrical. At $s = 0$, $\delta = s(t_{gL} - t_{gT})$ is also zero; so the fringe pattern does not have the characteristics of δ-fringes (Section 5.3.5). However, when $s \neq 0$, δ becomes nonzero and a further complexity is added to the boundary fringes. Even more complex fringes are expected for $g = 10\bar{1}2$ because an α-component must also be included.

For a more detailed discussion of the contrast of Dauphiné twin boundaries, the reader should consult the paper by McLaren and Phakey (1969), which illustrates the various types of contrast observed and compares

Figure 8.19. BF image of a Dauphiné twin boundary in a wedge-shaped crystal at $s = 0$ for $\mathbf{g} = 10\bar{1}1$. Note the change in spacing of the thickness fringes across the twin boundary.

them with fringe profiles calculated from the general theory of Gevers et al. (1965).

Dauphiné twin boundaries are sometimes parallel to low-index planes, but very often there appears to be no crystallographic control. This was certainly the case for the Dauphiné twin boundaries observed between the Brazil twin boundaries in amethyst quartz by McLaren and Phakey (1969).

Dauphiné twinning can result from external stress due to thermal shock, local pressure, bending, or torsion (Frondel 1962). It is not surprising, therefore, that Dauphiné twins are often observed in both experimentally and naturally deformed quartz crystals. Dauphiné twins are also produced when a crystal of hexagonal β quartz transforms on cooling at $T_c = 573\,^\circ$C to the trigonal α-form. It is probable that most of the Dauphiné twins observed in undeformed quartz crystals at normal temperatures have been formed in this way. The generation of Dauphiné twins at the phase transformation has been studied directly in a transmission electron microscope fitted with a specimen heating stage by van Tendeloo, Lunduyt, and Amelinckx (1976). They found that, in a crystal across which there was a temperature gradient ($T_c \pm \Delta T$), the boundary region between α and β structures was defined by a fine-scale array of Dauphiné twins, which develop in the α quartz near T_c and become progressively finer into the β-quartz region. The role of these twins in the transformation mechanism is discussed by Putnis and McConnell (1980). TEM observations of transformation twins in other minerals are discussed in the next section.

8.7 Transformation-induced twinning and modulated structures

Whenever a crystal transforms (usually on cooling) to a lower symmetry structure in which a number of orientational variants are possible, transformation twins are an almost invariable consequence if the new phase is nucleated at several different places in the original structure. The formation of Dauphiné twins at the β-to-α transformation in quartz is a classic example, which has already been mentioned briefly in Section 8.6. We now consider the twins and modulated structures associated with the monoclinic-to-triclinic transformation in the K feldspars, $KAlSi_3O_8$.

Sanidine is monoclinic (space group $C2/m$), and there is complete disorder in the occupation of the tetrahedral (T) sites by the Al and Si atoms. Over geological time, ordering takes place. In low (or maximum) microcline, the ordering is complete (all Al in T_1O sites), and the symmetry is reduced to triclinic ($C\bar{1}$). There are four main orientational variants in this structure: two orientations related by the albite twin law (rotation of 180° about b^*) and two orientations related by the pericline twin law (rotation of 180° about b). The composition planes of these two twins are, respectively, (010) and the rhombic section which is parallel to b and approximately normal to (001). Thus, the characteristic cross-hatched pattern observed in (001) sections between crossed-polarizers in the optical microscope has, for many years, been simply interpreted as intersecting sets of albite and pericline twin lamellae formed at the monoclinic-to-triclinic transformation. However, TEM observations indicate that this model is too simple. Because these observations, collectively, also constitute an excellent example of the application of the principal modes of operation of TEM to a specific mineralogical problem, we discuss them in some detail.

In most specimens, albite twinning predominates over pericline twinning. Generally, the albite twinning is on a very fine scale and occurs in domains separated by untwinned domains, arranged to form several types of overall microstructure (McLaren 1984).

Figure 8.20(a) is a DF micrograph ($g = 20\bar{1}$) showing part of a domain of albite twins and its boundary with an untwinned domain. Pericline twins,

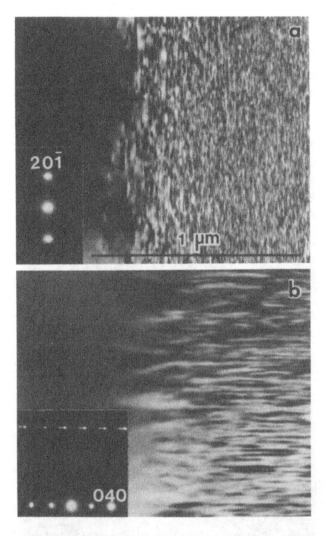

Figure 8.20. DF micrographs showing the boundary between a domain of fine-scale albite twinning (right) and an untwinned domain (left) in microcline. In (a), **g** = 20$\bar{1}$ and the albite twins are in contrast. In (b), **g** = 040 and the albite twins are out-of-contrast; however, fine-scale lamellae approximately parallel to **b** are now visible. The plane of the specimen is (001). (From McLaren 1978.)

if present, would be out-of-contrast for this operating reflection. Figure 8.20(b) is a DF micrograph with **g** = 040 of the same area. For this reflection, the albite twins are out-of-contrast, and the observed contrast suggests the existence of fine-scale pericline twins. However, there is no evidence in the SAD patterns for pericline twinning. In Figure 8.20(b), note

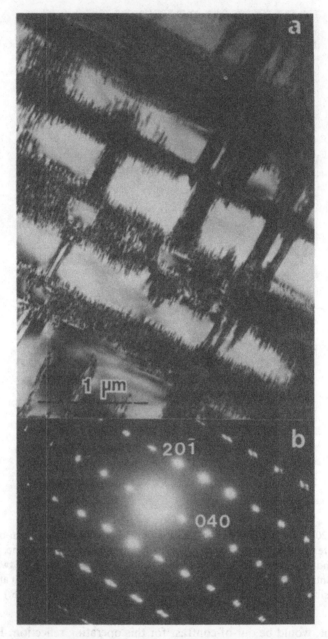

Figure 8.21. (a) BF micrograph ($g = 20\bar{1}$) showing a cross-hatched pattern in microcline. The plane of the specimen is (001). (b) Symmetrical electron diffraction pattern of area shown in (a). Note the streaking of the albite twin spots. (From McLaren 1978.)

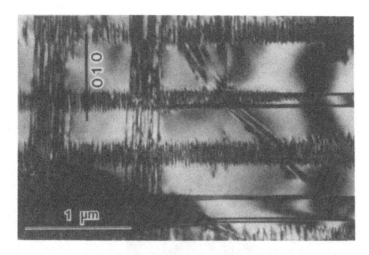

Figure 8.22. BF micrograph of microcline showing pericline twins partially
converted to fine albite twins. Other twins have been totally converted.
Remnant portions of the pericline composition planes are slightly inclined.
Beam direction near [001]. (From Fitz Gerald and McLaren 1982.)

that the reflections along the $0k0$ row are streaked normal to b^*, but there
is no splitting of these spots as would be expected if pericline twins were
present. Contrast of the type shown in Figure 8.20(b) was always observed with $g = 0k0$ in domains showing fine-scale albite twinning and is
clearly due to a modulated structure that is *qualitatively* similar to fine-
scale pericline-twin lamellae.

Figure 8.21(a) is a BF image of a type of microstructure that is commonly observed. It consists essentially of lamellar domains (approximately
parallel to b^*) that are alternately albite twinned on a fine scale and essentially untwinned. Some of these fine albite twins bridge the gaps between
the twinned lamellae, forming a cross-hatched pattern; also there are indications of a chessboard pattern. The SAD pattern, Figure 8.21(b), of
this area shows no evidence for pericline twinning, only albite twinning.
However, all the diffraction spots are streaked (and/or broken up into
discrete spots), indicating a narrow range of misorientations involving
rotations of up to $\pm 1°$ about the direction of the electron beam. In spite
of the absence of pericline twin spots, the nature of the microstructure
strongly suggests that it was derived from a simple microstructure of pericline twin lamellae – the domains of fine albite twins being formed by the
serration of the boundaries of the original pericline twins. A similar microstructure is shown in Figure 8.22, where one of the [010] lamellae is

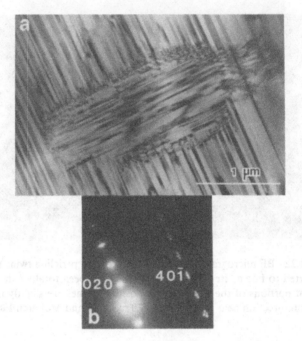

Figure 8.23. BF micrograph of microcline oriented with the electron beam near [104]. (a) Domain of albite twins completely enclosed in a matrix of pericline twins. (b) The SAD pattern corresponding to the area shown in (a). The streaks arise from the complex twin intersections in the domain boundary.
(From Fitz Gerald and McLaren 1982.)

seen to transform from a simple pericline twin lamella, to a lamella with one boundary serrated, and finally to a lamella consisting of fine, lens-shaped albite twins.

The observation that single pericline boundaries are commonly replaced by a large number of albite twin boundaries, possibly involving a 100-fold increase in total area of twin boundary per unit volume, suggests that the energy of a pericline twin boundary is greater than that of an albite twin boundary by a similar factor at normal temperatures (see later).

It was only after a long search of a number of microcline specimens that McLaren (1978) and Fitz Gerald and McLaren (1982) found regions that gave the *4-spot diffraction pattern* suggesting the coexistence of both albite and pericline twins (M-twinning). Akizuki (1972) found none of these regions, and they were apparently not common in the specimens studied by Tibballs and Olsen (1977). One such area is shown, with its associated selected area diffraction pattern, in Figure 8.23; it will be seen that within the field of view there is a domain of albite twins that is completely enclosed in a domain of pericline twins. The boundary between the domains is extremely complex.

When albite and pericline twin lamellae (A, B, and A', B') of comparable widths intersect, a chessboard, cross-hatched pattern frequently develops. This pattern is observed both on the optical microscope scale and on the TEM scale, as shown in Figure 8.24. DF observations show that the diagonally opposite, clear "window" areas (at the intersections of A and A' or B and B' twin lamellae) are related by the albite twin law operation plus a small misorientation that is qualitatively similar to that relating the albite and pericline twin laws. The areas AB' and A'B (at the intersections of A and B' and of A' and B twin lamellae) consist of fine-scale albite twin lamellae. The small misorientation is easily seen (i) by diffraction contrast in TEM (McLaren 1978), (ii) in corresponding SAD patterns, or (iii) by the misalignment of the fine albite twin lamellae at the adjacent corners of two diagonal windows, as shown in Fitz Gerald and McLaren (1982, fig. 8). The nature and origin of the chessboard pattern can be understood with the aid of Figure 8.25.

Consider a domain of albite twin lamellae ABABA... and a separate domain of pericline twin lamella A'B'A'B'A'.... Now consider a third domain in which these lamellae intersect so as to produce a chessboard pattern, as shown. The squares of this chessboard pattern are formally labeled AA', BA', AB', and BB'. Now, crystallographically (as distinct from morphologically) the only difference between albite and pericline twins is the twin axis (namely, b^* and b, respectively). A and A' are structurally identical but differ slightly in crystallographic orientation; similarly for B and B'. Thus, after some slight relaxation, regions AA' and BB' should become strain-free, and their mutual crystallographic orientation is expected to be somewhere between the two extremes defined by the albite and pericline twin laws.

The regions AA' and BB' correspond to the so-called windows in Figure 8.24, which are always found to be essentially albite-twin related. On the other hand, the regions AB' and BA' are formally required to be a mixture of the two twin orientations. The obvious way to achieve this is for these regions to twin on a finer scale. Figure 8.24 shows that the regions corresponding to AB' and BA' are albite twinned on a fine scale. Such regions are also modulated in a way that is qualitatively similar to fine-scale pericline-twin lamellae, as in Figure 8.20(b).

Partially ordered K-feldspars (such as orthoclase, adularia, and intermediate microcline) characteristically exhibit a tweed pattern consisting

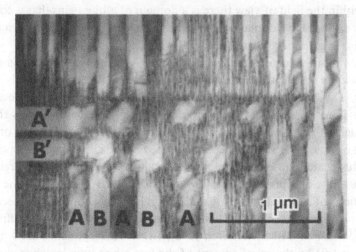

Figure 8.24. BF micrograph of microcline. The intersection of albite twins (A and B) with pericline twins (A′ and B′) has produced a chessboard pattern of finely twinned regions and untwinned window regions. The electron beam is near [104]. (From Fitz Gerald and McLaren 1982.)

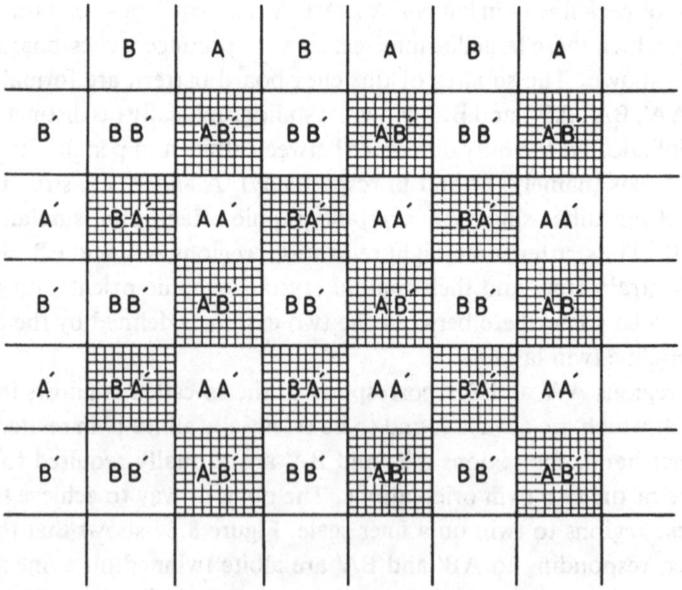

	B	A	B	A	B	A
B′	B B′	A B′	B B′	A B′	B B′	A B′
A′	B A′	A A′	B A′	A A′	B A′	A A′
B′	B B′	A B′	B B′	A B′	B B′	A B′
A′	B A′	A A′	B A′	A A′	B A′	A A′
B′	B B′	A B′	B B′	A B′	B B′	A B′

Figure 8.25. Diagram showing the idealized intersection of a set of albite-twin lamellae ABABA... with a set of pericline twin lamellae A′B′A′B′A′... to form the chessboard pattern of cross-hatched twinning. (From McLaren 1978.)

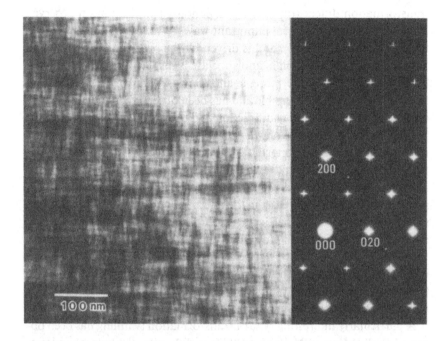

Figure 8.26. BF image (and its associated selected area diffraction pattern) showing the tweed microstructure in orthoclase. The shortest wavelength of the modulation is about 4.5 nm. (From McLaren and Fitz Gerald 1987.)

of two orthogonal sets of modulations (Figure 8.26), which is *qualitatively* similar to two intersecting sets of albite and pericline twin lamellae. The contrast arises because of the twinlike misorientation, and so each set can be put out-of-contrast separately by selecting an appropriate diffraction vector **g**. These tweed microstructures in K-feldspars were first described by McConnell (1965) and have since been studied by Nissen (1967), McLaren (1974, 1984), Bambauer, Krause, and Kroll (1989), and McLaren and Fitz Gerald (1987), who also used convergent beam electron diffraction (Section 3.10) and ALCHEMI (Section 7.4) to show that regions of tweed microstructure are partially ordered and that the average symmetry is triclinic (and not monoclinic as implied by the name orthoclase).

A transformation from monoclinic to triclinic on cooling from high temperatures is also observed in anorthoclase and gives rise to a cross-hatched microstructure, which is observed in **b*c*** sections. However, this transformation differs from the corresponding transformation observed in the K-feldspars in several important ways. Whereas in microcline, the transformation is diffusive and is driven by the ordering of Si and Al in T-sites, the transformation in anorthoclase is displacive and is not driven by Si/Al ordering. Furthermore, in anorthoclase (unlike the K-feldspars), the transformation is reversible and has been observed directly under laboratory conditions by x-ray diffraction and optical microscopy. Recently, Smith, McLaren, and O'Donnell (1987) have investigated the twin microstructure in a number of anorthoclase single crystals as a function of temperature (over the range 20°–950°C) in a transmission electron microscope fitted with a heating specimen stage. These observations showed that many points of similarity exist in the microstructures of anorthoclase and microcline. However, of particular importance were the observations that showed the relative stability of albite and pericline twins is a function of temperature, the pericline twin being more stable than the albite twin at high temperatures near the monoclinic-to-triclinic transformation, and the opposite at room temperature. This conclusion is consistent with the microstructures observed in microcline.

A particularly interesting type of transformation twinning has been observed by Reeder and Nakajima (1982) in dolomite; the twin boundaries resemble antiphase domain boundaries (see Section 8.8). Modulated structures are also observed in dolomites (van Tendeloo, Wenk, and Gronsky 1985; Wenk, Barber, and Reeder 1983).

8.8 Antiphase domain boundaries

When a disordered structure becomes ordered, antiphase domains can form within a given orientation variant if *translational variants* exist in the ordered structure. The concept of a translational variant and the process by which antiphase domains are formed can be easily understood by considering the ordering in a simple two-dimensional structure containing two types of atom (A and B) when there are only two possible translational variants. Figure 8.27(a) shows a structure in which all possible atom sites are occupied by either atom A or atom B with equal probability; the structure is disordered. Below a critical temperature T_c, the structure begins to order. Suppose that the ordering is nucleated at two places within the specimen: at the top left-hand corner and at the bottom right-hand corner. Figure 8.27(b) shows that when the ordering is complete, the specimen consists of two domains that are out of phase with one another and are joined along the solid line shown. These domains are called

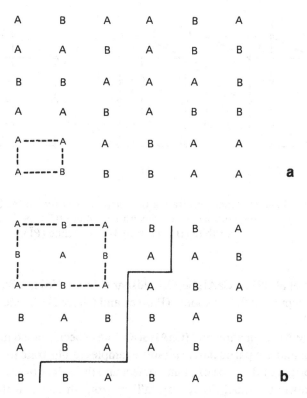

Figure 8.27. Diagrams illustrating the formation of an antiphase domain boundary during ordering of a crystal consisting of two types of atom, A and B.

antiphase domains, and the boundary between them is an *antiphase domain boundary* (APB). The unit cell edges of the ordered structure are double the length of those of the disordered structure (thus, the ordering produces a *superlattice*), and the domains are related by a displacement vector $\mathbf{R} = \frac{1}{2}\langle 10 \rangle$ in the ordered structure. Hence, the domains correspond to the two translational variants.

APBs are essentially stacking faults and have been observed by TEM in a number of minerals that undergo order–disorder transformations. Examples include omphacite (Carpenter 1979; Champness 1973; Phakey and Ghose 1973), pigeonite (Bailey et al. 1970; Carpenter 1979; Christie et al. 1971; Fujino, Furo, and Momoi 1988), calcic plagioclase feldspars (Christie et al. 1971; Czank et al. 1973; Czank, Schulz, and Laves 1972; Heuer et al. 1972; McLaren 1973; McLaren and Marshall 1974; Müller et al. 1973; Müller, Wenk, and Thomas 1972; Nord, Heuer, and Lally 1974;

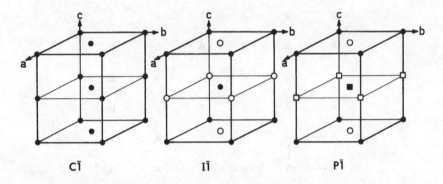

Figure 8.28. Diagrams illustrating the relationship between the lattices of the monoclinic feldspar structure with $c \approx 1.4$ nm ($C\bar{1}$), body-centered anorthite ($I\bar{1}$) and primitive anorthite ($P\bar{1}$).

Nord et al. 1973), $CaAl_2Ge_2O_8$-feldspar (Müller, Vojdan-Shemshadi, and Pentinghaus 1987), scapolite (Phakey and Ghose 1972), and wenkite (Lee 1976).

The APBs in anorthite ($CaAl_2Si_2O_8$) have been studied in considerable detail and are particularly suitable examples to illustrate the principles of the origin of this type of planar defect and the nature of the TEM images.

Anorthite, $CaAl_2Si_2O_8$, crystallizes from the melt in the high-albite structure (space group $C\bar{1}$), in which the Si and Al atoms are disordered among the T-sites. The diffraction pattern shows a-reflections ($h+k = 2n$, $l = 2n$) only. On cooling, this structure transforms to a body-centered structure ($I\bar{1}$) due to the regular alternation (ordering) of Si and Al. This leads to a doubling of the length of the c-axis and the addition of b-reflections ($h+k = 2n+1$, $l = 2n+1$) in the diffraction pattern. On further cooling, this structure transforms to a primitive structure ($P\bar{1}$), which is characterized by the addition of c-reflections ($h+k = 2n$, $l = 2n+1$) and d-reflections ($h+k = 2n+1$, $l = 2n$). The regular alternation of Si and Al atoms is retained, but the positions and vibrations of the Ca and framework atoms change. These three structural types are illustrated in Figure 8.28, and the reader is referred to Smith and Brown (1988) for a more detailed discussion. Consideration of these structures shows that at each transformation, APBs can be produced with the following possible fault vectors $\mathbf{R}$:

$$C\bar{1} \rightarrow I\bar{1}: \quad \mathbf{R}_1 = \tfrac{1}{2}[001]$$

$$\mathbf{R}_2 = \tfrac{1}{2}[110]$$

Table 8.5. *Values of* $\alpha = 2\pi \mathbf{g} \cdot \mathbf{R}$ *for the possible fault vectors of APBs in anorthite for b-, c-, and d-reflections*

Fault vector	Reflection	α
$\mathbf{R}_1$	b	π
	c	π
	d	0
$\mathbf{R}_2$	b	π
	c	0
	d	π
$\mathbf{R}_3$	b	0
	c	π
	d	π

$$I\bar{1} \rightarrow P\bar{1}: \quad \mathbf{R}_1$$
$$\mathbf{R}_2$$
$$\mathbf{R}_3 = \tfrac{1}{2}[11\bar{1}]$$

An APB is visible as an α-fringe pattern (Section 5.3.1) if

$$\alpha = 2\pi \mathbf{g} \cdot \mathbf{R} \neq 2\pi n$$

where n is an integer. Clearly, α is zero for all a-reflections, and the APBs are out-of-contrast in images formed with these reflections. Values of α for $\mathbf{R}_1$, $\mathbf{R}_2$, and $\mathbf{R}_3$ with b-, c-, and d-reflections are given in Table 8.5; note that all *visible* APBs give rise to $\alpha = \pi$ fringes (Section 5.3.3). At the exact Bragg angle ($s = 0$), these fringe patterns are symmetrical about the center of the foil, the fringe spacing is $t'_g/2$, and the central fringe is dark in DF. Values of t'_g for b-reflections in the range $-4 < h, k, l < 4$ vary from 1400 nm for the strongest reflections to over 60,000 nm for the weakest for 200-kV electrons (McLaren and Marshall 1974). Thus, even for the strongest b-reflections, the specimen thickness for good electron transmission is expected to be less than $t'_g/2$. It follows that only the central dark fringe of the π-fringe pattern is observed at $s = 0$. Using the theory of van Landuyt et al. (1964), McLaren and Marshall (1974) showed that the central fringe becomes bright on either side of $s = 0$. The DF image (using a b-reflection) of an APB crossing a bend contour shown in Figure 8.29 has the predicted characteristics. Additional fringes may be observed for larger deviations from $s = 0$ in thicker crystals.

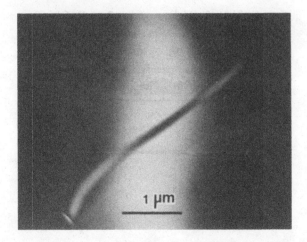

Figure 8.29. A single antiphase boundary crossing a bend contour in body-centered bytownite (An_{77}). DF image on a b-reflection. Note reversal of contrast with $s = 0$ and $s \neq 0$. (From McLaren and Marshall 1974.)

APBs imaged in DF with b-, c-, and d-reflections at $s = 0$ in the same region of a specimen of synthetic anorthite, which had been cooled from the melt, are shown in Figure 8.30, together with the SAD pattern in which all the reflections are sharp. There are clearly two coexisting sets of APBs: one set is in contrast for a b-reflection, and another set is in contrast for a c-reflection. Both sets are in contrast for a d-reflection. We see from Table 8.5 that the APBs imaged with a b-reflection in Figure 8.30(b) have a displacement vector $R_2 = \frac{1}{2}[110]$ and were formed at the $C\bar{1}$ to $P\bar{1}$ transformation, whereas the APBs imaged with a c-reflection in Figure 8.30(c) have a displacement vector $R_3 = \frac{1}{2}[11\bar{1}]$ and were formed at the $I\bar{1}$ to $P\bar{1}$ transformation. For some c-reflections (such as $11\bar{1}$ and $13\bar{1}$) t'_g is sufficiently short that the APB images show more than the central fringe at $s = 0$ (see, for example, Müller et al. 1973).

APBs can also be imaged by direct resolution of an appropriate set (or sets) of lattice planes. In Figure 8.31 an APB is revealed by the direct resolution of $(11\bar{1})$ planes ($d = 0.68$ nm). The displacement of $d/2$ is clearly seen at the boundary, which is slightly inclined to the foil.

8.9 Grain boundaries in mineral systems

The ideas about the structure of GBs and the methods for imaging them by TEM that were discussed in Section 5.3.8 evolved from studies of GBs

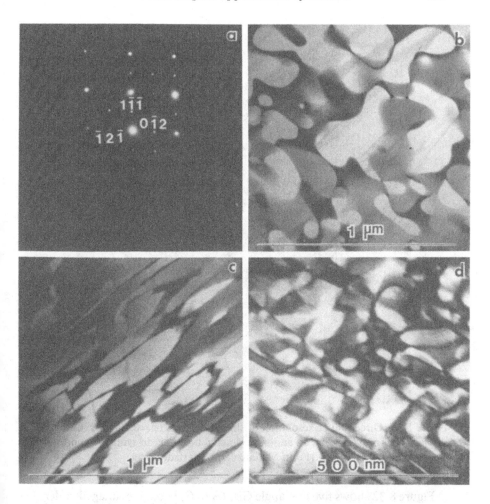

Figure 8.30. Antiphase domain boundaries in synthetic anorthite (An_{100}). (a) Diffraction pattern showing sharp *a*-, *b*-, *c*-, and *d*-reflections. (b), (c), and (d) Antiphase domain boundaries imaged in DF with *b*-, *c*-, and *d*-reflections, respectively. (From McLaren and Marshall 1974.)

in metals (Gleiter 1982) and ceramics (Duffy 1986). Comparatively little attention has been given to the GBs in rocks, and the few TEM studies that have been made have not revealed much detail about their structure. Therefore, this section is restricted to illustrating the several types of GB images that have been observed in some important monomineralic rocks, synthetic aggregates, and bicrystals.

Figure 8.31. Direct lattice resolution of the (11Ī) planes ($d = 0.68$nm) showing a displacement of $d/2$ across an antiphase domain boundary in synthetic anorthite (An_{100}). (From McLaren and Marshall 1974.)

Figure 8.32 shows five low-angle GBs (A to G) in olivine, imaged under slightly different diffracting conditions by Boland, McLaren, and Hobbs (1971). In the BF image, Figure 8.32(a), all the subgrains are diffracting essentially equally, and the changes of orientation across the GBs are not apparent. However, in the DF images (b) and (c), the change of orientation across each GB now shows up as a marked change of contrast. Several types of boundary image can be identified in these micrographs:

1 Thickness fringes, as at boundary A in Figure 8.32(a, b, c).
2 Thickness fringes on which is superimposed a fine structure consisting of an array of parallel lines, as at boundary B in Figure 8.32(a, c) and as at boundary C in Figure 8.32(a, b, c). At some places along B the individual lines of the fine structure are clearly identified as single

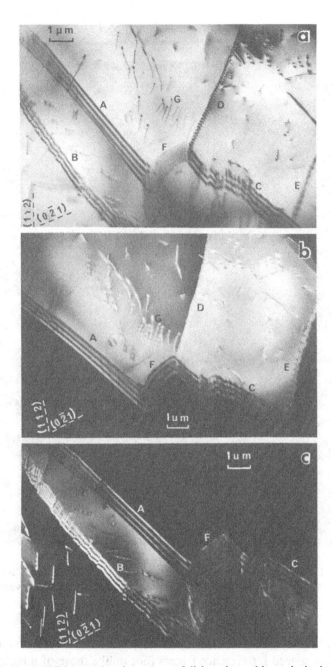

Figure 8.32. (a) BF image showing arrays of dislocation subboundaries in olivine. Boundaries A, B, C, and G are parallel to (100); D and E are parallel to (001); F is approximately parallel to ($\bar{1}$01). The changes in orientation across the subboundaries are shown by changes of contrast in the DF images (b) and (c). (From Boland et al. 1971.)

241

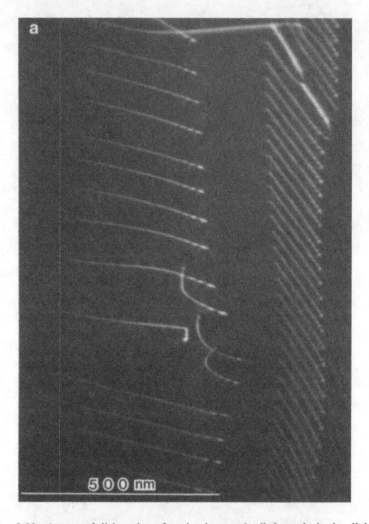

Figure 8.33. Arrays of dislocations forming low-angle tilt boundaries in olivine. (a) Dislocations in the two boundaries shown have the same Burgers vector and sign. (Courtesy J. D. Fitz Gerald.)

dislocations, whereas at other places the fine structure looks more like closely spaced interference fringes (not unlike those of a moiré pattern), which have a distinctly lower contrast than the clearly identified dislocations. Tholen (1970) has shown that if dislocations in a low-angle boundary are closer together than one extinction distance, they will give rise to a fringe system that is indistinguishable from moiré fringes. Moiré fringes can usually be identified because their orienta-

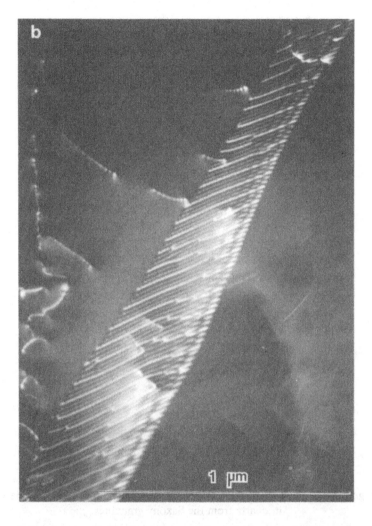

Figure 8.33. *Continued.* (b) Boundary consists of two sets of dislocations with different Burgers vectors. (From Karato et al. 1986.)

tion and spacing depend on the diffraction vector **g** (see Section 6.6). Ledges on boundaries A, B, and C are also clearly visible as kinks in the thickness fringes.

3 Boundaries D and E (and the very low-angle boundary G) are seen to be arrays or networks of dislocations.

Arrays of dislocations forming low-angle tilt boundaries in olivine are shown in Figure 8.33. The contrast of the dislocations in each of the two

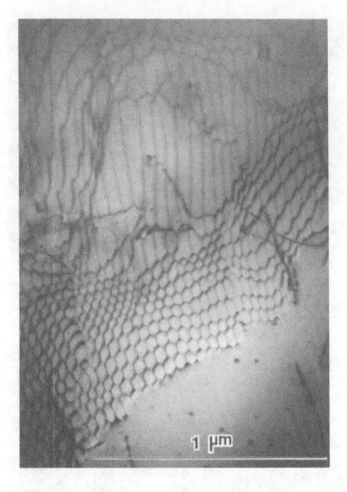

Figure 8.34. Network of dislocations forming a low-angle twist boundary
in quartz from the Saxony granulites.

boundaries shown in Figure 8.33(a) indicates that the dislocations are of
the same Burgers vector and sign. However, the boundary shown in Figure 8.33(b) is clearly made up of two sets of dislocations of different Burgers vectors. Figure 8.34 shows a well-developed network of dislocations
forming a low-angle twist boundary in quartz.

   The diffraction effects directly attributable to periodic arrays of dislocations that form low-angle boundaries in olivine have been studied by
Ricoult and Kohlstedt (1983). Figure 8.35 shows an edge-on tilt boundary
approximately parallel to (100) and its associated SAD pattern. The extra

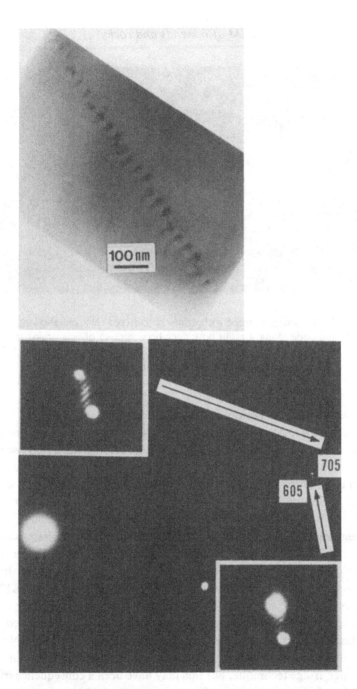

Figure 8.35. BF image of the dislocations forming a (100) tilt boundary in olivine viewed edge-on, and the electron diffraction pattern from the boundary region. Note the fine structure of the 605 and 705 reflections shown in the inserts. (From Ricoult and Kohlstedt 1983.)

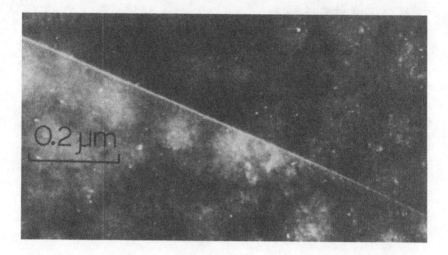

Figure 8.36. DF image formed with diffusely scattered electrons showing a band about 10 nm wide along a grain boundary between two olivine grains in a hot-pressed olivine-basalt aggregate annealed for 200 hours near 1250°C at 1 GPa. (From Vaughan and Kohlstedt 1982.)

reflections due to the periodic array of dislocations are clearly seen in the enlarged inserts. Also note that these extra reflections are streaked normal to the plane of the boundary. Ricoult and Kohlstedt (1983) used the length of the streaks to determine the boundary "thickness," which they found to be about one-third of the dislocation spacing. However, as discussed in Section 5.3.8, this thickness may have no simple quantitative relationship to the width of the distorted region at the boundary.

Vaughan and Kohlstedt (1982) used lattice-fringe images, phase-contrast images (Section 5.7.1), and DF diffuse-scattering images to study the distribution of the glass-phase in hot-pressed, olivine-basalt aggregates. In DF diffuse-scattering images, almost every GB appeared as a narrow, ≲10 nm wide, bright band, suggesting the presence of a glass phase (Figure 8.36). However, phase-contrast images of the same boundaries indicated that the glass phase, if present, had a maximum thickness of ≈1 nm. No evidence of a glass phase was found in most of the boundaries examined by the lattice-fringe technique, but this may have been a consequence of the difficulty of obtaining images of boundaries for which there was no overlap of the grains.

A lattice-fringe image of a high-angle (≈40°) boundary in a synthetic bicrystal of calcite is shown in Figure 8.37. No evidence for any second

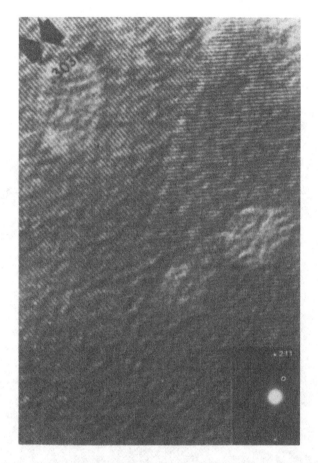

Figure 8.37. Lattice-fringe image showing a high-angle grain boundary in a bicrystal of "wet" synthetic calcite (From Hay and Evans 1988.)

phase is seen at the boundary, although it could be obscured by the slightly overlapping grains. Moiré fringes are not observed in the region of overlap because, with a 40° misorientation of 0.303-nm lattice fringes, the moiré–fringe spacing is only 0.43 nm (Eq. 6.8). Figure 8.38 shows the moiré fringes at the region of overlap when the misorientation is only ≈12°.

It is clear from the observations summarized in this section that there is still a great deal to learn about the structures of GBs in mineral aggregates and rocks. Although the currently available techniques will no doubt yield further useful information, significant progress may depend on the development of novel techniques of specimen preparation, diffraction, and imaging.

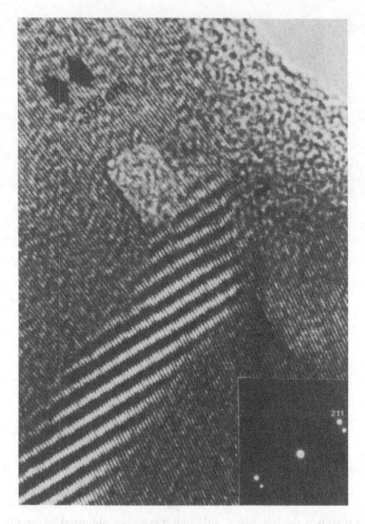

Figure 8.38. Lattice-fringe image of a grain boundary in Oak Hall limestone.
The boundary plane is inclined and gives rise to moiré fringes.
(From Hay and Evans 1988.)

8.10 Exsolution microstructures

In a number of important mineral systems, solid solution exists at elevated temperatures, but exsolution into chemically distinct phases takes place on cooling. In general, exsolution proceeds either by a nucleation and growth mechanism or by spinodal decomposition (as defined by Cahn

1961, 1968), depending on the cooling rate, the original composition, and the nature of the free energy versus composition curves. The thermodynamics of exsolution and the possible mechanisms have been reviewed by Champness and Lorimer (1976) and by Putnis and McConnell (1980) and need not concern us specifically here: our primary aim is to illustrate and interpret the TEM images of some of the most common types of microstructure that develop during exsolution in several important groups of rock-forming minerals.

8.10.1 Exsolution in the alkali feldspars. The exsolution microstructures in the alkali feldspars have been studied in considerable detail by TEM since the first observations of a cryptoperthite by Fleet and Ribbe (1963), who showed that the specimen consisted of alternating, irregular sheets of orthoclase and (finely twinned) low albite, approximately parallel to ($\bar{6}01$) and a few hundred nanometers thick. These original observations were of considerable mineralogical importance, but a more appropriate starting point for a general discussion of exsolution microstructures in the alkali feldspars is perhaps the TEM work of McConnell (1969a). He examined large alkali feldspar crystals ($Ab_{63}Or_{37}$) from a lava and observed a modulated microstructure, similar to that shown in Figure 8.39(a), with a periodicity of ≈ 10 nm approximately normal to ($\bar{6}01$), which he interpreted as exsolution of Na- and K-rich phases. Later, Owen and McConnell (1971, 1974), using previously homogenized specimens of the same crystals, regenerated the modulated microstructure by hydrothermal annealing at temperatures in the range 500°–600°C at 100 MPa for 4 days or less. The wavelength λ of the modulation increased with increasing temperature and annealing time. Yund, McLaren, and Hobbs (1974) observed a similar behavior in a homogenized specimen (Or_{33}) annealed in air at 600° and 700°C for periods up to 173 days. A DF image and its associated SAD pattern of the microstructure produced on annealing at 600°C for 4 hours are shown in Figure 8.39(a, b). The wavelength of the modulation in the image is 9 nm, which is in good agreement with the value of 8.5 nm determined from the spacing of the satellites to the Bragg reflections in the SAD pattern. The fact that there are only two satellites (one on each side of the Bragg reflection) indicates that the modulation is essentially sinusoidal, with only one Fourier component. Figure 8.39(c) shows the modulated microstructure after annealing at 600°C for 28 days; here we see that λ has increased to about 20 nm. Annealing at 700°C for 28 days produced a microstructure with $\lambda \approx 70$ nm and a square-wave modulation, as shown in Figure 8.39(d). When the wave-

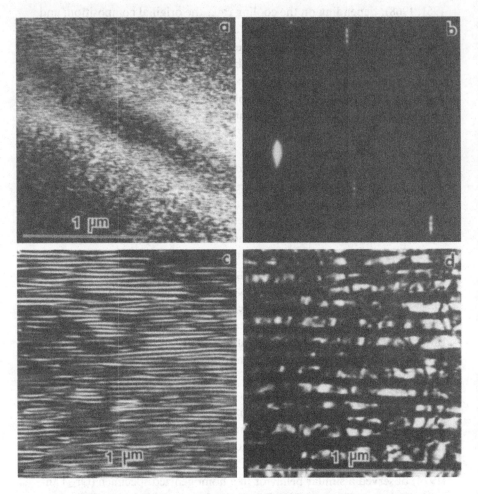

Figure 8.39. DF images of exsolution lamellae experimentally induced in an
alkali feldspar (Or_{33}). (a) Sample annealed at 600°C for 4 hours, $g = 400$.
(b) Diffraction pattern of (a). (c) Sample annealed at 600°C for 28 days, $g = 200$.
(d) Sample annealed at 700°C for 28 days, $g = 200$. (From Yund et al. 1974.)

length exceeds about 20 nm, it is greater than the diameter Δ of the coherence area of the electron beam; so satellite reflections are not observed
(see later). Instead, pairs of closely spaced Bragg reflections are seen, corresponding to the two exsolved phases with slightly different lattice parameters. The contrast in the images arises mainly because of the small
misorientation between adjacent lamellae. Owen and McConnell (1974)
found that when λ was of the order of 60 nm, the albite lamellae were

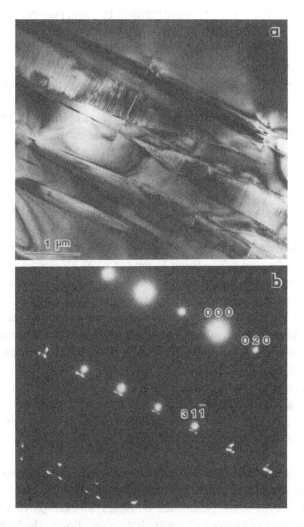

Figure 8.40. (a) BF micrograph showing albite-twinned albite exsolution lamellae in a cryptoperthite (Or_{72}) and (b) its associated SAD pattern. (From McLaren 1974.)

often albite-twinned, as observed in the cryptoperthite studied by Fleet and Ribbe (1963).

Figure 8.40 is a BF image of the exsolution microstructure commonly observed in cryptoperthites, together with its SAD pattern. The electron beam is approximately normal to (001). The single spots corresponding to the K-feldspar (monoclinic) and the pairs of spots corresponding to the two twin orientations in the albite lamellae (triclinic) are easily identified. Figure 8.40(a) shows that the width of the albite twins varies considerably. However, in some regions of the albite lamellae the width of the twins appears to be constant over distances of the order of 1 μm. Furthermore, the average width of a twin pair (the twin periodicity $\bar{\omega}$) within an albite lamella increases as the width t of the lamella increases. This suggests that the twinning occurs in order to minimize the stress at the interfaces between adjacent monoclinic and triclinic lamellae. On the basis of this suggestion, Willaime and Gandais (1972) and McLaren (1974) showed theoretically that

$$t = k\bar{\omega}^n \qquad (8.7)$$

where k is a constant and n lies somewhere between 2 and 3. Pairs of twin spots are observed in the SAD pattern only if $\bar{\omega}$ is significantly greater than the diameter Δ of the coherence area of the electron beam, which (as discussed in Section 1.6) is about 15 nm for a 100-kV electron microscope operating under normal conditions. If $\bar{\omega}$ is significantly less than Δ, superlattice reflections, whose spacing is inversely proportional to $\bar{\omega}$, will be observed (McLaren 1974; McLaren and MacKenzie 1976). If the SAD pattern is derived from an area of specimen containing a wide range of twin periodicities, streaking of the twin spots will be observed, as shown in Figure 8.40(b).

More complex exsolution microstructures are observed in cryptoperthites of similar composition when the K-feldspar lamellae become triclinic (see, e.g., Brown and Parsons 1984; Champness and Lorimer 1976; Willaime, Brown, and Gandais 1976).

8.10.2 Exsolution microstructures in the plagioclase feldspars. Modulated microstructures, similar to those observed in the alkali feldspars, are also observed in the plagioclase feldspars; an example in a bytownite (An$_{75}$) is shown in Figure 8.41. Note that, in addition to the high-contrast modulations parallel to (010), there is a second modulation, less regular and lower contrast, which crosses the first at approximately 90°. Consequently, these microstructures are often referred to as *tweed* textures. In addition to the tweed texture, several other distinct types of exsolution microstructure are observed in the plagioclase feldspars. Perhaps the simplest is the microstructure observed in the *peristerites,* whose bulk compositions are in the range An$_3$ to An$_{16}$.

The (001) SAD pattern of a typical peristerite is shown in Figure 8.42(a). By comparing this figure with (b), we can see that it is the superposition of the reciprocal lattice sections of albite (An$_{1\pm1}$) and oligoclase (An$_{25\pm5}$).

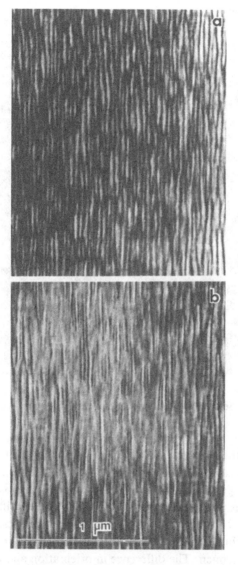

Figure 8.41. DF images ($g = 24\bar{2}$) of the modulated structure in bytownite (An₇₅).
(a) Average deviation from the exact Bragg angle $\bar{s} \neq 0$. (b) $\bar{s} = 0$. Note how the
contrast washes out in (b). (From McLaren 1974.)

Fleet and Ribbe (1965) were the first to show by TEM that the albite and
oligoclase phases were arranged as alternating lamellae parallel to planes
near (0$\bar{8}$2). Figure 8.43 shows the lamellae imaged in DF using the pair of
albite and oligoclase reflections marked in the corresponding SAD pattern.

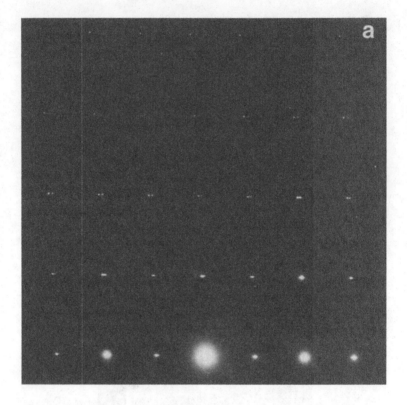

Figure 8.42. (a) Selected area diffraction pattern of a typical peristerite (An$_{10}$).

The stronger reflection is from oligoclase. Thus, in the micrograph, the bright lamellae correspond to oligoclase and the dark lamellae to albite. A BF image of peristerite lamellae in another specimen is shown in Figure 8.44. In this image, the two types of lamellae are diffracting almost equally and the contrast arises at the lamellar boundaries. Because these are slightly inclined, fringes are seen at the boundaries in the thicker regions of the specimen. The difference in orientation and lattice parameters across the boundaries appears to be accommodated without any pronounced elastic strain or additional defects, such as dislocations in the boundary.

A somewhat more complex exsolution microstructure involving albite twin lamellae in a peristerite of bulk composition Ab$_{91.3}$An$_{4.0}$Or$_{4.7}$ was studied by Gjonnes and Olsen (1974) using a TEM fitted with an energy dispersive x-ray spectrometer (Chapter 7). In this specimen, the *twin*

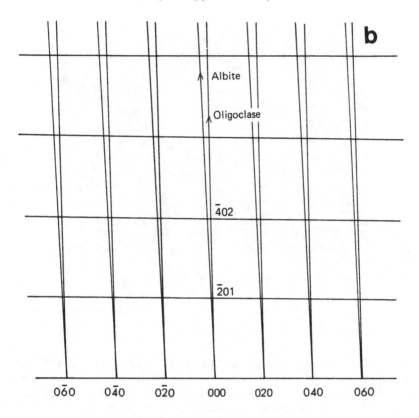

Figure 8.42. *Continued.* (b) Superimposed oligoclase and albite reciprocal lattice sections corresponding to the diffraction pattern shown in (a). (From McLaren 1974.)

lamellae parallel to (010) were approximately 1 μm wide and had compositions that were alternatively rich and poor in both Or and An. It was only in the Or- and An-rich lamellae that *finer-scale exsolution lamellae* (approximately parallel to the twin lamellae) were observed. The exsolution lamellae were too narrow to allow a direct determination of composition. However, from the average lattice constants obtained from x-ray powder patterns and the observed selected-area electron diffraction patterns, it was inferred that the two phases had compositions near An_0 and An_{25} (see also Cliff et al. 1976).

Plagioclase feldspars in the intermediate composition range An_{25} to An_{75} characteristically exhibit a diffraction pattern (Figure 8.45) in which the *e*- and *f*-reflections indicate the presence of a superlattice whose period

Figure 8.43. DF image of a peristerite (An$_{7.2}$) and its diffraction pattern. The image is formed with the arrowed pair of reflections. The brighter spot is from oligoclase; thus, in the micrograph, the bright lamellae correspond to oligoclase and the dark lamellae to albite. (From McLaren 1974.)

(2–8.5 nm) and orientation are functions of composition; see Smith and Brown (1988) for a review of these so-called e-plagioclases. Although there are no indications in the diffraction patterns for the presence of more than one phase, the existence of two kinds of alternating lamellae of width of the order of 100 nm is revealed in TEM images. Figure 8.46 is a pair of DF images formed with an a-reflection ($\mathbf{g} = 20\bar{2}$) operating in an (001) foil. It can be seen that the lamellae contrast is dependent on the orientation of the crystal with respect to the incident electron beam. This suggests that the contrast arises because of a slight misorientation between adjacent lamellae. Optimum contrast is obtained in DF when $\bar{s}$ (the *mean* deviation from the exact Bragg angle) is not zero, as shown in Figure 8.46(a). However, in Figure 8.46(b), the contrast becomes extremely weak when $\bar{s} = 0$. This behavior can be understood by referring to Figure 4.13. It will be seen that in DF at $s = 0$, two slightly misoriented lamellae diffract with almost equal intensity, but when $st_g = 0.75$, for example, the

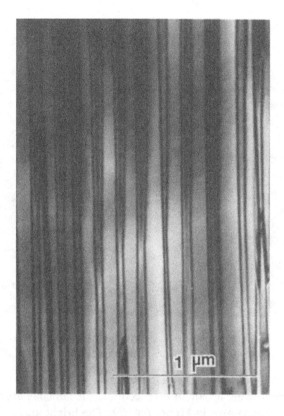

Figure 8.44. BF micrograph of exsolution lamellae in a peristerite (An$_{7.6}$) with the electron beam approximately normal to (001). The overlapping interfaces give rise to complex fringe patterns. (From McLaren 1974.)

diffracted intensities are markedly different. The contrast in Figure 8.46(a) implies that $t_g(s_1 - s_2) = t_g \Delta s$ is about 0.5. Taking $t_g = 100$ nm, we find that this value of $t_g \Delta s$ corresponds either to a change of orientation $\Delta \theta$ of about 3 minutes of arc, from Eq. (4.39), or to a fractional change of the d-spacing ($\Delta d/d$) of 0.1. A value of $\Delta d/d$ of this magnitude would be easily detected in the SAD pattern, whereas an orientation change of 3 minutes would not. Because there is no evidence for the lamellar structure in the diffraction patterns, it can be concluded that the observed contrast arises predominantly from a small change in orientation between adjacent lamellae (McLaren 1974).

The lamellar contrast in DF using an a-reflection is quite low, even under the optimum conditions ($\bar{s} \neq 0$) of Figure 8.46(a). However, in DF

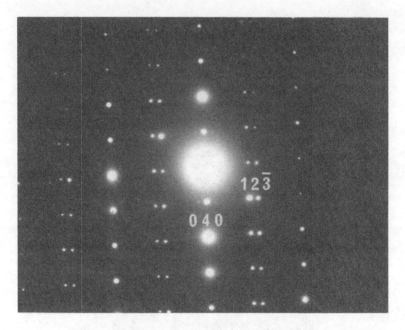

Figure 8.45. Diffraction pattern of a (100) foil of an intermediate plagioclase (An$_{52}$) showing a-, e-, and f-reflections. (From McLaren 1974.)

images formed using a pair of e-reflections, the lamellae are revealed in high contrast, as shown in Figure 8.47(a). The bright and dark lamellae may, for convenience, be designated A and B, respectively, because there is no reversal of contrast with the sign of $\bar{s}$. This characteristic indicates that the structure factors for the e-reflections, $F_e(A)$ and $F_e(B)$, are significantly different; in fact, it appears that $F_e(B) = 0$. It follows that the 3-nm superlattice associated with the e- and f-reflections in this specimen should be observed in the A-lamellae only. This has been found by direct resolution of the superlattice fringes (i) in DF using a pair of e-reflections, Figure 8.47(b), and (ii) in BF using $0\bar{2}0$, 000, 020, and the six associated f-reflections (see McLaren 1974, fig. 10). Because no b-reflections are observed in the diffraction patterns, it must be concluded that the B-lamellae have the high-temperature, disordered structure C$\bar{1}$, characterized by a-reflections only. The high contrast achieved in DF images using a pair of e-reflections reveals much more detail of the structure of the lamellae than is observed using a-reflections. Examination of Figure 8.47(b) indicates that the B-lamellae consist of walls of irregularly shaped domains of the C$\bar{1}$ structure. In a similar image, published by Wenk and Nakajima

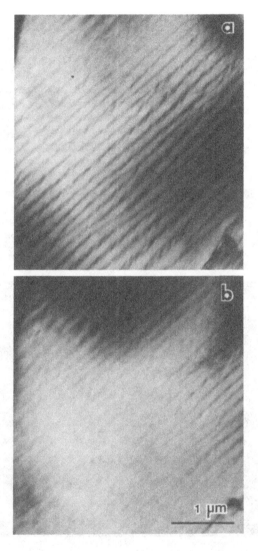

Figure 8.46. DF micrographs ($g = 20\bar{2}$) of the schiller lamellae in a labradorite (An$_{53}$). (a) $s \neq 0$ and (b) $s = 0$. Note the low contrast in (b). (From McLaren 1974.)

(1980), the B-lamellae are seen as a mixture of dark, irregularly shaped domains of C$\bar{1}$ structure and bright domains exhibiting the e-superlattice fringes. In addition, the orientation of the fringes in the A and B domains differs by up to about 15°, and the fringe spacing differs by a factor of about 1.25. These observations were interpreted as due to two distinct

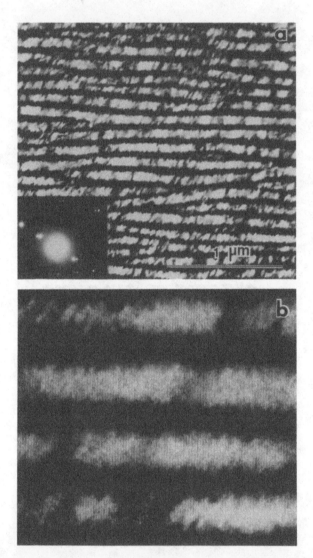

Figure 8.47. (a) The schiller lamellae in labradorite (An$_{52}$) imaged in DF with the 123 pair of *e*-reflections. The plane of the specimen is (100). The lamella-period is approximately 120 nm. (b) Higher magnification micrograph imaged under the same conditions showing the 3-nm superlattice associated with the *e*- and *f*-reflections. (From McLaren and Marshall 1974.)

e-plagioclase structures, although the presence of two structures is not apparent in the associated SAD pattern. Lattice fringes are not always easy to interpret since their position bears no simple relation to the position of the atomic planes and since changes in the visibility, orientation, and spacing of the fringes can occur due to local variations in crystal thickness and/or deviation from the exact Bragg angle; see Sections 6.2.1 and 6.5. Thus, the changes in orientation and spacing of the superlattice fringes observed on passing from A-lamellae to B-lamellae (as well as the "dislocations" observed in the image near the boundaries with the CĪ structure) might arise in a *single* *e*-plagioclase phase because of local changes in *s* (corresponding to $\Delta\theta$ of only a few minutes of arc) and/or because of local changes of thickness of the *e*-plagioclase phase due to the presence of the domains of the CĪ phase.

Nissen et al. (1973) and Cliff et al. (1976) found that the composition of the two types of lamellae in a labradorite (An_{54}) differed by at least 12 percent An.

8.10.3 Exsolution in the pyroxenes. The variety of exsolution microstructures that develop in these minerals have been studied in great detail by TEM, and excellent reviews have been given by Champness and Lorimer (1976), Champness (1977), and Buseck, Nord, and Veblen (1980). Many of the exsolution features observed in the pyroxenes, such as the tweed texture, are essentially the same as those already discussed and, therefore, need not be considered further. However, some important characteristics of pyroxene exsolution microstructures warrant special mention.

Figure 8.48 shows exsolution lamellae of pigeonite on (001) and (100) in augite. Note that a tweed structure, in which the (001) component dominates, has developed in the augite but not in those regions adjacent to the pigeonite lamellae. Champness and Copley (1976) observed (100) lamellae characterized by α-fringes (Section 5.3.1) within the (001) pigeonite lamellae and showed that the fault vector $\mathbf{R} = \frac{1}{6}[001]$. These fringes may be due to lamellae of orthopyroxene one unit cell wide. McLaren and Etheridge (1976) observed α-fringes with the same fault vector due to lamellae of clinopyroxene one unit cell wide in deformed orthopyroxene.

Figure 8.49 shows a pair of (100) augite lamellae in an orthopyroxene matrix observed by Champness and Lorimer (1973). Using HRTEM, electron diffraction, and analytical electron microscopy, they showed that the augite is primitive monoclinic with an **a** lattice parameter of 0.96 nm and that the augite is richer in calcium than the matrix. Between the augite lamellae is a fine distribution of small, coherent (100) platelets less than 0.25 μm in diameter and 1.8 nm (one matrix lattice parameter) wide. The platelet-free zone adjacent to the augite lamellae was found to have a lower calcium concentration than other areas of the matrix. This observation indicates that the growth of the augite lamellae drained calcium from the matrix and suppressed later nucleation of the Ca-rich platelets.

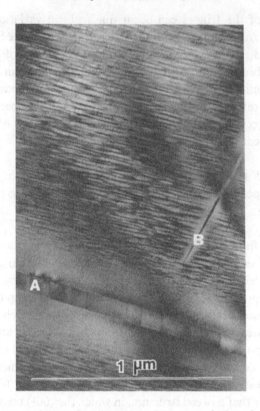

Figure 8.48. (001) and (100) lamellae (A and B, respectively) of pigeonite in an augite from the Whin Sill, northern England. Between the large (001) lamellae is a tweed structure in which the (001) component is the more prominent. Note the absence of the tweed structure adjacent to the pigeonite lamellae and the anti-phase domain boundaries (formed during the C to P transition) in A.
(From Champness 1977.)

Ross and Huebner (1979) found that dislocations were often present in the augite-orthopyroxene interface, presumably to relieve stress due to the difference in the unit cell sizes of the two phases.

A number of observations show that augite lamellae nucleate on other defects such as grain boundaries (Champness and Lorimer 1973), APBs (Carpenter 1978), and dislocations (Nord, Heuer, and Lally 1976).

8.11 Chemically distinct intergrowths observed by HRTEM

The crystal structures of many minerals can be visualized as ordered intergrowths of two or more structurally distinct planar units. Thompson

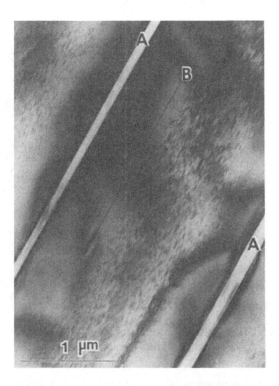

Figure 8.49. Microstructure of an orthopyroxene from the Sillwater Complex showing large (100) augite lamellae A, intermediate phase B, and a fine distribution of platelets that nucleated homogeneously. (From Champness 1977.)

(1978) called such structures *polysomatic*. Although most polysomatic structures are well-ordered, many cases have been documented of structures in which the planar units are arranged in a less-ordered (perhaps even random) manner. The basic type of planar defect involved in polysomatic structures is the Wadsley defect (see Hyde and Bursill 1970) which, unlike the simple stacking fault, is characterized by a fault vector with a component normal to the plane of the defect. Thus, the structure in the immediate vicinity of the Wadsley defect is chemically distinct from the structure of the matrix, and the presence of these defects results in nonstoichiometry in the host crystal. Wadsley defects are formed during crystal growth or as the result of a chemical reaction involving diffusion.

Over the past decade, HRTEM has made its greatest contributions to the understanding of this class of planar defect; a review of this work as applied to silicate minerals is given by Veblen (1985b). The studies of

Figure 8.50. HRTEM image showing terminating triple-chain lamellae (3) in augite. (From Veblen 1985b.)

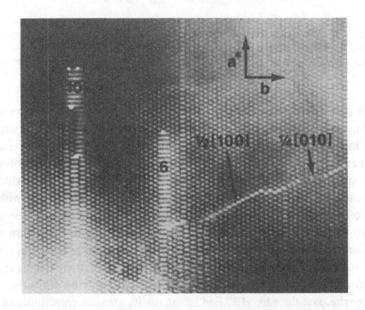

Figure 8.51. HRTEM image showing 3-, 4-, 5-, 6-, and 10-chain defects in the amphibole (2-chain) structure. (From Veblen 1985b).

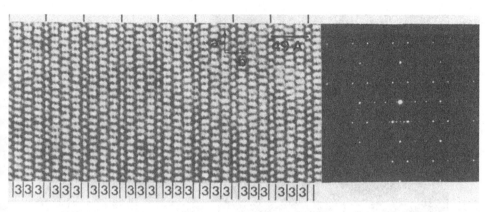

Figure 8.52. HRTEM image of an ordered chain silicate having the chain sequence (2333). The 2-chain slabs are not labeled. (From Veblen 1985b).

chain-width defects and disorder in pyriboles (pyroxenes, amphiboles, and other closely related structures) provide striking examples of the power of HRTEM in this field of mineralogy.

HRTEM has shown that the pyroxenes, which ideally contain single silicate chains parallel to [100], often contain intergrowths of amphibole-type (double) chains, jimthompsonite-type (triple) chains, or even wider chains. These chain-width errors can be visualized as Wadsley defects parallel to (010). The most common chain-width defects in pyroxenes are triple (3) and double (2) chains, and Figure 8.50 shows a number of terminating 3-chain defects (1.35 nm wide) in an augite (Px) crystal. Figure 8.51 shows an amphibole crystal that has suffered chemical alteration: 3-, 4-, 5-, 6-, and 10-chain defects have grown into the amphibole (2-chain) structure. Eggleton and Boland (1982) have used HRTEM to follow the weathering of enstatite to talc through a sequence of transitional phases.

Wadsley defects can be ordered over large distances. For example, chesterite consists of alternating double and triple chains. Figure 8.52 shows a structure in which the sequence of chain widths is (2333). Only by the use of HRTEM can structures of this type be discovered and determined, since they consist of intergrowths (on the unit-cell scale) with other minerals in crystals that are usually too small to be studied by x-ray diffraction techniques. In the examples given, the images require virtually no interpretation in terms of electron diffraction because the intergrowths of different chain widths can be identified directly. However, for more complex structures, this simple approach may not be adequate, and images

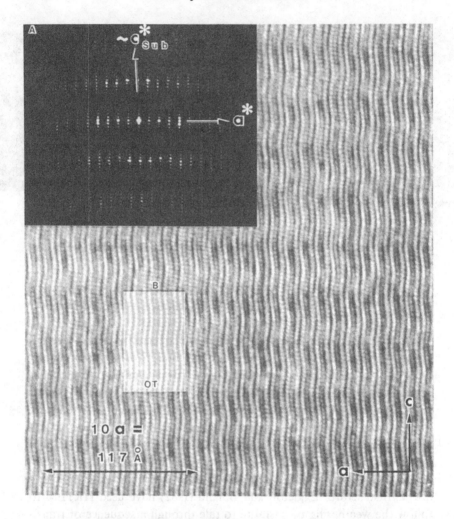

Figure 8.53. Typical [010] image of cylindrite, with corresponding diffraction pattern (A). The inset B is a simulated image (foil thickness $t = 2$ nm, defect of focus $\Delta f = 40$ nm). (From Williams and Hyde 1988.)

observed under different experimental conditions must be compared with simulated images computed for different possible structural models in order to determine which model best fits the observations (see Section 6.4). An example of observed and simulated images (with the corresponding diffraction pattern) in cylindrite is shown in Figure 8.53 (Williams and Hyde 1988).

HRTEM has been applied to a very large number of minerals; some examples recently studied include carlosturanite (Mellini, Ferraris, and Compagnoni 1985), orientite (Mellini, Merlino, and Pasero 1986), zinkenite (Smith 1986), eastonite (Livi and Veblen 1987), andalusite (Ahn, Burt, and Buseck 1988), biotites and muscovites (Konings et al. 1988), and sapphirine (Christy and Putnis 1988).

8.12 Determination of site-occupancies in silicates using ALCHEMI

Many minerals contain minor elements whose distribution among possible sites can be ordered or disordered. Olivine, for example, in addition to Mg, Si, and O usually contains Fe and smaller amounts of Mn and Ni, which substitute for Mg atoms in the two nonequivalent M1 and M2 sites. The distribution of Mg, Fe, Mn, and Ni in M1 and M2 sites in a specimen of San Carlos olivine was determined by Tafto and Spence (1982) using ALCHEMI, and more recently, McCormick, Smyth, and Lofgren (1987) have used the same technique to determine the site-occupancies of these minor elements in synthetic olivines as a function of equilibration temperature. Recall from Section 7.4 that ALCHEMI is not fully quantitative for the determination of site-occupancy in some structures because of the so-called localization effect. However, because these studies of olivine indicate that planar ALCHEMI can be used with confidence in this important mineral, it is therefore appropriate to consider this particular application of the technique as an example.

If the olivine structure (space group Pbnm) is viewed parallel to [100], it will be seen that the Si and M2 sites line up on (002) planes that alternate with planes containing M1 sites only (see Brown 1982, fig. 1). Thus, the channeling orientation corresponds to $g = 002$. X-ray spectra were recorded for two channeling orientations ($\theta < \theta_{002}$ and $\theta > \theta_{002}$) and for a third orientation in which all strong reflections were avoided. The x-ray emission from the Si atoms provides the reference signal proportional to the thickness-averaged electron intensity on the planes containing the M2 sites, and the proportions of Mg, Fe, Mn, and Ni on M1 sites were calculated from the x-ray emissions of these atoms using the equations derived in Section 7.4. The results are summarized in Figure 8.54.

The results of McCormick et al. (1987) indicate that the degrees of ordering at different equilibration temperatures differ by less than the analytical errors, although there may be significant differences in ordering between olivines that have been quenched from temperatures of 300°C (or higher)

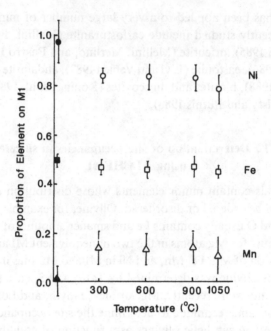

Figure 8.54. Proportions of Fe, Mn, and Ni in M1 sites of San Carlos olivine (solid symbols, Tafto and Spence 1982) and in synthetic olivine as a function of equilibration temperature (open symbols, McCormick et al. 1987). Error bars are ±1 e.s.d.

and those that have cooled in a normal lava flow. The ALCHEMI results on the San Carlos olivine are consistent with those obtained from x-ray structure refinement, suggesting that the electron channeling technique may be capable of identifying olivines that have been quenched very rapidly.

ALCHEMI has also been successfully used for determining Al/Si ordering in tetrahedral sites in K-feldspars by Tafto and Buseck (1983) and McLaren and Fitz Gerald (1987).

8.13 Radiation-induced defects

Many minerals are unstable when exposed to the beam in the electron microscope and transform from a crystalline to an amorphous (or glassy) state at a rate that varies widely from one mineral to another and as a function of a number of variables such as beam intensity, accelerating voltage, temperature, and even impurity content. A similar transformation can be induced by neutron irradiation, and TEM has been used to follow the process in quartz, in particular. Furthermore, some minerals, notably zircon, are often found in an amorphous (or *metamict*) state, which is generally considered to be the result of radiation damage from α-particles emitted by U and/or Th impurities over geological periods of time. In this section, we consider some examples of the application of TEM to studies of electron and neutron radiation damage and to meta-mictization.

8.13.1 Electron radiation damage.   Minerals that are known to become damaged in the beam of the electron microscope include topaz (Hampar 1971), albite and nepheline (McConnell 1969b), and quartz, which has been studied extensively since the first observations of McLaren and Phakey (1965a). They found that within a 1-minute exposure to a focused 100-kV electron beam, crushed fragments of amethyst quartz developed small damage centers, which were most easily imaged by structure factor contrast (Section 5.7.1), as shown in Figure 8.55(a). The observed contrast-reversal of the damage centers across the thickness contours is the same as for voids (see Figure 5.20) and indicates that the structure factor F_{gi} of the damage centers is less than that of the matrix F_g (i.e., $t_{gi} > t_g$). As the irradiation continued, the damage centers developed into strain loops, Figure 8.55(b), which grew and eventually coalesced, resulting in the complex contrast shown in Figure 8.55(c). After about 10 minutes of irradiation, the contrast began to disappear, as shown in Figure 8.55(d). Eventually, all contrast disappeared, and the SAD pattern confirmed that the specimen was amorphous. Specimens of synthetic quartz of unknown purity and of naturally occurring clear, smoky, and rose quartz all behaved in a similar way. However, the density of damage centers was significantly lower in these crystals than in amethyst quartz, and so the strain loops were able to grow much larger before coalescing. These strain loops were typical of the contrast expected from a spherical inclusion, as discussed in Section 5.7.2. In particular, they showed a *line of no contrast,* which was always normal to **g**. These observations are consistent with the damage centers being essentially small amorphous spheres in a crystalline matrix.

Das and Mitchell (1974) studied the radiation damage as a function of voltage. Black spot damage centers – essentially the same as the contrast shown in Figure 8.55(a) – were observed after a few minutes at 50–100 kV and after a few seconds at 350–650 kV. The size and density of the damage centers increased with increasing irradiation time at all voltages in the early stages. In the range 50–100 kV, the damage centers grew in size and appeared to be dislocation loops with an average diameter of about 70 μm.

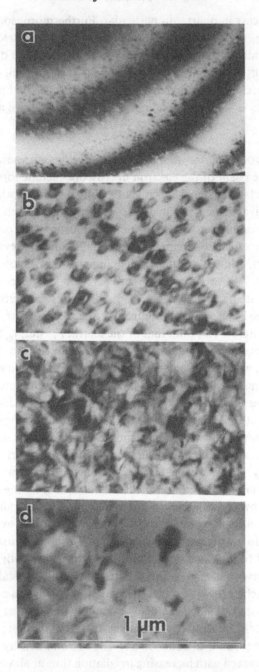

Figure 8.55. See caption on facing page.

However, at higher voltages (350–650 kV), the rate of nucleation of damage centers was higher, but there seemed to be little growth once they had formed. Additional irradiation at all voltages gradually induced the crystalline-to-amorphous transformation, as already described, and the electron dose required to produce the amorphous transformation increased approximately linearly with increasing voltage. On the basis of the different behavior with increasing voltage, Das and Mitchell (1974) suggested that two types of damage processes occur: (1) displacement damage, which produces the black spots and dislocation loops, and (2) ionization damage, which produces the amorphous transformation.

More recently, Pascucci, Hutchison, and Hobbs (1983) studied the electron radiation damage of quartz by HRTEM and its temperature dependence from 20 to 500 K. The HRTEM images indicate that the heterogeneously nucleated damage centers are disordered (amorphous) inclusions, rather than dislocation loops. At temperatures of about 450 K, the inclusions still form, but at a reduced density, and they grow to a larger size. Above about 500 K, no inclusions are observed, although an apparently homogeneous transformation to an amorphous state still takes place at about the same rate as at room temperature. Van Tendeloo et al. (1976) observed no damage near the α to β transformation at 846 K but did not mention the homogeneous amorphous transformation. Both the rate of growth of inclusions and the overall rate of amorphous transformation are retarded below room temperature. From these observations, Pascucci et al. (1983) suggest that the damage process is radiolytic (i.e., involves electron–electron interactions) and proceeds in two stages. The first stage is characterized by the heterogeneous nucleation of discrete strained inclusions (the black spots and strain loops), which are substantially disordered. The second stage is an independent, slower, homogeneous loss of long-range crystalline order in the surrounding matrix. There is some evidence to suggest that the damage centers may be nucleated at sites associated with hydrogen-related defects, although the evidence is far from conclusive. For example, Pascucci et al. (1983) found that the density of

---

Figure 8.55. *Facing page.* Electron radiation damage in quartz. (a) Damage centers produced within a 1-minute exposure to 100-kV electrons, imaged in BF by structure factor contrast. (b) As the damage centers grow during irradiation, they become visible by the strain they produce in the surrounding matrix. (c) On continued irradiation, the strain fields overlap. (d) Eventually, the crystal becomes amorphous, and the strain contrast disappears. (From McLaren and Phakey 1965a.)

damage centers in lunar quartz (which is, presumably, extremely "dry") was considerably lower than in synthetic quartz, which may contain as much as 3000 $H/10^6Si$ (Godbeer and Wilkins 1977; Paterson 1982). Recent studies of hydrogen-related defects in quartz (see Paterson 1989 for a review) provide strong evidence that the broad 3-μm infrared absorption band characteristic of "wet" synthetic quartz is due to aggregated water (Aines and Rossman 1984). TEM observations by McLaren et al. (1983) and Gerretsen, Paterson, and McLaren (1989) indicate that the water is present in high-pressure clusters whose sizes range from about 60 nm to about 10 nm, the limit of detection (see Section 5.7.2). Smaller clusters could, therefore, be present. Annealing at temperatures above about 550°C causes the clusters to evolve into strain-free bubbles, and McLaren and Phakey (1965a) found that crystals annealed at 700°C developed fewer damage centers than unannealed crystals. These observations suggest that the high-pressure clusters can act as nuclei for the radiation-damage centers, but a one-to-one correlation has not been demonstrated.

Optically clear crystals of amethyst quartz often exhibit the same 3-μm infrared absorption band as "wet" synthetic quartz and have comparable values of $H/10^6Si$ (Kekulawala, Paterson, and Boland 1981; Paterson and Kekulawala 1979). TEM observations (McLaren et al. 1983) indicate that the hydrogen is also present in high-pressure water clusters, which lie predominantly in the boundaries of the Brazil twins that are characteristic of this variety of quartz (see Section 8.2). McLaren and Phakey (1965a) found that the electron radiation damage occurs preferentially at the twin boundaries and that the transformation to the amorphous state starts at the twin boundaries and spreads out into the crystalline matrix, forming an amorphous slab between the right-handed and left-handed domains, as shown in Figure 8.56. The preferential damage at the twin boundaries may be due to the segregation of water. However, Comer (1972) found that Dauphiné twin boundaries (produced by cooling through the β to α transformation at 573°C) also damaged preferentially during irradiation, although there was no evidence that water-related defects (or any other impurities) had segregated at these boundaries.

Figure 8.56. *Facing page.* Preferential electron radiation damage at Brazil twin boundaries in amethyst quartz. (a) DF image showing the impurity segregation (probably high-pressure clusters of water) at the twin boundaries, viewed edge-on. (b) Damage begins at the boundaries and (c) spreads out into the matrix leaving lamellae of amorphous material. (d) Electron diffraction pattern of (c) showing the diffuse scattering due to the amorphous regions.

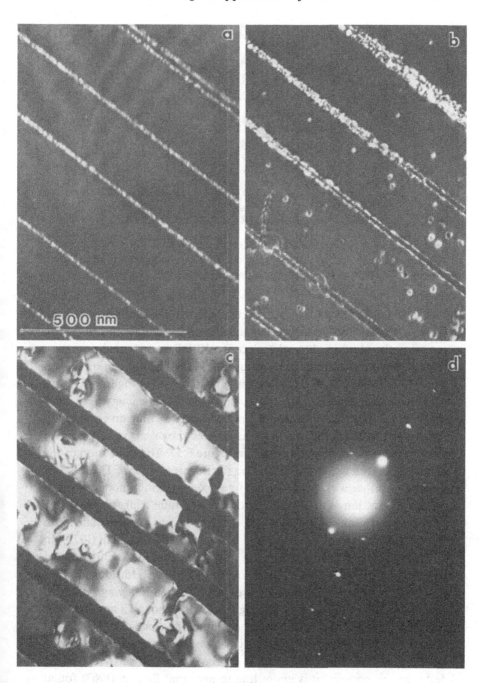

Figure 8.56. See caption on facing page.

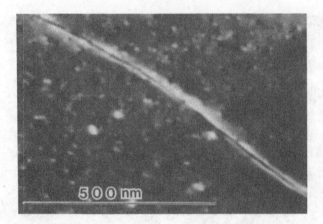

Figure 8.57. WBDF image of a dislocation in electron radiation damaged quartz. The double image of the dislocation is due to the development of amorphous material along the core.

Electron radiation damage also occurs preferentially along dislocations. Cherns, Jenkins, and White (1980) and Carter and Kohlstedt (1981) have recorded the changes that take place in WBDF images of dislocations (with $g \cdot b = n = 1$) as a function of time. Initially, the dislocation images were characterized by a single intensity maximum (as in Figure 5.17), but with increasing exposure the images split into two maxima whose separation continued to increase with time (without reaching an equilibrium separation) at about the same rate as isolated damage centers grow. An example of such an image is shown in Figure 8.57. The separation of the maxima does not appear to depend on the Burgers vector b. These observations, together with the absence of any constriction at nodes of intersecting dislocations and the absence of stacking fault fringes between the image maxima, make it unlikely that the dislocations are dissociated into two partial dislocations as suggested by Trepied and Doukhan (1978); in fact, these observations support the suggestion of Ardell, Christie, and McCormick (1974) that the material in the vicinity of the dislocation core has been transformed to glass by the electron beam in a way similar to that observed at twin boundaries (Figure 8.56). Although it appears to be unnecessary to invoke the idea that the preferential damage at the line and plane defects is dependent on impurity segregation, the isolated damage centers may have nucleated at local concentrations of water-related defects. In this context, it is interesting to note that Phakey (1967) found that the hydrous minerals topaz, brazilianite, and hydroxyapatite were readily damaged in the electron beam but that the anhydrous minerals fluoroapatite, diopside, and garnet were not. However, the role of water-related defects in the mechanism of electron radiation damage of quartz and other minerals is still unclear.

8.13.2 Neutron radiation damage. Quartz appears to be the only important rock-forming mineral in which neutron radiation damage has been studied by TEM. The first detailed study was made by Weissmann and Nakajima (1963) who irradiated (0001) and ($11\bar{2}0$) slices of both natural and synthetic crystals of optical quality with fast neutrons, with doses from 1×10^{19} to 1.5×10^{20} nvt.* Specimens thin enough for TEM were prepared by etching in a 1:1 mixture of HF (48%) and concentrated HNO_3. The irradiated specimens exhibited dark spot images characteristic of defect clusters whose diameter ranged from 6 to 20 nm. Both the size and the density of clusters increased with increasing dose. The SAD patterns were remarkably sharp and relatively free from the influence of lattice distortion even for doses of 10^{20} nvt. This was in marked contrast to the corresponding x-ray diffraction patterns, which for this dose exhibited only a single diffuse halo with a completely ill-defined line profile. For a comparable radiation dose, the values of size and volume fraction of the clusters were both larger for Z-cut crystals than for X-cut crystals. This dependence appeared to be related to the ease of diffusion of displaced silicon atoms in directions normal to the *c*-axis channel.

Later, Phakey (1967) studied the radiation damage in several varieties of quartz after exposure to fast neutrons of doses ranging from 10^{18} to 2.1×10^{20} nvt. Examples of the black spot damage that was observed are shown in Figure 8.58. The contrast was found to be tilt-dependent, which indicates that it is due to strain around defect clusters. Thus, the observed size of the black spots does not necessarily correspond to the size of the clusters themselves (see Section 5.7.2). Although the nature of the observed damage was the same for all varieties of quartz, the radiation dose D_0 for which the damage was *first observed* was different for each. This behavior is shown in Table 8.6; in general, D_0 is smaller for the relatively purer varieties, such as optical-quality natural and synthetic, than for the varieties, such as amethyst, that contain large concentrations of impurities.

When first observed, the damage centers were of the order of 5 nm in diameter. As the neutron dose was increased from D_0, the damage centers

* nvt = neutron number density (N/cm^3) × neutron velocity (cm/sec) × time (sec) = number of neutrons per unit area (N/cm^{-2}).

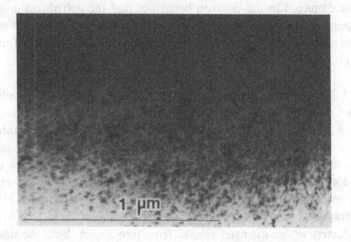

Figure 8.58. BF micrograph showing neutron radiation damage in synthetic
quartz (dose 3×10^{19} nvt, followed by annealing at 900°C for 16 hours).
(From Phakey 1967.)

showed a progressive increase in size to a maximum diameter of about
15 nm. At higher doses, some varieties became amorphous, judged from
both the images and the associated SAD patterns. The dose D_m at which
this occurred is also shown in Table 8.6.

In addition to the black spot contrast, most crystals of neutron-irradiated amethyst quartz showed black lines, which were subsequently identified as the tracks of particles produced by the fission of uranium impurity
atoms (see Section 8.13.3).

Phakey (1967) also studied the effects of annealing neutron-irradiated
quartz. Crystals that had been damaged with a dose of D_0 showed no
black spot damage after annealing at temperatures and times that varied
from one variety of quartz to another. These annealing conditions are
also given in Table 8.6. Specimens of optical-quality natural and synthetic
quartz that had been damaged with doses in excess of D_0 could not be
completely recovered even by annealing at 1,250°C for 16 hours.

The rate at which electron radiation damage occurred in neutron-irradiated specimens also varied from one variety of quartz to another.
Whereas synthetic quartz (dose D_0) became amorphous after a few seconds of exposure to a focused electron beam, amethyst quartz (dose <
D_0) was more stable in the electron beam than the sample before neutron
irradiation.

Table 8.6. *Characteristics of neutron-irradiation damage in quartz*

Quartz variety	Major impurity	Dose D_0 for which damage is first observed (nvt)	D_m required to produce the amorphous state (nvt)	Annealing conditions necessary to recover specimens damaged by a dose D_0
Synthetic	?	1×10^{19}	7×10^{19}	900°C for 16 hours
Natural colorless	Al, fairly low	3×10^{19}	–	1100°C for 15 hours
Smoky	Al, high	7×10^{19}	1×10^{20}	1100°C for 15 hours
Citrine				
(a) light color	Fe, OH	3×10^{19}	–	700°C for 3 hours
(b) deep color		1×10^{20}	–	700°C for 15 hours
Amethyst				
(a) light color	Fe, OH	1×10^{20}	–	700°C for 15 hours
(b) deep color		$>2.1 \times 10^{20}$	–	–

Source: From Phakey 1967.

The predominant mechanism of creation of defects by neutron irradiation is direct atomic displacement by collision. In an essentially covalent crystal such as quartz, fast neutrons are expected to produce broken Si—O bonds, as well as interstitial atoms and vacancies. The broken bonds give rise to disordered regions in the crystal. When the local concentration of these defects is sufficiently high, they constitute complex damage centers that strain the surrounding crystal elastically, producing the observed black spot contrast in TEM images. The observations summarized in Table 8.6 indicate that the neutron dose D_0 required to form clusters that are large enough to be observed by their strain contrast is larger in the more impure crystals. Phakey (1967) suggested that this indicates that the impurities do not act as preferential damage sites and proposed that the displaced atoms diffuse through the crystal and are trapped at impurity sites. On the basis of this proposal, it follows that if the number of trapping sites is comparable to the number of diffusing atoms, large clusters will not form. Thus, the critical dose D_0 required to produce damage centers large enough to be observed increases with increasing number of trapping sites.

It will also be noted in Table 8.6 that for a dose D_0, crystals of amethyst and citrine can be recovered by annealing at 700°C, whereas the purer crystals (e.g., optical-quality natural and synthetic quartz) require a temperature of 900° to 1,110°C. The reasons for this behavior, and for the influence of neutron irradiation on subsequent electron radiation damage, are not clear.

8.13.3 Fission fragment damage. As mentioned in the previous section, specimens of amethyst quartz that have been irradiated with fast neutrons usually exhibit black lines. Similar lines are also observed after irradiation with thermal neutrons, which suggests that the lines are the tracks of particles resulting from the fission of uranium atoms. Using the procedure of Price and Walker (1963), Phakey (1967) determined the concentration of uranium in amethyst crystals from measurements of the number of tracks intersecting a unit area of crystal, together with their average range. He found a concentration of the order of 5×10^{14} cm^{-3}. A typical track about 6.6 μm long is shown in Figure 8.59. The contrast of fission-fragment tracks has been considered in detail by Chadderton (1964). He assumed that the damage produced by a fission fragment in an initially perfect crystal consists of a central cylinder of disturbed material surrounded by a radial strain field. This is mathematically very similar to the model invoked by Ashby and Brown (1963a) to determine the contrast due to spherically symmetrical coherency strain from precipitates (see

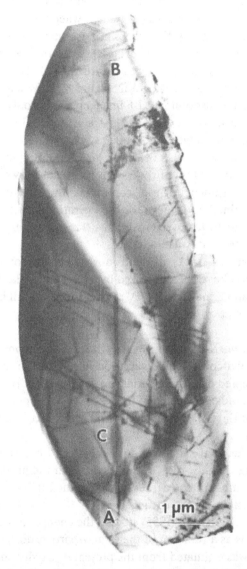

Figure 8.59. Fission-fragment tracks in neutron-irradiated amethyst quartz. Note the double image of the track AB and the long-range strain effects in the matrix at C. (From Phakey 1967.)

Section 5.7.2). Chadderton (1964) used this model in conjunction with the two-beam dynamical theory of electron diffraction, Eqs. (5.5a, b), to calculate contrast profiles for tracks at different positions in the crystal and for different values of the deviation s from the exact Bragg angle and the absorption length t_g''. Since the displacements **R** around the track are radial, the tracks are invisible, or very weak, when they are aligned parallel to the diffraction vector **g**. For tracks in the center of the crystal foil (and more or less parallel to the foil surfaces), a central line of no contrast is observed when $s \approx 0$; compare with Figure 5.22(b). Thus, the track gives rise to a double image, as shown in Figure 8.59. When $s \neq 0$, one dark line appears to be the dominant BF image, provided only one reflection is operating. Intensity oscillations of the track image with depth for tracks inclined to the foil surface are frequently observed, as with dislocations (see Figure 5.16). Sometimes the tracks are discontinuous because of the mechanism of energy loss of the fission fragment. However, if the separation between segments of a discontinuous track is about the same as the effective extinction distance for the operating reflection, then proper precautions must be taken to ensure that the discontinuous nature of the track image is not a diffraction effect.

Fission-fragment tracks in neutron-irradiated zircon and in natural metamict zircon have been observed by Vance and Boland (1975) and by Yada, Tanji, and Sunagawa (1981), respectively, and will be considered in the following section.

8.13.4 Metamictization of zircon $(ZrSiO_4)$.

Zircons from different localities show a considerable variation in physical properties, and it is generally accepted that these variations are the result of structural damage due to the radiations emitted by U and/or Th impurities over geological periods of time. The extensive literature on zircon has been reviewed by Speer (1982).

Although the total radiation dose consists of contributions from α-particles, β-particles, γ-rays, recoil nuclei, and the products of spontaneous fission, most of the earlier work assumed that the α-particles are chiefly responsible for the damage.

Holland and Gottfried (1955) studied the changes in density and unit-cell dimensions as a function of the total α-particle dose D per milligram of zircon. D was estimated from the present α-activity and the Th/U ratio, together with the age T of the sample, on the assumption that equilibrium in the uranium and thorium decay series had been established in a time that was short compared with T. T was determined from the present α-activity and lead content (see Faure 1977). It was found that for $D < 10^{15}$, the density was approximately 4.7 g/cm^3. As the dose was increased, the density decreased slowly at first and then more rapidly, and finally approached asymptotically a value of approximately 3.95 g/cm^3, for $D > 1.2 \times 10^{16}$. Zircons at the extremes of this density range are commonly referred to as *high* and *low* (or metamict), respectively. The variation in unit-cell dimensions with α-dose was determined from x-ray powder diffraction patterns using the 101, 200, and 112 reflections. The observations indicated that the transformation from high to metamict zircon could be considered as a three-stage process:

Stage I: $10^{14} < D < 2 \times 10^{15}$ α/mg. The 112 reflection shifts from $2\theta = 35.63°$ to $35.4°$ and decreases in intensity. The density decreases to 4.6 g/cm^3.

Stage II: $2 \times 10^{15} < D < 4.6 \times 10^{15}$ α/mg. The 112 reflection shifts further towards $2\theta = 35°$ and disappears, accompanied by its reappearance at $2\theta = 35.6°$. The density drops to 4.35 g/cm^3.

Stage III: $4.6 \times 10^{15} < D < 1.26 \times 10^{16}$ α/mg. The 112 reflection at $2\theta = 35.6°$ decreases in intensity and disappears. Zircons with this type of x-ray diffraction pattern are usually referred to as *x-ray amorphous*. The density falls to 3.95 g/cm^3.

Specimens with x-ray diffraction patterns characteristic of these three stages were examined by electron microscopy and diffraction by Bursill and McLaren (1966). All specimens of a sample characteristic of Stage I exhibited single-crystal SAD patterns with sharp reflections corresponding to high zircon, and there was no evidence of radiation damage (or fission-fragment tracks) in BF images. However, black spot contrast, typical of radiation damage, was observed in images of all specimens prepared from a sample characteristic of Stage II. Thickness contours were also observed, indicating that the Stage II specimens were still essentially single crystals; for most specimens, the single-crystal SAD patterns exhibited sharp reflections corresponding to high zircon. The density of black spots in the TEM images of Stage II zircons varied somewhat from specimen to specimen; and in specimens with a relatively high density, each reflection in the associated SAD pattern was accompanied by a weak diffuse reflection displaced slightly toward the origin of reciprocal space, indicating an expanded lattice. The specimens from samples characteristic of Stage III also exhibited basically single-crystal SAD patterns. However, the reflections were quite diffuse compared with those observed for specimens of Stages I and II. In addition, the reflections were slightly streaked along circular arcs about the origin. This observation, together with the fact that no thickness extinction contours were observed, suggests that these metamict zircons are composed of many small crystallites that are very slightly misoriented with respect to each other. The black spot image contrast suggests that the crystallites are of the order of 10 nm in size. Recent HRTEM observations (see, e.g., Chakoumakos et al. 1987) are in general agreement with this interpretation. Such crystallites are expected to produce particle-size broadening of the x-ray reflections; thus, a microstructure of this kind is not inconsistent with the x-ray amorphous character of the metamict zircons. The SAD patterns of the metamict zircons also exhibited a diffuse ring and some extra reflections which indicate the presence of ZrO_2. One metamict zircon also contained large numbers of irregularly shaped bubbles or voids 50–100 nm in diameter. Bubbles of helium have been frequently observed in α-irradiated materials after annealing (see, e.g., Nelson, Mazey, and Barnes 1965), but it seems unlikely that the bubbles observed in this zircon were produced in this way because no bubbles were observed in another x-ray amorphous zircon. It is more likely that the bubbles contain water, which is not an uncommon impurity in zircon (Aines and Rossman 1984). This suggestion is supported by the observation that the metamict zircon with bubbles almost completely recrystallized on annealing at 1,250°C for 16 hours, whereas annealing at 1,450°C for a similar time was required to recrystallize a metamict zircon without bubbles. Frondel and Collette (1957) found previously that, when annealed in water vapor at 100 MPa pressure, metamict zircons recrystallize at a temperature below that needed for recrystallization in dry air. Bursill and McLaren (1966) found that the recrystallization of a metamict zircon without bubbles was accompanied by the nucleation and growth of prismatic dislocation loops (up to 100 nm in diameter) on the close-packed {010} planes. This observation suggests that recrystallization involves the condensation of point defects produced by the irradiation.

Bursill and McLaren (1966) observed no fission-fragment tracks in any of the specimens examined. However, in a more recent HRTEM study of high, intermediate, and low zircons, Yada et al. (1981) concluded that metamictization proceeds principally by the formation of fission tracks. In high zircon, fission-fragment tracks in an otherwise perfect structure were observed. The tracks were 2–3 nm wide and about 100 nm long and terminated in elliptical areas with nearly amorphous structure at both ends. Within the tracks, the lattice was almost completely transformed to an amorphous state. Point defects, assumed to correspond to atomic displacements by recoil nuclei, were also seen adjacent to the tracks. Intermediate zircon consisted of amorphous regions (low zircon) and regions consisting of slightly misoriented domains of near-perfect structure about 10 nm in diameter separated by amorphous boundaries. The amorphous regions showed a high density of white lines, which were also interpreted as fission tracks. On the basis of these observations, Yada et al. (1981) proposed that the metamictization of zircon is due primarily to the destruction of the structure along the path of nuclear particles that form the fission tracks. They considered the breakdown of the structure by α-irradiation and recoil nuclei to be of secondary importance, contrary to the view of most other researchers in this field (see Chakoumakos et al. 1987, for example).

Optical microscope observations have shown that zircons often exhibit extremely complex microstructures (on the scale of 1–100 μm), in which changes of birefringence correlate with the distribution of U and Th (Chakoumakos et al. 1987). However, no attempts appear to have been made to relate these microstructures to other impurities (such as water-related species) and crystal defects, both of which may significantly influence the processes of metamictization and recrystallization. Such a study involving TEM might also provide important information about the diffusion and leaching of radioactive impurities (and the products of their decay), processes that have important implications for ceramic nuclear-waste disposal and for techniques of age determination based on measurements of Pb/U isotopic ratios.

9

Mineralogical applications of TEM
II. Dislocations and microstructures associated with deformation

Although the role of dislocations in the plastic deformation of crystalline materials was appreciated by physical metallurgists by 1950, structural geologists concerned with the deformation of rocks were relatively slow to respond to the new concepts. For example, dislocations are mentioned only briefly by Turner and Weiss (1963) in their classic book *Structural Analysis of Metamorphic Tectonites*. John Christie and David Griggs of the University of California at Los Angeles were probably the first geologists to use dislocation concepts to interpret the deformation characteristics of an important rock-forming mineral (Wenk 1979). In 1964, Christie, Griggs, and Carter (1964) succeeded, after many earlier failures, in plastically deforming high-quality single crystals of natural quartz and attempted to explain the deformation lamellae observed in the optical microscope in terms of *en echelon* arrays of edge dislocations lying in the slip planes. TEM observations of replicas of etched surfaces of the deformed crystals provided some supporting evidence for the proposed model. However, the first direct observations of the dislocations and other defects in thin foils of these deformed specimens were made by McLaren et al. (1967), following an earlier TEM investigation of dislocations in thin foils of milky vein quartz by McLaren and Phakey (1965b).

Since then, TEM has been used to study dislocation microstructures in a wide range of naturally and experimentally deformed minerals and rocks. In general, the aim of the experimental studies is to determine the deformation mechanisms by relating the evolution of the observed microstructures to the macroscopic deformational behavior observed under varying conditions of temperature, confining pressure, chemical environment, strain-rate, stress, and total strain, and then to use this knowledge to interpret the microstructures observed in naturally deformed specimens and hence to determine their deformational history.

The aim of this chapter is to illustrate and interpret the TEM images of the various types of dislocations and dislocation microstructures that are observed in deformed minerals and then to relate these observations to the macroscopic deformational behavior. It is not the intention to review the now very large literature devoted to the deformation and associated microstructures of a wide range of minerals and rocks. In fact, to illustrate the TEM images of dislocations and the evolution of dislocation microstructures in deformed minerals as directly and simply as possible, the examples have been chosen largely from experimentally deformed single crystals of a relatively small number of important rock-forming minerals and their synthetic equivalents. In general, the microstructures observed within the grains of experimentally deformed polycrystalline specimens and in naturally deformed crystals tend to be more complex than those observed in experimentally deformed single crystals, but they are composed of the same basic microstructural elements. Collectively, the examples chosen illustrate most of the basic characteristics of the microstructures that have so far been observed in both naturally and experimentally deformed minerals. However, because most experimental deformation studies of minerals aim to understand the conditions and processes of deformation of rocks within the Earth, some discussion of the microstructures observed in naturally deformed minerals, and their tectonic significance, is given in Section 9.11.

Before discussing these microstructures and their evolution during deformation, it will be helpful to consider briefly some fundamental aspects of the deformation processes in crystalline solids (in particular, the role of dislocations) and the basic types of deformation experiments.*

9.1 Slip and the need for dislocations

If a solid is stressed beyond its elastic limit, it will acquire a permanent deformation. The deformation can be either brittle or ductile depending on (i) the material, (ii) the hydrostatic pressure, (iii) the temperature, and (iv) the strain rate. In general, a solid is more likely to deform in a brittle manner at low hydrostatic pressures, low temperatures, and at high strain-rates. Convesely, high hydrostatic pressures and temperatures and low strain-rates favor ductile deformation.

Most crystalline materials which can undergo a large permanent strain without fracture deform in a complex manner that is neither viscous nor perfectly plastic. At low temperatures, such materials deform by a process

---

* Readers familiar with rheological principles and elementary dislocation theory may wish to omit Sections 9.1, 9.2, and 9.3.

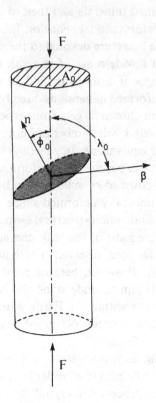

Figure 9.1. Diagram illustrating the concept of the resolved shear stress τ on a slip plane T. It is easily shown that

$$\tau = \frac{F\cos\lambda_0}{A_0/\cos\varphi_0} = \sigma\cos\varphi_0\cos\lambda_0$$

of *slip* in which one part of a crystal is translated with respect to an adjacent part of the crystal without any change of volume. Slip usually takes place on a specific crystallographic plane (the *slip plane T*) in a specific crystallographic direction (the *slip direction t*). The *slip system* is defined by T and t. In many materials (particularly metals), the slip planes are parallel to the closest packed planes, and the slip directions are parallel to the shortest translation vectors of the Bravais lattice. The ease with which a given slip system operates may be a function of temperature and strain-rate. Clearly, a slip system can operate only if there is a stress in T along t. This leads to the concept of a *resolved shear stress,* which is explained in Figure 9.1. It is usually found that when a crystal is subjected to an increasing load, slip occurs first on the slip system for which the resolved shear stress is greatest.

For crystals of reasonably pure, well-annealed metals at a given temperature, slip begins when the resolved shear stress reaches a certain critical value, which is characteristic of each metal. In the case of aluminum, for example, the observed critical shear stress σ_{c0} is usually about 4×10^5 N/m^2 (≈ 4 bars $= 0.4$ MPa). Theoretically, for a perfect crystal, the resolved shear stress is expected to vary periodically as the lattice planes slide over each other and to have a maximum value that is simply related to the elastic shear modulus μ.* This was first pointed out in 1926 by Frenkel who, on the basis of a simple model, estimated that the critical resolved shear stress was approximately equal to $\mu/2\pi$ (see Kittel 1968).[†] In the case of aluminum (which is approximately elastically isotropic), $\mu = c_{44} = 2.7 \times 10^{10}$ N/m^2, so the theoretical critical resolved shear stress is about $10^4 \sigma_{c0}$ for the slip system $\langle 100 \rangle \{100\}$.

An explanation of the tendency for crystalline solids to deform plastically at stresses that are so much smaller than the calculated critical resolved shear stress was first given in 1934 independently by Taylor, Orowan, and Polanyi. They introduced the concept of the dislocation into physics and showed that the motion of dislocations is responsible for the deformation of metals and other crystalline solids. At low temperatures, where atomic diffusion is low, dislocations move almost exclusively by slip.

Dislocation movement requires only a small stress compared with that required for the simultaneous movement of one atomic plane over another because only a few atoms are directly involved in the slip process at any instant (see Figure 9.2). However, at higher temperatures, edge dislocations can move out of their slip planes by a process called *climb*, in which atoms (or vacancies) diffuse to, or away from, the dislocation core (Figure 9.3). The climb of dislocations is, therefore, an important process in high-temperature deformation. In some materials, deformation twinning may be important, especially at low temperatures.

The theory of dislocations – what they are, what they do, and how they produce macroscopic effects that can be observed and measured – was highly developed (Read 1953) before individual dislocations were routinely observed directly by transmission electron microscopy. Dislocations and

* For slip in a $\langle 100 \rangle$ direction on a $\{100\}$ plane in an elastically isotropic cubic crystal, μ is the elastic stiffness component c_{44}.

† Later calculations (MacKenzie 1949) show that the theoretical shear strength is probably about $\mu/15$.

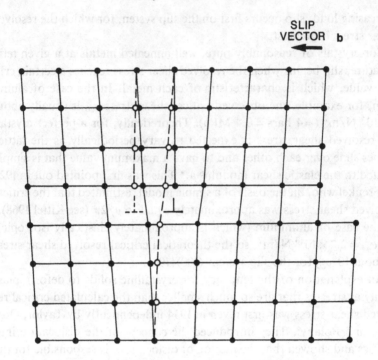

Figure 9.2. Diagram illustrating the movement of an edge dislocation by one lattice spacing $|\mathbf{b}|$ to the left. The solid and open circles represent the atom positions before and after the movement, respectively.

other extended defects were first studied in detail by this technique in the late 1950s (see Hirsch et al. 1960). These and subsequent observations have contributed greatly to dislocation theory and to our understanding of the role of dislocations in deformation and in many other solid state processes. Many excellent books on dislocations have appeared since Read's classic work, and the reader is referred to that of Hull and Bacon (1984) for a modern and concise introduction to the topic. An account of the deformation processes in metals, ceramics, and minerals has been given by Poirier (1985), and more geologically oriented accounts can be found in Nicolas and Poirier (1976), Hobbs, Means, and Williams (1976), and Wenk (1985).

9.2 Experimental deformation techniques

In most deformation experiments, the specimens have been deformed in compression at a *constant strain-rate* (10^{-3}–$10^{-6}\,\mathrm{s}^{-1}$) under conditions of

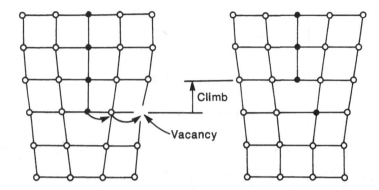

Figure 9.3. Diagrams illustrating the climb of an edge dislocation due to the diffusion of atoms away from the dislocation core or by the movement of vacancies in the opposite direction.

high temperature ($\approx 500°$–$1,400°$C) and confining pressure (0.3–1.5 GPa). The constant strain-rate is achieved by loading the specimen with a piston that advances at a constant speed. The specimen first deforms *elastically* before reaching the *initial yield stress,* after which the stress is no longer proportional to strain. The stress may actually decrease for a short time after the yield point (*yield drop*), before becoming constant or beginning to rise again with increasing stress. Stress-strain (σ-ϵ) curves of this type are shown in Figure 9.4. This rise in stress with increasing strain is called *strain* (or *work*) *hardening.* The rate of work hardening (or the *work coefficient*) at any point (σ, ϵ) on the stress-strain curve is specified by the slope ($d\sigma/d\epsilon$) at that point. Strain softening can also occur. The *initial yield stress* and the *flow stress at a given strain* are both used to define strength; but, as will be seen, these stresses are associated with different dislocation processes and should not be confused. A variant of the constant strain-rate experiment is the *stress-relaxation experiment.* Here the specimen is first deformed at a constant strain-rate beyond the elastic limit into the plastic regime until the flow stress reaches some particular value. The forward motion of the piston (which maintained the constant strain-rate) is then stopped, and the decrease in stress is measured as a function of time as the specimen relaxes, perhaps at different temperatures.

Much important information about deformation mechanisms has come from *creep experiments,* in which the stress is kept constant and the strain is measured as a function of time, producing a *creep curve* (Figure 9.5). In general, the creep curve exhibits three stages of deformation: (i) *transient creep,* in which the strain-rate changes with time (this stage may be

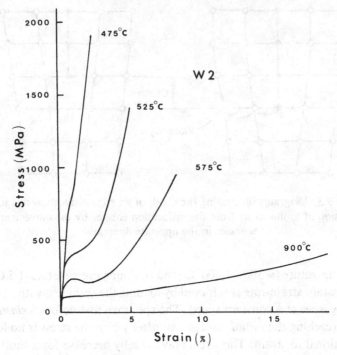

Figure 9.4. Stress-strain curves observed by Morrison-Smith et al. (1976) at the temperatures shown with specimens cut from the z-growth region of a single crystal of synthetic quartz (W2 with a water content of 1,600–200H/10^6Si). Stress normal to ($10\bar{1}1$); confining pressure 300 MPa; strain-rate 10^{-5} s^{-1}.

preceded by an *incubation period,* or *delay time,* during which no strain is detected); (ii) *steady-state-creep,* in which strain-rate is constant; and (iii) *accelerating creep,* which is often associated with fracture. Creep experiments, particularly on synthetic quartz and olivine single crystals, have been carried out at atmospheric pressure with relatively low compressive loads of only a few hundred megapascal at temperatures up to about 1,000°C. These experiments are particularly useful for measurements of strain-rate as a function of temperature and stress in the low-stress range.

The effects of very high stresses and strain-rates have been investigated in *microhardness experiments.* In these experiments, loads of 50–500 g (corresponding to stresses as high as 2 GPa) are exerted by a diamond or sapphire Vickers indenter for about 20 seconds at temperatures up to 1,000°C. Clearly, steady-state flow is never achieved; but such experiments have provided important information about the dislocations involved in the deformation of olivine, for example.

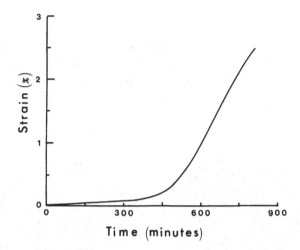

Figure 9.5. A typical creep curve for a single crystal of synthetic quartz (X-507 with 365H/10^6Si) with a stress of 140 MPa applied normal to (10$\bar{1}$1) at 510°C. Note the incubation period followed by a region of constant strain-rate. (After Linker et al. 1984.)

9.3 Dislocation processes involved in deformation

In this section, the basic dislocation processes involved in the progressive deformation of a crystalline solid are discussed briefly to provide background for the detailed discussion of the deformation microstructures observed by TEM in specific minerals to follow. Particular attention is given to relating the nucleation, glide, climb, multiplication, and interaction of dislocations to the various stages of the creep and stress-strain curves. More discussion can be found in the texts referred to in Section 9.1.

9.3.1 Dislocation nucleation. Dislocations are present in most crystals although there are some notable exceptions. For example, the dislocation density ρ in most common metals of reasonable purity (e.g., 99.99%) is usually at least 10^6 cm^{-2}, even after annealing for long periods at temperatures near the melting point (Young and Savage 1964). Many observations have shown that when the applied stress reaches the critical resolved shear stress for a particular slip system in such a crystal, a rapid multiplication of dislocations is followed by a progressive increase in dislocation density with increasing strain. Extensive plastic deformation would not be possible without this dislocation multiplication because mobile grown-in dislocations can annihilate each other or can reach the

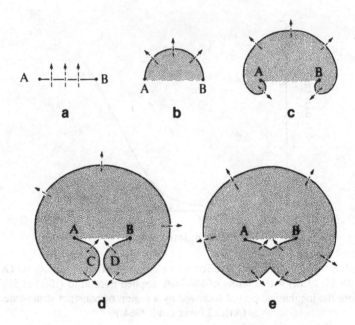

Figure 9.6. Diagrams illustrating the operation of a Frank–Read source. (a) *AB* is a short segment (of length *L*) of dislocation lying in its slip plane, which is the plane of the diagram. The continuations of the dislocation to the left of *A* and to the right of *B* do not lie in the slip plane and therefore are not shown. The applied stress produces a normal force σ on *AB*, which then bows out. If σ is increased beyond a critical value $\sigma_c \approx \mu b/2L$, corresponding to (b) where the curvature is a maximum, the dislocation becomes unstable and expands indefinitely. The expanding loop doubles back on itself, as shown in (c) and (d). The approaching parts *C* and *D* of the expanding loop are of opposite sign; so when they meet, they annihilate each other, producing a closed loop and regenerating the segment *AB*, as shown in (e). The process then begins again. Slip occurs in the shaded area produced by the expanding loop. Note that the stress needed to activate the source is inversely proportional to the length of the segment *AB*. (After Read 1953.)

surfaces of the crystal, where they are lost for further deformation. There is strong evidence that the dislocation multiplication is due mainly to mechanisms such as that of the *Frank–Read source,* whose operation is shown in Figure 9.6. Such multiplication mechanisms depend on the existence in the undeformed crystal of segments of *potentially mobile* dislocations. In most common metals, this is not a problem. However, in many rock-forming minerals and their manmade equivalents, the density of potentially mobile dislocation segments, which could act as Frank–Read type sources, can be effectively zero; fresh dislocations must be nucleated in some way before deformation can be initiated at stresses well below the theoretical strength. Dislocation nucleation is often associated with cracks and local stress concentrations around small clusters of impurity atoms.

Of course, many mineral crystals have much higher dislocation densities, in the range $10^6 < \rho < 10^9$ cm^{-2}. This is probably the result of extensive straining, generally at temperatures sufficiently high for recrystallization to accompany the deformation (Green 1976). Whether segments of these dislocations can glide easily and act as Frank–Read type sources in a laboratory deformation experiment will depend on a number of factors. It may be that during recovery and recrystallization in nature many of the dislocations climbed into stable arrays and networks and that they are not lying in slip planes suitably oriented with respect to the direction of the applied stress. Further, dislocations that have existed over geological periods of time could be pinned by the segregation of impurities in the dislocation cores (Knipe 1980). In such crystals, the nucleation of fresh dislocations of appropriate Burgers vectors and orientations may be necessary before deformation can begin.

The dislocation nucleation just discussed is a preyield phenomenon; in any deformation experiment, it may occur (i) during any preconditioning treatment at temperature and pressure before the shear stress is applied, (ii) during the incubation period in a creep test, or (iii) during the nominally elastic region in a constant strain-rate experiment. Thus, the microstructure of the crystal immediately prior to the onset of deformation may not be the same as the microstructure of the as-grown crystal.

9.3.2 The yield point, work-hardening, and recovery. The yield stress, whether in a creep or a constant strain-rate experiment, is determined by the onset of dislocation mobility, usually glide. The subsequent deformation depends on the density ρ_m of mobile dislocations and their speed v. Provided the dislocations are distributed reasonably homogeneously in the specimen, the deformation is described by the Orowan equation

$$\dot{\epsilon} = \bar{\rho}_m b \bar{v} \qquad (9.1)$$

where $\dot{\epsilon}$ is the strain rate, $\bar{\rho}_m$ the average density of mobile dislocations of Burgers vector **b**, and $\bar{v}$ their average speed. $\bar{\rho}_m$ obviously depends on the number of sources that can be activated at a given stress, and $\bar{v}$ depends on the number of obstacles of various types that provide a resistance to the mobility of the dislocations. By formulating expressions for $\bar{\rho}_m$ and $\bar{v}$ in terms of appropriate variables, the Orowan equation can be used to calculate creep and stress-strain curves. Such microdynamical calculations have been carried out for lithium fluoride (Johnson 1962), for crystals with the diamond structure (Alexander and Haasen 1968), and for quartz (Griggs 1974).

In a constant strain-rate experiment, the rapid multiplication of dislocations following the yield point can produce more mobile dislocations than are necessary to maintain the imposed strain-rate and consequently the stress drops. The deformation will continue at a constant stress provided any decrease in $\bar{v}$ is compensated by an increase in $\bar{\rho}_m$, or vice versa. However, in general, the stress rises with increasing strain. The slope ($d\sigma/d\epsilon$) of the stress-strain curve is determined by the competition between two dislocation processes: namely, *work-hardening* and *recovery*, which we now consider briefly.

Although there is as yet no unified theory of work-hardening, the process is essentially due to the mutual interaction between the dislocations, which produces barriers to continuing dislocation motion. Some of the important interaction processes include:

1   Dislocations emitted from a source pile up against an obstacle, producing a back stress σ_i that opposes the applied stress σ. If the effective stress ($\sigma - \sigma_i$) acting on the source becomes less than the critical stress $\sigma_c \approx \mu b/2L$ to activate the source (see Figure 9.6), the source will stop emitting dislocations. To maintain the imposed strain-rate, the applied stress σ increases to drive the blocked dislocations over the obstacle, to reactivate the source, or to activate other sources with smaller L.

2   When a moving dislocation of Burgers vector $\mathbf{b}_1$ intersects a stationary dislocation of Burgers vector $\mathbf{b}_2$, which crosses the slip plane, two distinct types of interaction can occur: (a) The two dislocations may interact to form a short segment of dislocation of Burgers vector $\mathbf{b}_3 = \mathbf{b}_1 + \mathbf{b}_2$ between two three-fold nodes. If this new segment is sessile, there is a drag on the moving dislocation, and it bows out in its slip plane on each side of the sessile segment. (b) If the moving dislocation cuts through the stationary dislocation, the moving dislocation may become jogged. A *jog* in an edge dislocation can glide along with the dislocation without exerting any drag. However, this is not true in general. For example, the jog in a moving screw dislocation that has cut through another screw dislocation can move along with the moving dislocation only by climb. Thus, at low temperatures, where diffusion is low, the jog produces a drag on the dislocation.

3   When the dislocation density increases to about 10^9 cm^{-2}, the dislocations tend to become *tangled*. This occurs particularly when the

dislocations of one slip system interact in the ways just described with the dislocations of secondary slip systems. Because dislocation tangles also act as barriers to the glide of other dislocations, they tend to grow quickly once they have been initiated. Thus, tangles are probably the most obvious microstructural characteristic of a work-hardened specimen and often indicate that several slip systems have been simultaneously active during deformation.

When a work-hardened crystal containing a high density of tangled dislocations is statically annealed at temperatures $T \gtrsim 0.3T_m$, there is a general lowering of the dislocation density, and the dislocations tend to become rearranged into low-energy dislocation walls and networks forming boundaries (with short-range stress fields) between slightly misoriented subgrains with relatively low densities of dislocations. This *recovery* of the crystal from its high-energy work-hardened state is due to the *climb* of dislocations. As shown in Figure 9.3, climb is diffusion-controlled and therefore becomes more important as the temperature increases. During high-temperature deformation, dislocations move by climb as well as by glide; as a result, recovery processes occur that tend to counteract the effects of work-hardening. Thus, the slope $(d\sigma/d\epsilon)$ of the stress-strain curve depends on the competition between work-hardening and recovery. At low temperatures, work-hardening (mainly due to dislocation tangling) dominates, leading to a rapid rise in stress with increasing strain. However, at higher temperatures, recovery processes make dislocation glide possible at lower stresses, and hence the stress does not rise so rapidly with increasing strain.

The main recovery processes due to dislocation climb can be summarized as follows:

1 Dislocations of the same Burgers vector, but of opposite sign, climb toward one another and annihilate each other, thereby reducing the dislocation density.
2 Dislocations of the same Burgers vector climb into stable arrays of parallel dislocations, forming low-angle boundaries between subgrains of relatively low dislocation density.
3 Dislocations with different Burgers vectors climb and interact to form three-fold nodes and, subsequently, low-angle grain boundaries consisting of networks of dislocations.
4 Dislocations climb over obstacles in the slip plane. This process lessens hardening by reducing the drag on moving dislocations and by relieving the back stress associated with a pileup of dislocations, thereby allowing the dislocation source to continue to emit dislocations.

The most obvious microstructural characteristics of recovery are probably subgrain boundaries consisting of arrays of parallel dislocations or dislocation networks.

During annealing at high temperatures, new grains free of dislocations can form in the deformed or partially recovered structure. This process is called *static recrystallization* and results in the production of high-angle grain boundaries. The migration of these boundaries causes grains of low dislocation density to consume the grains of higher dislocation density, leading to a new undeformed state. Recrystallization can occur during deformation at high temperatures, in which case it is called *dynamic* (or *syntectonic*) *recrystallization*. In either case, recrystallization tends to reduce hardening.

9.4 Deformation of "wet" synthetic quartz

Quartz is one of the most abundant minerals in rocks of the Earth's crust. Many quartz-rich rocks show clear evidence of plastic deformation on a large scale, and the individual quartz grains of these deformed rocks usually exhibit deformation features such as deformation lamellae, deformation bands, and subgrains, which have been studied in great detail in thin sections by optical microscopy. In spite of this clear evidence of quartz deformation in nature, it was not until 1964 that Griggs and his co-workers succeeded, after many failures, in plastically deforming quartz crystals in the laboratory. These and later experiments have shown that although plastic deformation can be achieved, most optically clear single crystals of natural quartz are exceedingly strong and have strengths comparable to the theoretical strength (see Section 9.5). However, Griggs and Blacic (1964) observed that such crystals became very weak when deformed in the presence of water at temperatures of 800°C and above. Later, Griggs and Blacic (1965) found that a single crystal of synthetic quartz (which, according to its 3-μm infrared absorption band, contained grown-in water-related defects in a concentration corresponding to one H for every 130 Si) had a comparable strength at temperatures as low as 400°C. The geological implications of these observations are obvious; as a consequence, many rheological and microstructural studies have been made of "wet" synthetic quartz and of natural quartz deformed in both "wet" and "dry" environments in order to elucidate the mechanism of the *water* (or *hydrolytic*) *weakening*. Most of these studies have implicitly assumed that the same mechanism, involving the same water-related defects, operates in both "wet" synthetic quartz and water-weakened natural quartz. Furthermore, most explanations have evolved from the original hypothesis of Griggs and Blacic (1965), which in its most advanced form (Griggs 1974) states the following:

1　The water responsible for the weakening effect is present as isolated water-related point defects, which diffuse readily at temperatures near 500°C.
2　Under the influence of the applied stress, a dislocation loop can grow by glide only as sufficient HOH diffuses to the growing segment to saturate the newly created core and develop a cloud of hydrolyzed Si−O bonds in the neighborhood of the dislocation in order to reduce the Peierls stress (the fundamental "friction" to the glide of a dislocation in a perfect crystal) to a very low value.

The idea that water-related defects might also enhance dislocation climb was suggested by McLaren and Retchford (1969) and included by Griggs (1974) in his microdynamical calculations of the stress-strain curves based on his original hypothesis.

However, the Griggs hypothesis has been challenged recently in several ways:

1　The observed uptake of water by quartz under conditions of high temperature, pressure, and water fugacity indicates that the diffusivity and/or solubility of water-related point defects in solid solution are much lower than are required by the Griggs model (Gerretsen et al. 1989; Kronenberg, Kirby, and Rossman 1986; Rovetta, Holloway, and Blacic 1986).
2　The 3-μm infrared absorption band in "wet" synthetic quartz is due to high-pressure clusters of molecular water rather than point defects (Aines and Rossman 1984; Gerretsen et al. 1989).
3　The *initial* density of mobile dislocations is at least five orders of magnitude less than that required to initiate deformation in the Griggs (1974) microdynamical calculations (McLaren, Fitz Gerald, and Gerretsen 1989).

It would be inappropriate in the present context to attempt to review in detail the tortuous path followed by the many investigations of water-weakening that have been made since the initial discovery of the effect.[*] Suffice to say that there is, as yet, no direct independent evidence for an

----

[*] Blacic and Christie (1984) have provided an excellent review of research to that date.

influence of water on the pure glide of dislocations. However, there is good evidence that water can influence the nucleation of potentially mobile dislocations in both synthetic quartz (McLaren et al. 1989) and natural quartz (Fitz Gerald et al. 1990), but in different ways, and that dislocation climb is enhanced in wetter crystals (Cordier and Doukhan 1989; Fitz Gerald et al. 1990; Hobbs 1968; Hobbs, McLaren, and Paterson 1972; McLaren et al. 1989; McLaren and Retchford 1969; Tullis and Yund 1989).

The following subsections are concerned with the grown-in defects in synthetic quartz and the dislocation microstructures associated with the various stages of the creep and stress-strain curves. The microstructural evolution in natural quartz deformed under various conditions of water fugacity, and other experimental variables, will be considered in Section 9.5.

9.4.1 Grown-in defects. The density of grown-in dislocations in commercial synthetic quartz crystals is typically of the order of 10^3 cm^{-2} (McLaren et al. 1971). The dislocations are predominantly of edge character with $\mathbf{b} = \frac{1}{3}\langle 11\bar{2}0 \rangle$ and are aligned approximately parallel to [0001] in the boundaries of the growth cells that are characteristic of these crystals. These dislocations are present in both strong "dry" crystals and weak "wet" crystals, and there is no evidence that they play an active role in the deformation of "wet" crystals. "Wet" crystals characteristically contain high-pressure clusters of molecular water, which (see Section 9.4.2) act as highly efficient sources of the dislocations involved in the deformation process. These high-pressure clusters are imaged by the strain they produce in the surrounding crystal matrix (Section 5.7.2). Typical examples in a (11$\bar{2}$0) crystal plate are shown in Figure 9.7(a, b). Note that in the "drier" region of this crystal, where the water content corresponds to about 200H/10^6Si, the clusters appear as black spots, which are sometimes not easily distinguishable from electron beam damage (Section 8.13.1). However, in the "wetter" regions, Figure 9.7(b), where the water content corresponds to about 1,600H/10^6Si, the clusters are significantly more numerous and larger, and their strain fields are clearly seen. Approximately one-third of the clusters have a very characteristic appearance, and one example is circled. This image consists of two sets of lobes symmetrically placed about a line of no contrast, similar to that shown in Figure 5.22. Thus, the image could be due to strain around a small spherical inclusion, in which case the line of no contrast would always be normal to $\mathbf{g}$. However, in the present case, the line of no contrast is always parallel to [0001], independently of $\mathbf{g}$, which suggests that the contrast is due to a

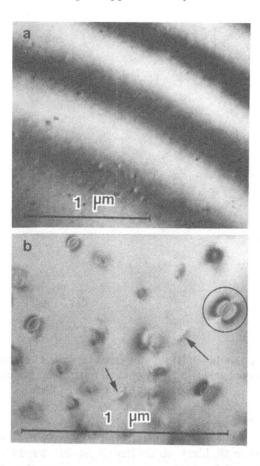

Figure 9.7. BF micrographs showing the strain associated with high-pressure clusters of molecular water in as-grown "wet" synthetic quartz (crystal W2). (a) Region with a water content corresponding to 200H/10^6Si. (b) Region with 1,600H/10^6Si. No strain is associated with the arrowed clusters because they intersect the foil surface; the water escapes, and the stress is relaxed. The strain field of the circled cluster is characteristic of a lens-shaped inclusion whose plane is normal to the foil surface. (From McLaren et al. 1983.)

small lens-shaped inclusion normal to the plane of the specimen: that is, parallel to ($\bar{1}100$). Two observations appear to confirm this suggestion. First, if an inclusion intersects one of the surfaces of the thin foil, the water will escape, and the strain in the surrounding crystal vanishes. Features that are consistent with this interpretation are commonly observed; two examples are indicated by arrows in Figure 9.7(b). Second, in DF

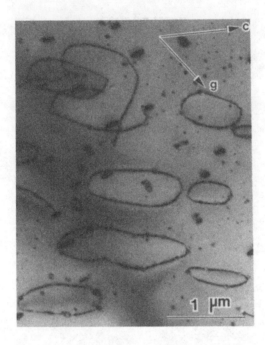

Figure 9.8. BF image ($\mathbf{g} = 10\bar{1}1$) showing prismatic dislocation loops ($\mathbf{b} = \langle 11\bar{2}0 \rangle$) and strain-free bubbles in "wet" synthetic quartz after heating at 550°C for 2 hours. (From McLaren et al. 1989.)

images with $\mathbf{g} = 0003$ at $s = 0$, the lobes are out of contrast and the inclusions appear as small, black, strain-free ellipsoidal-shaped features whose long axes are parallel to [0001]. It appears, therefore, that the other more common strain features seen in Figure 9.7(b) are due to lens-shaped inclusions lying parallel to the inclined planes ($0\bar{1}10$) and ($10\bar{1}0$).

Gerretsen et al. (1989) measured the image width (w/t_g), radius r_l, and thickness a of about 70 clusters in two crystals of synthetic quartz. Using the analysis of the strain contrast around platelike precipitates given in Section 5.7.2, they estimated the average strain ϵ to be in the range 2 ± 1 percent. The pressure P in the clusters was then estimated from this strain to be 0.4 ± 0.15 GPa, using an expression for an oblate spheroidal cavity of aspect ratio 4:1 from Neuber (1958). The water content estimated from the observed size and number density of clusters, assuming they are all filled with H_2O at 0.4 GPa, was consistent with the value determined from infrared absorption.

On heating, these clusters evolve into strain-free bubbles and associated dislocation loops (Figure 9.8). The following mechanism of growth was proposed by McLaren et al. (1983). At any temperature T, there is a critical bubble diameter above which the "steam" pressure P exceeds the pressure p for a spherical bubble in mechanical equilibrium. If P becomes greater than p, the bubble will increase in volume until $P = p$, the increase in volume being achieved by the pipe diffusion of Si and O away from the bubble site into a linked edge (prismatic) dislocation loop. More recent observations (Gerretsen, McLaren, and Paterson 1987; Gerretsen et al. 1989) of the evolution of the clusters indicate that the critical size above which a bubble will grow at a temperature T is increased by an increase in the external hydrostatic pressure. Thus, dislocation nucleation tends to be inhibited by an increase in pressure.

9.4.2 Creep deformation. The deformation of single crystals of "wet" synthetic quartz under a constant compressive load at atmospheric pressure has been studied by McCormick (1977), Kirby and McCormick (1979), and Linker et al. (1984). The creep curves for all samples loaded almost immediately after the test temperature had been reached are characterized by an incubation period that decreased rapidly with increasing axial stress and increasing temperature. A typical creep curve is shown in Figure 9.5.

Recent TEM observations (McLaren et al. 1989) have shown that during the incubation period the high-pressure clusters of molecular water evolve to form a microstructure consisting mainly of prismatic dislocation loops on which there are strain-free water bubbles, as described in the previous section. Furthermore, the incubation period appears to be the time required to nucleate the dislocation loops plus the time required for the loops to expand by climb until segments of the loops lying in potential slip planes are long enough for them to act as Frank–Read sources under the given applied stress (Figure 9.6). Because the loops are also produced when the crystals are heated without an applied load, appropriate preheating can reduce the incubation period to essentially zero. However, very long heat treatment (e.g., >250 hours) can so reduce the density of potentially mobile dislocation segments – and destroy the sources of dislocations (i.e., the high-pressure clusters) – that the incubation period is greatly increased from the original value (Kirby and McCormick 1979). Some examples of the dislocation loops produced during the incubation period (or a preheat of similar duration) are shown in Figure 9.8.

The Burgers vectors of the loops are easily determined using the $\mathbf{g} \cdot \mathbf{b} = 0$ and $\mathbf{g} \cdot \mathbf{b} \times \mathbf{u}$ invisibility criteria (Section 5.6). We consider some examples in foils normal to an **a**-axis. First, most of the dislocations are out-of-contrast in DF images with $\mathbf{g} = 0003$ at $s = 0$, as shown in Figure 9.9(a), indicating that $\mathbf{b} = \frac{1}{3}\langle 11\bar{2}0\rangle$. Second, dislocation loops with these Burgers

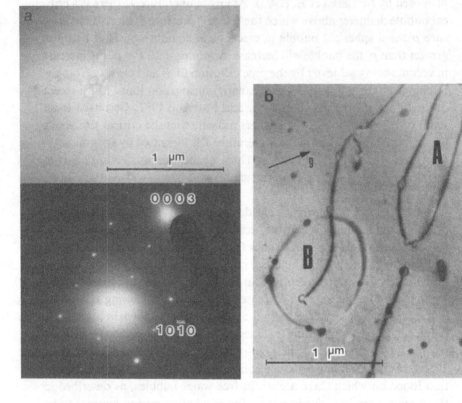

Figure 9.9. (a) DF micrograph and its associated selected area diffraction pattern, showing dislocations of $\mathbf{b} = \langle 11\bar{2}0 \rangle$ in heated "wet" synthetic quartz effectively out-of-contrast for $\mathbf{g} = 0003$. (b) BF micrograph showing a prismatic dislocation loop A for which $\mathbf{g} \cdot \mathbf{b} = \pm 1$ and a prismatic loop B for which $\mathbf{g} \cdot \mathbf{b} = 0$, with $\mathbf{g} = (\bar{1}101)$. Note that those segments of loop B that also satisfy the equation $\mathbf{g} \cdot \mathbf{b} \times \mathbf{u} = 0$ are out-of-contrast. (From McLaren et al. 1989.)

vectors give rise to two distinct types of image contrast with all the $\mathbf{g} = \{10\bar{1}1\}$ accessible in a normal-to-**a** foil. In Figure 9.9(b), for example, the elongated loop A (like those in Figure 9.8) appears as a single line of high contrast. On the other hand, the contrast of the near-circular loop B is distinctly different: those parts of the loop that are approximately parallel to **g** are out-of-contrast, whereas those that are approximately normal to **g** exhibit a double image. These features are characteristic of a pure edge-dislocation loop lying in the plane of the foil; **b** is normal to the foil, and hence $\mathbf{g} \cdot \mathbf{b} = 0$ (see Section 5.4.2). For similar prismatic loops with the other crystallographically equivalent Burgers vectors, $\mathbf{g} \cdot \mathbf{b} = \pm 1$ for all

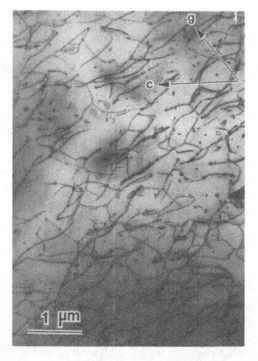

Figure 9.10. BF image ($g = 10\bar{1}1$) of the dislocation microstructure in a specimen of "wet" synthetic quartz deformed about 0.3 percent in creep with an axial stress of 100 MPa normal to ($10\bar{1}1$) at 550°C for 30 minutes, after a preheat of 2 hours without a load. (From McLaren et al. 1989.)

accessible $g = \{10\bar{1}1\}$, and the dislocation exhibits essentially the same contrast all round the loop, as for the loop A. Dislocation loops that are in strong contrast for $g = 0003$ were also nucleated (apparently via microcracks) during the incubation period (or preheat); they obviously have a Burgers vector with a large c-component. The contrast of dislocations whose Burgers vectors have a large c-component is considered later.

The onset of deformation at the end of the incubation period is accompanied by an abrupt change in the dislocation microstructure, with a sudden increase in the dislocation density of about an order of magnitude, as shown in Figure 9.10. The dislocation density is about 2×10^9 cm^{-2}, and about one-tenth of those dislocations tend to be aligned along [0001] and have a Burgers vector with a large c-component. Kirby and McCormick (1979) found that the dislocation density was effectively independent of strain beyond about 1 percent. Some examples of Burgers vector determinations in deformed crystals are illustrated in Figures 9.11 and 9.12.

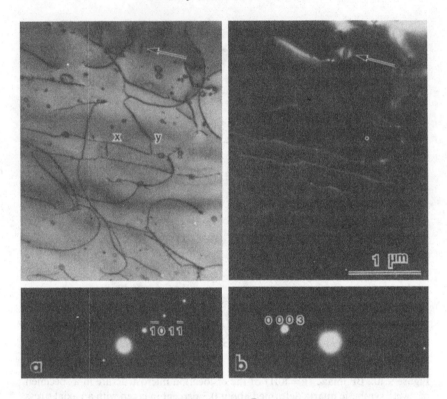

Figure 9.11. (a) Kinematical BF image ($g = 10\bar{1}1$) and (b) WBDF image ($g = 0003$) of dislocations in "wet" synthetic quartz after deformation in creep with an axial load of 100 MPa normal to ($10\bar{1}1$) at 550°C for 3.5 hours.
(From McLaren et al. 1989.)

Figures 9.11(a, b) are a kinematical BF and WBDF pair of an area of a normal-to-**a** foil with $g = 10\bar{1}1$ and $g = 0003$, respectively. By applying the $\mathbf{g} \cdot \mathbf{b} = 0$ invisibility criterion, we find that the dislocations that are in contrast in (a) and out-of-contrast in (b) have $\mathbf{b} = \frac{1}{3}\langle 11\bar{2}0 \rangle$, while those that are in contrast in both have $\mathbf{b} = [0001]$ or $\frac{1}{3}\langle 11\bar{2}3 \rangle$. Close inspection of the original negative of (b) reveals that these dislocations have the double-image contrast characteristic of $\mathbf{g} \cdot \mathbf{b} = 3$ (see Section 5.5.2 and Figure 5.17). The long dislocation marked x has interacted with a dislocation (marked y) that has $\mathbf{b} = \frac{1}{3}\langle 11\bar{2}0 \rangle$ because it is out-of-contrast for $g = 0003$. This interaction has produced a short segment of dislocation that is out-of-contrast for $g = 10\bar{1}1$ and in contrast for $g = 0003$; therefore, it must have $\mathbf{b} = \frac{1}{3}\langle 11\bar{2}3 \rangle$. Thus, dislocation x must have $\mathbf{b} = [0001]$; because it is approximately parallel to [0001], it is essentially in screw orientation.

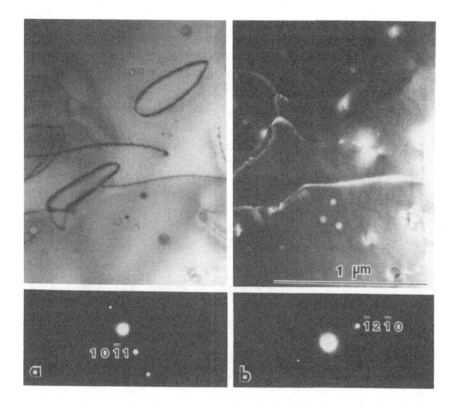

Figure 9.12. (a) Kinematical BF ($\mathbf{g} = 10\bar{1}1$) and (b) WBDF image ($\mathbf{g} = \bar{1}2\bar{1}0$) of the dislocations in "wet" synthetic quartz deformed under the same conditions as the specimen shown in Figure 9.11. (From McLaren et al. 1989.)

Figures 9.12(a, b) are a kinematical BF and a WBDF pair of an area of a foil that is parallel to $\mathbf{a}_2$ ($= [\bar{1}2\bar{1}0]$) and normal to ($10\bar{1}1$), with $\mathbf{g} = 10\bar{1}1$ and $\mathbf{g} = \bar{1}2\bar{1}0$, respectively. As can be seen, two dislocation loops, and the dislocation passing between them, are in strong contrast for $\mathbf{g} = 10\bar{1}1$ and completely out-of-contrast for $\mathbf{g} = \bar{1}2\bar{1}0$. All these dislocations must, therefore, have $\mathbf{b} = [0001]$. The orientation and shape of the loops suggest that they are lying nearly parallel to (0001).

Small prismatic loops viewed edge-on may give rise to a characteristic contrast of two symmetrical lobes, particularly when $\mathbf{g}$ is normal to the plane of the loop, as discussed in Section 5.7.2. The contrast and orientation of the (arrowed) dislocation loop viewed edge-on in Figure 9.11(a, b) are consistent with its being a prismatic loop of $\mathbf{b} = [0001]$.

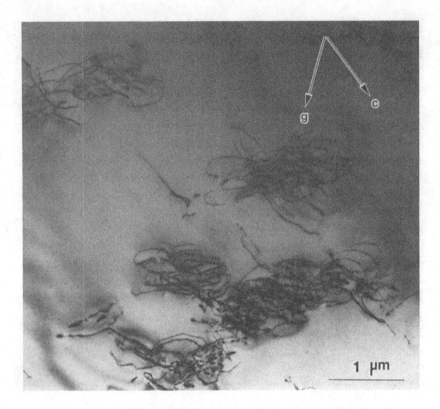

Figure 9.13. BF image ($g = 10\bar{1}1$) of the dislocation tangles generated at the high-pressure water clusters in "wet" synthetic quartz deformed at a constant strain-rate at 475°C. (From McLaren et al. 1989.)

9.4.3 Deformation at a constant strain-rate. In a constant strain-rate experiment, the applied stress increases linearly with time during the elastic regime. At high temperatures (>500°C), dislocation loops (and associated strain-free bubbles) are nucleated and grow by climb during the elastic regime, just as they do during the incubation period of a creep test. Clearly, the time (and hence the stress) at which plastic flow begins is determined by the rate at which the stress increases and the rate at which the dislocation loops grow. The rate at which the stress increases during the elastic regime is determined by the imposed strain-rate, and the rate at which the loops grow is determined mainly by the temperature. Thus, the initial flow stress (at a given strain-rate) decreases with increasing temperature, as observed in Figure 9.4. Likewise, at a given temperature, an increase in strain-rate causes an increase in the initial flow stress, as is also

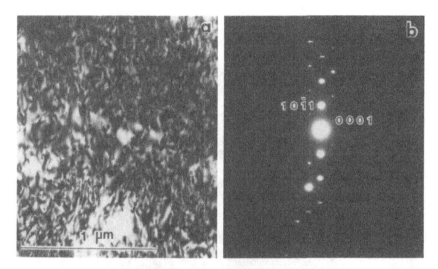

Figure 9.14. (a) BF image ($\mathbf{g} = 10\bar{1}1$) showing the very high density of dislocations generated in a region containing a high density of large high-pressure clusters of water in a "wet" synthetic quartz crystal deformed at a constant strain-rate at 475°C. (b) The associated selected area diffraction pattern showing radial streaking of the spots due to the presence of domains misoriented by up to 4°.
(From McLaren et al. 1989.)

observed (Hobbs et al., 1972). Most constant strain-rate experiments have been carried out with a hydrostatic confining pressure of at least 300 MPa, which tends to decrease the rate at which dislocation loops are nucleated compared with experiments at atmospheric pressure. Thus, at low temperatures ($< 500°C$) the stress can rise to quite high values (≈ 750 MPa) before prismatic dislocation loops are nucleated. Under these conditions, tangles of dislocation loops are generated at the high-pressure clusters, apparently directly by the applied stress since no clusters of strain-free bubbles are subsequently seen. This was first observed by Morrison-Smith, Paterson, and Hobbs (1976),* and examples of dislocations nucleated in this way are shown in Figure 9.13. Particularly when the density of clusters is high, these tangles lead to rapid work-hardening at low temperatures (see Figure 9.4). The dislocation density becomes so high that it is very difficult to resolve individual dislocations (see Figure 9.14), even in

* Since Morrison-Smith et al. (1976) associated the strain features observed in these specimens (Figure 9.7) with acmite inclusions and not with high-pressure clusters of molecular water, they did not relate dislocation nucleation with the water in "wet" synthetic quartz.

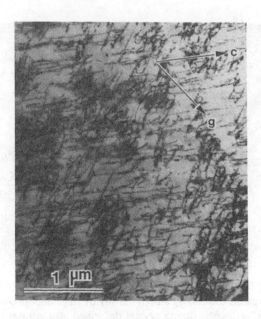

Figure 9.15. BF image ($g = 10\bar{1}1$) of the dislocation microstructure in a region containing a lower density of large high-pressure water clusters than in the region shown in Figure 9.14(a), after deformation at a constant strain-rate at 475°C. (From McLaren et al. 1989.)

WBDF images. Figure 9.15 shows the dislocations in a deformed region in which the original density of clusters was rather lower; it is clear that there are two sets of dislocations. Contrast experiments, such as those described in Section 9.4.2, have shown that the dislocations that are approximately parallel to [0001] have a Burgers vector with a c-component (probably $\mathbf{b} = [0001]$), while the other curved (and somewhat tangled) dislocations have $\mathbf{b} = \frac{1}{3}\langle 11\bar{2}0\rangle$. An interesting feature of the microstructures of specimens deformed at low temperatures is that there appear to be no clusters or strain-free bubbles. This observation suggests that the water (originally in the clusters) is now distributed in the dislocation cores, assuming that it has not diffused out of the crystal. The ragged fine-structure of these dislocation images suggests the presence of many pinning points (associated with the water) along the dislocations, which actually produce a drag on the dislocation glide. Very similar dislocation images were observed in specimens deformed under comparable conditions by Morrison-Smith et al. (1976). When these deformed crystals are annealed in air at atmospheric pressure for 2 hours at 600°C, significant microstructural

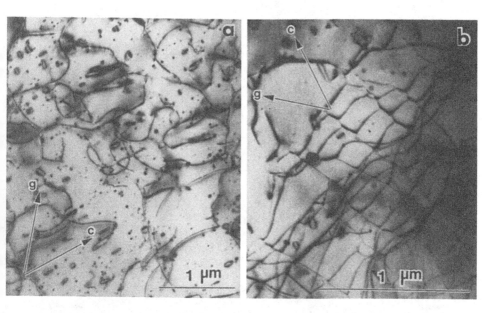

Figure 9.16. BF images (g = 10Ī1) showing the dislocation microstructure in "wet" synthetic quartz deformed at 475°C and subsequently annealed at atmospheric pressure at 600°C for 2 hours. Note the bubbles in (a) and the dislocation networks in (b). This microstructure should be compared with that shown in Figure 9.15. (From McLaren et al. 1989.)

changes occur, as shown in Figure 9.16. In general, the dislocations are smoothly curved, and many have interacted to form networks; also there is much debris of small dislocation loops. These features together with a noticeable decrease in dislocation density indicate that considerable re-covery, due to dislocation climb, has occurred during annealing. Another feature of the annealed specimens is that there are now many bubbles, both isolated and linked by dislocations. These bubbles, particularly the smaller ones, are more easily identified in out-of-focus phase-contrast im-ages (see Section 5.7.1 and Figure 5.21) in which the dislocations are out-of-contrast, as illustrated in Figure 9.17. The presence of these bubbles in the annealed specimens appears to confirm the suggestion made earlier that in the specimens deformed at low temperatures (475°C) the water originally in the clusters is distributed in the dislocation cores.

Specimens deformed at $T \geq 600°C$ exhibit microstructures that are very similar to those observed in specimens deformed at 475°C and subse-quently annealed at the higher temperature. The lower work-hardening

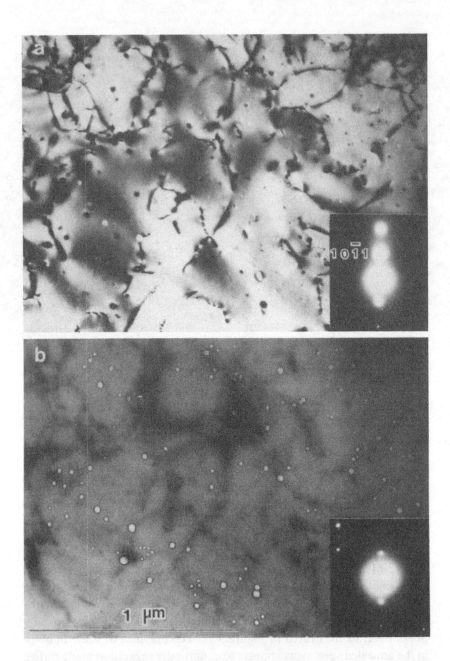

Figure 9.17. Out-of-focus phase contrast images of small bubbles in deformed and annealed "wet" synthetic quartz. (a) Normal BF image ($g = 10\bar{1}1$), in which the dislocations are in strong contrast. Close inspection reveals the presence of small bubbles, but often it is difficult to distinguish them from small dislocation loops. (b) BF out-of-focus phase contrast image of the same area. The dislocations are effectively out-of-contrast, but the small bubbles are visible in strong contrast.

rates observed in specimens deformed at $T \geq 600°C$ are clearly associated with the recovery mechanisms revealed in the microstructure (Morrison-Smith et al. 1976). The stress relaxation observed by Griggs (see McLaren and Retchford 1969) is also associated with recovery processes. As already mentioned, abundant evidence exists that the rate of recovery is positively correlated with the bulk water content of synthetic quartz crystals, although the mechanism of water-enhanced climb is not clear. However, since recovery characteristics always appear to be associated with the presence of strain-free bubbles on dislocations, McLaren et al. (1989) suggested that these bubbles promote climb by acting as sources and sinks for Si and O vacancies, which diffuse along the dislocation cores to and from jogs. Thus, enhanced climb may not be due to a direct interaction of dislocations and water-related point defects. A similar view has been expressed by Doukhan and Trepied (1985) and Cordier and Doukhan (1989).

In specimens deformed to several percent strain (or more) at low to intermediate temperatures and stresses, where neither work-hardening nor recovery processes predominate, dislocations tend to tangle into localized walls (Kirby and McCormick 1979; McCormick 1977; McLaren et al. 1970; Morrison-Smith et al. 1976). These walls behave as optical phase objects and give rise to the *deformation lamellae* that are commonly observed in deformed crystals by optical microscopy (see Section 1.3 and McLaren et al. 1970). Similar walls of tangled dislocations develop in metals in the *power-law-breakdown* creep regime where both recovery-controlled and glide-controlled deformation mechanisms are operative (see, e.g., Drury and Humphreys 1986).

It is clear from the preceding discussion that both the nucleation and the climb of dislocations are associated with the grown-in water in "wet" synthetic quartz. McLaren et al. (1989) have argued that these facts are sufficient to rationalize all the main features of the stress-strain curves (as well as the observation that the initial yield stress is essentially independent of the average water content) without postulating an influence of water on the *pure glide* of dislocations. They also suggest that the low initial yield stress (and hence the low *strength*) of "wet" synthetic quartz compared with that of "dry" natural (and synthetic) quartz is due fundamentally to the ease with which dislocations are nucleated at the high-pressure clusters of molecular water. However, as noted in Section 9.2, *strength* is sometimes defined as the stress at a given percent strain. In this case the strength depends on the slope of the (σ-ϵ) curve following the yield point and clearly is lower for "wetter" crystals in which the rate of recovery is high.

9.5 Experimental deformation of single crystals of natural quartz

Large, clear single crystals of natural quartz, such as those used for piezo-electric and optical applications, are essentially free of dislocations and twins. Such crystals are exceedingly strong; even at confining pressures in the region of 2–3 GPa, they fracture in a brittle manner without any evidence of plastic flow at room temperature (Christie, Heard, and La Mori 1964). However, Christie, Griggs, and Carter (1964) and Carter, Christie, and Griggs (1964) obtained clear evidence of plastic flow in specimens deformed in compression at high temperature (500°–900°C) and confining pressure (1.5–2 GPa). The observed yield stress at 700°C was about 3 GPa. With a shear modulus of 47 GPa (Poirier 1985), this yield stress corresponds to a shear stress on the slip plane (0001) of about half the theoretical strength. Thin lamellae of deformation-induced Brazil twins (which could account for some of the strain; see Section 8.2 and Figure 8.6) and regions of dislocations were observed in these specimens by McLaren et al. (1967). Because the dislocations were not obviously associated with any other deformation-induced feature or with any grown-in defects, it appeared that the dislocations were nucleated directly by the high applied stress. The fact that the dislocations were straight and parallel to low-index directions indicated that the Peierls stress was high. However, at 900°C the strength fell to about 2 GPa, and TEM observations revealed dense tangles of dislocations in which the dislocation density was estimated to be about 10^{12} cm^{-2} (see Griggs 1974).

As mentioned in Section 9.4, Griggs and Blacic (1964) observed a dramatic weakening of these crystals when deformed at 1.5 GPa confining pressure in the presence of water at $T \geq 800°$C. However, Mackwell and Paterson (1985) were unable to reproduce this result at a confining pressure of 300 MPa. They interpreted the absence of weakening as due to a decrease in the diffusion coefficient of the water-related defects involved in the Griggs (1974) mechanism, which was reviewed briefly in Section 9.4.

Hobbs (1985) proposed an alternative model of hydrolytic weakening in terms of the influence that charged H or OH defects have on the concentrations of other charged defects – such as kinks and jogs on dislocations or vacancies and interstitials – and hence on the deformation rate. In this model, the diffusion of water-related defects need not play a role in facilitating deformation. To test these ideas, Ord and Hobbs (1986) conducted a series of experiments to investigate the influence of the chemical environment on the strength of single crystals of natural quartz loaded

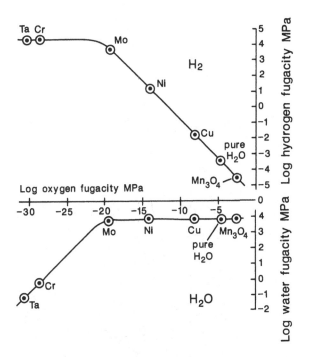

Figure 9.18. Oxygen, water, and hydrogen fugacity relationships at 1.65 GPa and 800°C. (From Ord and Hobbs 1986.)

normal to (10$\bar{1}$0), at 800°C, with a confining pressure of 1.64 GPa and a strain-rate of 10^{-5} s^{-1}. The environment of the crystal was controlled during a 20-hour preheat and during deformation by use of a solid-oxygen buffer and excess water encapsulated with the quartz in a silver jacket. The fO_2, fH_2O, and fH_2 relationships for the buffers used are shown in Figure 9.18, and the observed stress-strain curves are shown in Figure 9.19.

Ord and Hobbs (1986) used the stress at 5 percent strain as a measure of strength and found that the Ta-buffered specimens (corresponding to low H_2O fugacity) have a high strength compared with the Mn_3O_4-buffered specimens (corresponding to high H_2O fugacity). For each of these two buffering conditions, the microstructures of these deformed specimens (which exhibit the extremes of strength) and of specimens that have been (i) pre-heated without an applied load and (ii) pre-heated and loaded only to a peak stress corresponding to the expected onset of plastic deformation have been examined in detail by Fitz Gerald et al. (1990), using both optical microscopy and TEM. These observations have revealed important

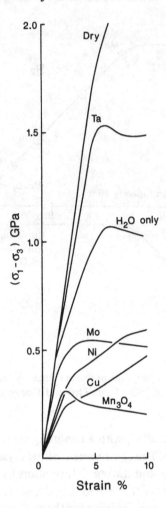

Figure 9.19. Stress-strain curves for single crystals of natural quartz deformed in buffered assemblies at 1.64 GPa confining pressure, 800°C, and strain-rate of 10^{-5} s^{-1} after 20 hours preconditioning. (From Ord and Hobbs 1986.)

information about the nucleation of dislocations in initially dry natural quartz, the mobility of these dislocations, the interaction of brittle and ductile processes, and the influence of the chemical environment on these processes.

9.5.1 Microstructural development during the preheat and (nominally) elastic regimes. The Ord and Hobbs (1986) experiments were carried out with specimens cut from a single crystal of natural quartz that exhibited very few extended crystal defects. The main imperfections were rare planes of isolated dislocations and small bubbles. The overall dislocation density was low, $\approx 10^4 \text{ cm}^{-2}$. The preheat was intended to bring the specimen into equilibrium with its environment (and hence produce an equilibrium concentration of the point defects, which, on the Hobbs model, is expected to control the subsequent deformational behavior), without any major microstructural change. However, the optical and TEM observations of Fitz Gerald et al. (1990) have shown that many microstructural modifications occur during the preheat and the nominally elastic regimes. These modifications are expected to influence the subsequent deformation. It was found that although the same *type* of microstructural features develop in all specimens, there was a distinct *quantitative* difference between the Mn_3O_4-buffered and Ta-buffered specimens. The most conspicuous microstructural features are cracks, narrow bands of recrystallized quartz, and a unique feature (termed a *microcrack-ladder*) whose development involves both brittle and ductile processes (see Section 9.5.2). These features all tend to be oriented near the long axis of the specimens and are associated with solid, fluid, or mixed solid and fluid inclusions, which x-ray microanalysis in the electron microscope (Chapter 7) has shown are derived from the specimen environment (e.g., buffer and jacket material). In general, these features are more numerous in the Mn_3O_4-buffered specimens and tend to be more common near one end of the specimen. Thus, immediately prior to the onset of deformation, the microstructure of the specimen is markedly different from the original microstructure. Furthermore, there is a clear quantitative difference between the Mn_3O_4- and Ta-buffered specimens, which will presumably have an important influence on their mechanical strengths.

9.5.2 Microstructures in specimens deformed under conditions of high water fugacity (Mn_3O_4-*buffer*). The distribution of strain and the types of deformation features generally observed by optical microscopy in these specimens are illustrated schematically in Figure 9.20. Macroscopically, the strain is accommodated mainly at one end in a narrow zone 1–2 mm thick. A band of completely recrystallized quartz extends the full width of the sample, and within this band the grain and subgrain boundaries are decorated with very small inclusions of opaque material. Adjacent to the fully recrystallized zone is a band in which the deformation is heterogeneous; recovery features such as subgrains dominate, but there are also patches of recrystallization. Further into the specimen are the so-called microcrack-ladders, shown schematically in Figure 9.21. A microcrack-ladder consists of a longitudinally oriented planar central zone up to 20 μm

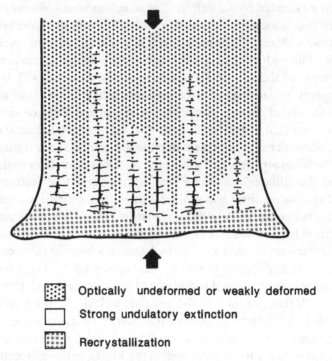

Optically   undeformed or weakly deformed

Strong undulatory extinction

Recrystallization

Figure 9.20. Schematic view of one end of a single crystal of natural quartz after deformation under conditions of high water fugacity (Mn_3O_4-buffer) showing the spatial distribution of recrystallization and microcrack-ladders.
(From Fitzgerald et al. 1990.)

wide, together with a regular set of microcracks at right-angles to the central zone (i.e., parallel to the specimen ends) that extend away from the central zone on both sides, giving a total width of up to 0.5 mm. The central zone contains solid, fluid, or mixed solid and fluid inclusions that occupy about 1 percent of its volume. The regions of quartz adjacent to the array of inclusions are undulatory when viewed between crossed polarizers; regions outside the microcracked zone of the microcrack-ladder appear to be undeformed. The microcrack-ladder structure is most clearly seen in the optical microscope when the planar central zone is perpendicular to the plane of the thin section.

Because the microcrack-ladders are observed in the preheat and in the peak-loaded specimens as well as in the weakly deformed central regions of the deformed specimens, they are likely to contain much information about the initial stages of deformation. Such information is completely

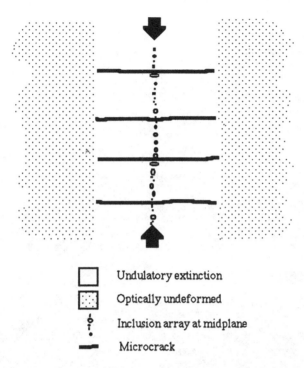

Undulatory extinction

Optically undeformed

Inclusion array at midplane

Microcrack

Figure 9.21. Schematic diagram of a microcrack-ladder showing the longitudinal array of inclusions parallel to the applied stress, the horizontal unloading fractures within the plastically deformed zone, and the surrounding undeformed crystal. (From Fitz Gerald et al. 1990.)

eliminated from the fully recrystallized zones. Therefore, Fitz Gerald et al. (1990) used TEM to examine the structure of microcrack-ladders in detail.

The dislocation microstructure revealed by TEM at the margin of a microcrack ladder is shown in Figure 9.22(a). All the dislocations have $\mathbf{b} = \frac{1}{3}\langle11\bar{2}0\rangle$, with short, leading screw segments and long, trailing edge segments. This dislocation geometry suggests that the deformation is due essentially to the glide of the screw segments, although their cuspate shapes clearly indicate some strong pinning action that hinders their motion. There are regions with high density of tangled dislocations separated by regions with a lower density. Closer to the central zone of the microcrack-ladder, the density of dislocations with $\mathbf{b} = \frac{1}{3}\langle11\bar{2}0\rangle$ becomes more uniform, with lower maximum densities. Figure 9.22(b) shows that closer still to the central zone the microstructure changes to (i) looping dislocations

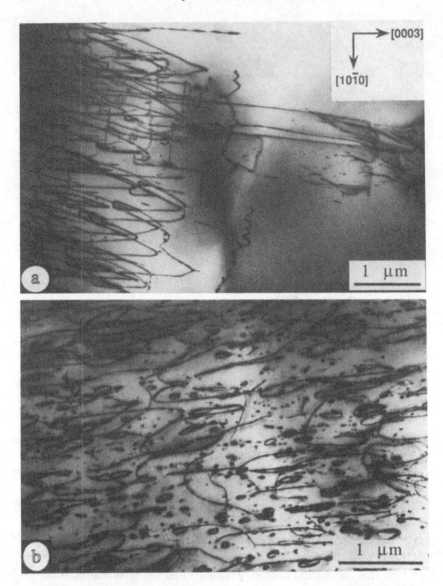

Figure 9.22. Dislocation microstructures observed at a microcrack-ladder in a single crystal of natural quartz experimentally deformed under conditions of high water fugacity (Mn₃O₄-buffer). In all micrographs, the electron beam is parallel to [$\bar{1}2\bar{1}0$] and $\mathbf{g} = 10\bar{1}1$. The loading direction [$10\bar{1}0$] and [0001] are marked.
(a) At the microcrack-ladder margin, BF image.
(b) Closer to the central zone, BF image.

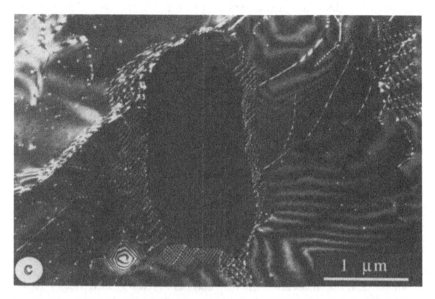

Figure 9.22. *Continued.* (c) Within about 10 μm of the central zone, DF image. (From Fitz Gerald et al. 1990.)

with $\mathbf{b} = \frac{1}{3}\langle 11\bar{2}0\rangle$, (ii) complete prismatic loops with $\mathbf{b} = \frac{1}{3}\langle 11\bar{2}0\rangle$, elongated parallel to [0001], and (iii) some small unidentified debris (perhaps small prismatic loops collapsing by climb). Overall, there is a decrease in the density of edge dislocations of $\mathbf{b} = \frac{1}{3}\langle 11\bar{2}0\rangle$ as the plane of the central zone is approached. Within 10 μm or so of the central zone, the dislocation microstructure changes completely. First, the dislocations have Burgers vectors $\mathbf{b} = [0001]$ and $\mathbf{b} = \frac{1}{3}\langle 11\bar{2}3\rangle$ as well as $\mathbf{b} = \frac{1}{3}\langle 11\bar{2}0\rangle$. Second, the microstructures show clear evidence of recovery, with well-organized networks of dislocations and small subgrains, as shown in Figure 9.22(c).

The trace of the planar central zone of the microcrack-ladder is marked by the presence of negative crystals and inclusions. This characteristic indicates that microcrack-ladders must be related to *healed* longitudinal fractures. The formation of unhealed microcracks normal to the loading direction (i.e., the "rungs" of the ladders) can be understood as follows. Undulatory extinction and dislocation microstructures observed within the microcrack-ladders indicate that plasticity is restricted to a narrow zone (≈ 100 μm wide) within the otherwise-undeformed host crystal. Therefore, the host crystal responded elastically while the load was applied; the *elastic* strain $\Delta l/l$ being about 0.5 percent. Now, in the plastic zone of the microcrack-ladder, the *plastic* strain ϵ produced at the same time is given by $\epsilon = \bar{\rho} b \bar{x}$, where $\bar{\rho}$ is the average dislocation density, b the magnitude of the "average" Burgers vector, and $\bar{x}$ the average distance the dislocations have moved. A plastic strain of 0.5 percent (equal to the elastic strain just estimated) requires $\bar{x}$ to be 10 μm, which is in reasonable order-of-magnitude agreement with the half-width of the regions of undulatory extinction and dislocation activity. It is clear that the stress in the specimen varies from place to place. In the microcrack-ladder zones, the internal stress is relaxed by plastic deformation while the surrounding crystal remains elastically stressed. When the load is removed at the end of the experiment, the undeformed host relaxes elastically, thereby stressing the microcrack-ladder zones. These tensile stresses cause multiple microcracks to propagate parallel to the specimen ends, but only within the plastically deformed zone of each microcrack-ladder.

Longitudinal recrystallized zones away from the extensively deformed end zones are all associated with healed fractures.

In general, the dislocation microstructures develop symmetrically about the healed fractures; and in the outer parts of the plastic zones, the cusp-shaped screw dislocation segments indicate that the direction of glide is overwhelmingly away from the fractured region. The dislocation density in the margins of the plastic zones also tends to increase toward the central fracture. Taken together, these observations indicate that the dislocation sources are near the healed fractures, that is, the planar central zones of microcrack-ladders. Unfortunately, the dislocation microstructures observed close to the healed fractures are extensively recovered, leaving no direct evidence as to the nature of the primary sources. However, mobile dislocations could be nucleated at or near a healed fracture (i) by segments of misfit dislocations in networks at the healed fracture acting as Frank–Read sources, (ii) by local stress concentrations at inclusions in the healed fractures being high enough to punch out new dislocations, and (iii) by plastic relaxation of stress at the crack tip (see Smith 1979).

In more uniformly deformed regions (which exhibit subgrains visible in the optical microscope), the dislocation microstructure is similar to that observed about 10 μm from the central zones of microcrack-ladders and does not vary significantly over distances of several hundred micrometers. The dislocation density inside large subgrains is low, but there is much fine debris. The dislocations have Burgers vectors $b = \frac{1}{3}\langle 11\bar{2}0 \rangle$ and $b = [0001]$ or $\frac{1}{3}\langle 11\bar{2}3 \rangle$.

9.5.3 Microstructures in specimens deformed under conditions of low water fugacity (Ta-*buffer*).   Although microcrack-ladders were observed in specimens preheated and peak-loaded under conditions of low water fugacity, they were rare compared with the Mn_3O_4-buffered specimens. Furthermore, none were observed in the one specimen that was deformed to about 10 percent strain (Figure 9.19). Longitudinal zones of recrystallization were also rare in this specimen. The most conspicuous deformation features observed optically were narrow lamellae, ≈ 5 μm wide and parallel to $\{10\bar{1}1\}$. TEM observations show that in the regions containing these lamellar structures, the microstructure (Figure 9.23a) consists of dense tangles of dislocations all with the same Burgers vector (one of the three possible $\frac{1}{3}\langle 11\bar{2}0\rangle$). The tangles were separated by regions of essentially zero dislocation density. The dislocation sources could not be positively identified within the tangles though there was an occasional indication of cracks within the tangles. At the margins of each tangle, screw dislocations have run out for a short distance, leaving behind long near-edge segments approximately parallel to [0001]. The screw dislocations exhibit many cusps, and trails of fine debris accompany most apices (Figure 9.23b). This geometry, like that observed at the margins of microcrack-ladders (see Section 9.5.2), suggests that the deformation is due essentially to the hindered glide of the screw segments away from dislocation sources deep within the tangle of dislocations.

9.5.4 Microstructural interpretation of water weakening of natural quartz.   The stress-strain curves (Figure 9.19) clearly demonstrate a correlation between the mechanical strength and the chemical environment imposed on initially strong (and essentially defect-free) crystal. The observed difference in the stress required to deform these specimens must be related to the nature, density, and behavior of the various types of crystal defects – in other words the microstructures – introduced during the preheating and deformation regimes. Overall, the *style* of the microstructure in longitudinal zones of plasticity and in recrystallized regions appears to be independent of the imposed chemical environment. However, the *volume fraction* of these deformed structures is related to the chemical environment, being large in the Mn_3O_4-buffered specimens (high water fugacity) and very small in the Ta-buffered specimens (low water fugacity). If we postulate that all recrystallization was initiated at plastic zones surrounding healed fractures, then it follows that one effect of high water fugacity is enhancement of the number and growth rate of the longitudinal

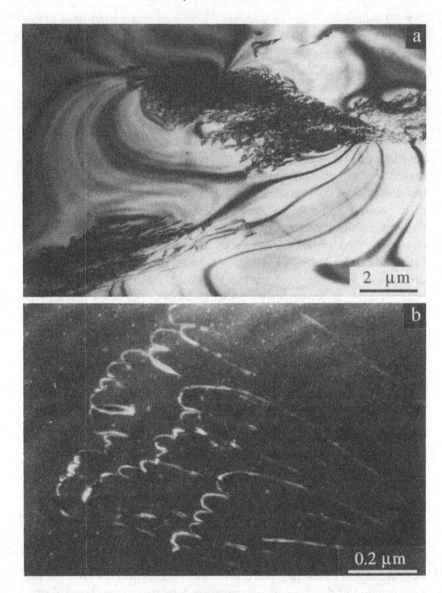

Figure 9.23. (a) BF image showing dense, localized tangles of dislocations in a single crystal of natural quartz experimentally deformed under conditions of low water fugacity (Ta-buffered). (b) Higher magnification WBDF image showing the dislocation arrangement at the edge of such a tangle.
(From Fitz Gerald et al. 1990.)

fractures that develop in the specimens, particularly at the ends where extensive recrystallization is observed (Figure 9.20). Thus, some *processing* of the lower ends of the specimens has occurred during the preheat and nominally elastic regimes, resulting in an easily deformable, weaker region. Even where recrystallization has not developed in the processed zone, the initially near-perfect crystal has been replaced by microstructures, consisting of recovery features (such as subgrains and dislocation networks) and dislocation loop structures, in which the dislocation density may be as high as 10^9 cm^{-2}. Thus, it appears that dynamic processes, including cracking, dislocation generation, dislocation glide, recovery, and recrystallization, were active during the preheat and nominally elastic regimes.

No such pervasive processing was observed in the Ta-buffered specimens although a few isolated microcrack-ladders and recrystallized bands were produced.

The microcrack-ladders in the Ta-buffered specimens were very similar to those observed in the Mn$_3$O$_4$-buffered specimens, suggesting that the microcrack-ladders were formed in the Ta-buffered specimens in the very early stages of the preheat before the buffer became fully active (i.e., under conditions of relatively high water fugacity). Thus, the dislocation configurations in microcrack-ladders in Mn$_3$O$_4$- and Ta-buffered specimens do not reveal any information about the relative mobility of dislocations under wet and dry conditions. However, the dislocations associated with the {10$\bar{1}$1} lamellae in the deformed Ta-buffered specimen (Section 9.5.3) were probbly nucleated under conditions of low water fugacity, since there is no evidence that they are associated with healed fractures. It is perhaps significant that these dislocations have not moved more than a few micrometers from their sources within the tangles, whereas the dislocations associated with the microcrack-ladders have moved 100 μm or so from their sources near the central zone. In neither case is it clear what limits the glide distance of the dislocations from their sources. It is conceivable that the glide distance could be related to a concentration gradient of water-related point defects (Griggs mechanism, Section 9.4), arising either from bulk diffusion in the quartz matrix or from pipe diffusion along dislocation cores from fluid inclusions at the healed fractures. However, lack of knowledge of the critical concentration of water-related defects that might control dislocation mobility, and of the precise values of the relevant diffusion parameters, make further speculation pointless. Nevertheless, it is clear that dislocation climb is enhanced near the fluid inclusions in the central zone of a microcrack-ladder and that at the margins of a microcrack-ladder the dislocations move predominantly by glide.

The microstructural observations described here highlight the complex interaction of brittle and ductile deformation processes that lead to the bulk mechanical weakening of initially strong, single crystals of natural quartz under conditions of high water fugacity. Perhaps the most important single aspect to arise from these observations is the nature of the processing of the specimen that occurs during the preheat and nominally elastic regimes prior to the onset of macroscopic deformation. The processing begins with longitudinal fracturing, which is subsequently overprinted by crack healing, the creation of a population of active dislocation sources, and numerous dislocations that can act as secondary sources of dislocation multiplication, microcrack-ladder structures, and zones of recrystallization. Whereas the processing is pervasive in the Mn_3O_4-buffered specimens (due to a high density of fractures), it is restricted to small localized regions in Ta-buffered specimens. Thus, in the former specimens, the bulk strength is determined by the relatively low strength of the large volume fraction of processed material in which there is a high density of dislocation sources and relatively mobile dislocations. On the other hand, the bulk strength of the Ta-buffered specimens is similar to that of the original crystal because there are only a few localized weak regions. The main effect of the high water fugacity appears to be in nucleating mobile dislocations via fractures and crack-healing, rather than in increasing the concentration of charged point defects as in the Hobbs (1985) model.

Although these microstructural observations provide no new information about the possible effect of water fugacity on dislocation glide, the observed enhancement of dislocation climb in the neighborhood of the fluid inclusions in the central zones of the microcrack-ladders is consistent with the earlier suggestion (Section 9.4.3) that bubbles in "wet" crystals promote climb by acting as sources and sinks for Si and O. Thus, in specimens with a reasonable density of mobile dislocations ($\approx 10^8$ cm^{-2}), the strength (at a given percent strain beyond the yield point) is expected to be reduced by the presence of water because of the enhancement of climb (Tullis and Yund 1989).

9.6 Deformation of feldspars

Because the feldspars are the most abundant minerals of the Earth's crust, their mechanical properties have been investigated in some detail. A review has been provided by Gandais and Willaime (1984). Thus, only a brief summary need be given here. However, the dislocations observed in deformed feldspars have some special characteristics and warrant some discussion.

The macroscopic properties of feldspars have been deduced from constant strain-rate experiments with both single crystals and polycrystalline specimens. The mechanical behavior is similar for all feldspars. Single crystals have several soft orientations, and (with confining pressures in the range 0.5–1.5 GPa and strain-rates from 10^{-5} to 10^{-6} s^{-1}) there is a brittle-to-ductile transition at 700°–800°C. There is some evidence for a decrease in strength when the deformation is carried out in the presence of water, but the mechanism is not as yet understood. The recent observations of Fitz Gerald et al. (1990), which indicate the role of fractures in the mechanism of the water-weakening of natural quartz (see Section 9.5), may also be relevant to the feldspars. All experiments show a trend for polycrystalline specimens to be stronger than their single crystal equivalents.

TEM observations of experimentally deformed specimens indicate that the dislocation arrangements can be broadly classified into two groups, depending on the deformation temperature. In the low temperature range (600° < T < 800°C), the dislocations tend to be concentrated into long narrow bands or walls separated by areas free of dislocations, as shown in Figure 9.24(a). At higher temperatures (T > 800°C), the dislocation distribution is more homogeneous and recovery features develop, manifesting the importance of climb processes. In highly strained specimens, recrystallization occurs at the high temperatures.

Burgers vectors and slip systems have been determined from TEM observations and structural considerations. In specimens deformed at low temperatures (700°–800°C), each dislocation is generally associated with an extended planar defect, see Figure 9.24(b). Marshall and McLaren (1977a, b) identified two kinds of fault. Faults of the first kind were out-of-contrast for all **g** parallel to the plane of the fault and in contrast (although always weak) for all other **g**. This implies that the fault vector **R** is normal to the fault plane and the fault is due to dilation or compression across the fault plane. The low contrast observed suggests that |**R**| is a small fraction of a lattice parameter. More detailed analysis of the fringe contrast of such faults by Gandais et al. (1987) has shown that **R** = 0.015[100] for dislocations belonging to the slip system (010)[101], for example. The second kind of fault is characterized by a fault vector that lies in the plane of the fault. The faults of this type studied by Marshall and McLaren (1977a, b) were always parallel to ($h0l$) with **R** = (1/x)[010], where x is not an integer. However, Gandais et al. (1987) observed faults

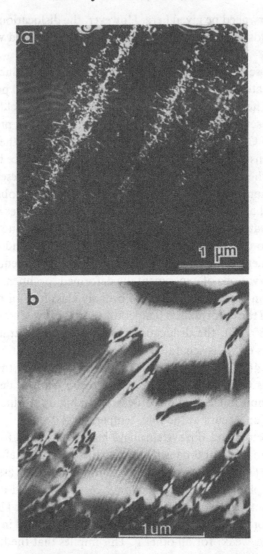

Figure 9.24. (a) WBDF image ($g = 004$; $s = 0.04$ nm^{-1}) showing the dislocations arranged in narrow bands or walls whose orientation is about 15° from (001) in deformed plagioclase feldspar (An$_{50}$; disordered C$\bar{1}$ structure). (b) DF image ($g = 004$; $s = 0$) showing dislocations and their associated slip plane faults in deformed plagioclase feldspar (An$_{95}$). (From Marshall and McLaren 1977a, b.)

of this type on other planes and showed, for example, that dislocations of $\mathbf{b} = [101]$ can be associated with faults on (12$\bar{1}$) with $\mathbf{R} = 0.015[101]$. It appears that both types of fault (regardless of the detailed structure) do define the slip plane of their associated dislocation. Thus, these faults can be used for determining the slip plane of an associated dislocation whose Burgers vector can be determined by the usual contrast criteria. The faults are probably associated with small configurational changes of the $(Si, Al)O_4$ tetrahedra near the slip plane. The observation that the faults are less conspicuous in specimens deformed at higher temperatures (900°–1,100°C) suggests that thermal vibrations help the tetrahedra to return to their normal positions after the passage of a dislocation.

At least a dozen slip systems have been identified by TEM in experimentally and naturally deformed feldspars (see Gandais and Willaime 1984). In many cases, the dislocations are dissociated, though the separation of the partial dislocations is usually small (≤ 50 nm). The dissociation of dislocations of $\mathbf{b} = [100]$ gliding in (010) in experimentally deformed sanidine was first observed by Kovacs and Gandais (1980), who suggested the following reactions:

$$[100] = \tfrac{1}{2}[100] + \tfrac{1}{2}[100]$$

or

$$[100] = \tfrac{1}{2}[101] + \tfrac{1}{2}[10\bar{1}]$$

Later, Gandais and Strunk (1983) observed that dislocations of $\mathbf{b} = [100]$ and $\mathbf{b} = [101]$ gliding in (010) tended to interact according to the reaction

$$[100] + [101] = [201]$$

followed by the dissociation

$$[201] = \tfrac{1}{2}[201] + \tfrac{1}{2}[201]$$

Olsen and Kohlstedt (1984) analyzed the dislocations in some naturally deformed intermediate plagioclase feldspars. All the known Burgers vectors except $\mathbf{b} = [100]$ were identified, and most, perhaps all, dislocations were dissociated by up to 20 nm. The microstructure was dominated by screw dislocations of $\mathbf{b} = [001]$, which had dissociated in (010) probably according to the reaction

$$[001] = \tfrac{1}{2}[001] + \tfrac{1}{2}[001]$$

first suggested by Kovacks and Gandais (1980).

More recently, Montardi and Mainprice (1987) made a detailed TEM study of dislocations in naturally deformed calcic plagioclases (An_{68-70}). As in the specimens studied by Olsen and Kohlstedt (1984), the microstructure was dominated by the slip system (010)[001]. The [001] dislocations dissociated according to the reaction just given, the separation being about 50 nm. The pair of gliding partial dislocations left behind a fault characterized by fringes of low contrast, as previously discussed. Image calculations with $\mathbf{R} = \frac{1}{30}[001]$ confirm that $|\mathbf{R}|$ must be very small. It is not surprising, therefore, that the third partial dislocation (which, presumably, must exist in order to restore the crystal structure at the opposite end of the fault from the two closely spaced leading partials) is not seen because its Burgers vector is also very small.

Finally, a brief mention should be made of the recent attempt to model the core structure of $(010)[001]$ dislocations in alkali feldspars using the experimental observation (Zheng and Gandais 1987a) that these dislocations are dissociated in the manner already described. Zheng and Gandais (1987b) showed that if the associated stacking fault on (010) with $\mathbf{R} = \frac{1}{2}[001]$ was placed within the double crankshaft of the $(Si,Al)O_4$ tetrahedra, tetrahedral entity is preserved and the $(Si,Al)-O$ bonds can be reconstructed in the core structures of the partial dislocations. As an extension of this work, Heggie and Zheng (1987) investigated the possible dislocation and planar defect structures by atomistic computer modeling and concluded that the model proposed by Zheng and Gandais (1987b) was the most energetically favorable dislocation structure.

9.7 Deformation of carbonates

Calcite and dolomite form large parts of the sedimentary continental crust. Consequently, their mechanical properties have been studied in some detail, and Wenk et al. (1983) have reviewed the rheology and associated microstructural development. Deformation takes place by both twinning and dislocation glide.

9.7.1 Microstructures in deformed calcite.   Single crystals of calcite compressed at room temperature without a confining pressure usually fail by fracture. However, Barber and Wenk (1979) observed that with a confining pressure of 1 GPa, high twin densities are developed in suitably oriented specimens. At low strains (≈ 2 percent), the deformed (sheared) twin lamellae are normally planar and parallel, having a thickness (≈ 1 μm) which is comparable to the untwinned matrix lamellae. The twin boundaries commonly contain closely spaced arrays of twinning dislocations, and the twinned lamellae usually contain numerous glide dislocations generated by the twinning. The untwinned matrix may contain only a low density of preexisting dislocations. These features are shown in Figure 9.25. However, deformation twinning can generate glide dislocations in both the twin lamellae and the matrix. The generation of dislocations by twinning is due to the need to relax the intense stresses produced by the

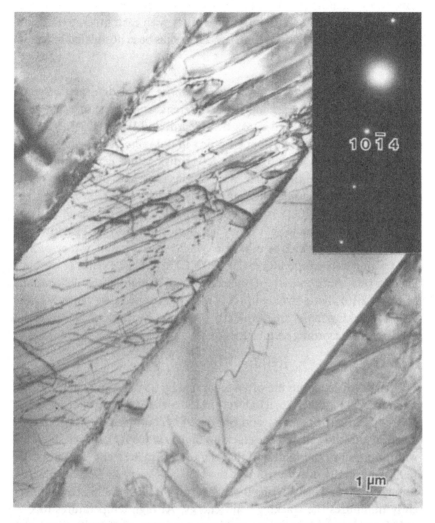

Figure 9.25. Glide dislocations and dislocation dipoles within a twin lamella in calcite compressed normal to (10$\bar{1}$0) to 10 percent strain at 25°C with 1 GPa confining pressure. (From Barber and Wenk 1979.)

shape change at the surfaces near the tip of a moving twin lamella. Multiple twinning (twinning on more than one plane) is common in calcite deformed at high temperatures (600°C), and high densities of tangled dislocations are generated by the passage of one twin lamella through another.

The slip systems {10$\bar{1}$4}⟨$\bar{2}$021⟩ have been identified in calcite over the temperature range 300°–500°C. TEM observations (Wenk et al. 1983) have shown that the dislocations are commonly curved, do not lie in well-defined planes, and tend to interact strongly with each other, even at the lower temperatures. Dislocation dissociation has been postulated but never observed.

9.7.2 Microstructures in deformed dolomite. The deformation characteristics of dolomite are markedly different from those of calcite and have been studied in detail by Barber, Heard, and Wenk (1981). Not only are the twin laws different, but twinning in dolomite occurs only at temperatures above about 250°C. The lower dislocation densities observed in twinned dolomite and at twin intersections is perhaps due to the greater ease of stress relaxation at the higher temperatures required for twinning.

Whereas $\{10\bar{1}4\}\langle\bar{2}021\rangle$ is the preferred slip system in calcite, it is so uncommon in dolomite that it has not been studied by TEM. However, $(0001)\langle2\bar{1}\bar{1}0\rangle$ slip, which is minor in calcite, is the major slip system at low-to-moderate temperatures (20°–300°C) in dolomite. The corresponding microstructure consists of numerous long, straight, screw dislocations of $\mathbf{b} = \frac{1}{3}\langle2\bar{1}\bar{1}0\rangle$, lying strictly in (0001). In the temperature range $\approx 225°$–500°C, stacking faults are frequently generated by the dissociation of the dislocations according to the reaction

$$\tfrac{1}{3}[2\bar{1}\bar{1}0] = \tfrac{1}{3}[1\bar{1}00] + \tfrac{1}{3}[10\bar{1}0]$$

The partial dislocations are often separated by up to several micrometers, as shown in Figure 9.26. High-resolution structure images observed at 500 kV (Barber, Freeman, and Smith 1983) confirmed this dissociation reaction and showed that the stacking faults on (0001) have considerable width normal to the fault and that the total displacement $|\mathbf{R}|$ across the fault is normally distributed over several (0006) layers. Delocalizing the displacement and distributing it over several sheets of the structure apparently minimizes the stacking fault energy. At high temperatures (600°–800°C), climbing dislocations and loop debris associated with the (0001) slip are observed, as shown in Figure 9.27. Cross slip of screw dislocations from (0001) to $\{\bar{1}012\}$ is also observed.

Slip on $\{\bar{1}012\}$ planes was first recognized in dolomite by Barber (1977) and was found to operate over a wide temperature range (200°–600°C). At temperatures near the middle of the range, slip on these planes creates long planar faults (Figure 9.28) which, at first sight, appear to be stacking faults and were described as such by Barber et al. (1981). However, the high-voltage high-resolution images of these faults later obtained by Barber et al. (1983) showed that their characteristics are novel: they are not

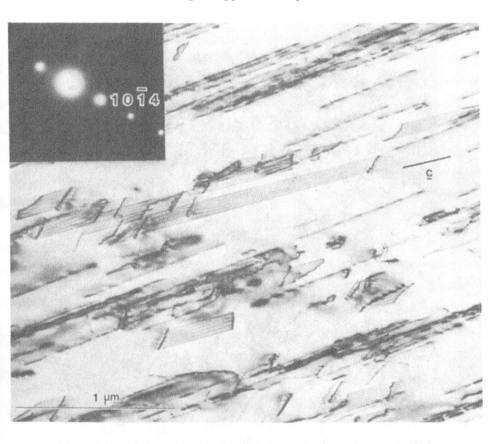

Figure 9.26. Stacking faults on (0001) induced by (0001) slip in dolomite deformed at 420°C. (From Barber et al. 1981.)

stacking faults as normally defined. The particular characteristics that lead to this conclusion are (1) the faults are rarely seen to terminate at partial dislocations; (2) the faults commonly do not terminate abruptly but attenuate over large distances and usually have no beginning or end; and (3) no displacements of lattice planes are observed at nearly all faults. What then are the faults, and what makes them visible in HVHRTEM images? Barber et al. (1983) suggested that the contrast shown by the faults is due to rotational disorder of the CO_3^- groups and that there is no antiphase relationship across the faults in either the cation or anion sublattices. The fault fringes are often very weak, and the faults are interpreted as relics or "memories" of planes over which slip has occurred. The following mechanism was proposed. The passage of a partial dislocation of

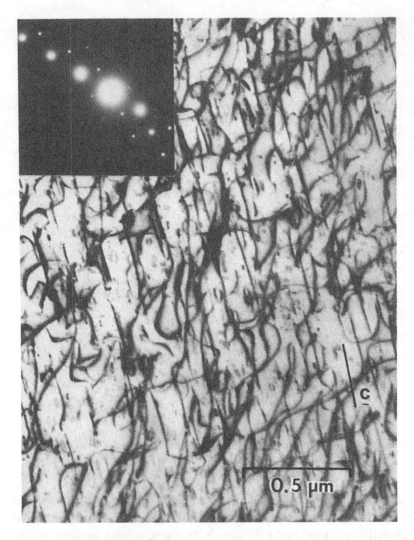

Figure 9.27. Climbing dislocations and loop debris associated with (0001) slip in dolomite deformed at 800°C. (From Barber et al. 1981.)

$\mathbf{b} = \frac{1}{6}[241] \equiv \frac{1}{6}[02\bar{2}1]$ along the slip plane $(\bar{1}02) \equiv (\bar{1}012)$ produces an APB (Reeder and Nakajima 1982), across which there is a rotation of 180° of the CO_3^- groups. As far as the cation sublattice only is concerned, the fault has created a thin layer of twin of low energy. However, across the fault, the CO_3^- groups are in a very high-energy configuration. The CO_3^- groups along the fault, and in several neighboring $(\bar{1}012)$ planes on either

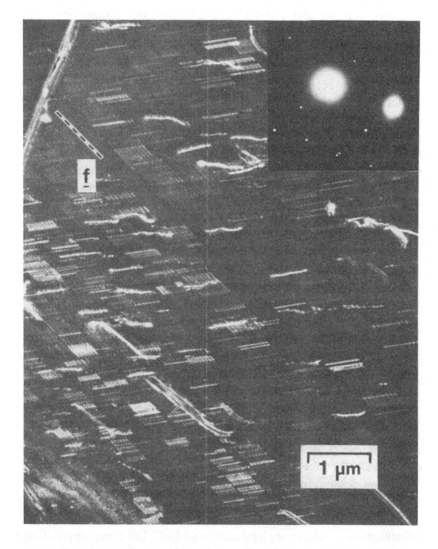

Figure 9.28. Fringe patterns due to faults associated with slip on ($\bar{1}012$) in dolomite deformed at 400°C. (From Barber et al. 1981.)

side, rotate away from their proper orientations. If some of the CO_3^- groups remain in the wrong orientations, then a relatively low-energy fault is formed, which marks the slip plane of the original dislocation. These faults are very similar to the slip plane faults observed behind gliding dislocations in deformed feldspars (Section 9.6).

At temperatures above 500°C, dislocation climb becomes apparent, and extensive cross slip of undissociated screw dislocations from {$\bar{1}012$} to (0001) takes place.

9.7.3 Microstructure and rheology. Perhaps the most interesting rheological characteristic of dolomite is the *increase* in strength (initial yield stress) with increasing temperature for (0001) slip. The rate of work-hardening also tends to increase with temperature for this slip system. Barber et al. (1981) have discussed these effects and the geological implications in detail. The origin of the work-hardening effect is not clear, but they suggest that the increase in strength with temperature is associated with the increase in thermal motion (translation and rotation) which increases the repulsion between passing CO_3^- groups and obstructs dislocation glide.

9.8 Deformation of olivine

Because olivine is one of the major components of the Earth's upper mantle, studies of the mechanical properties of this mineral are likely to provide important information about the dynamical processes taking place in the lithosphere and asthenosphere. Consequently, many rheological and associated microstructural investigations have been made of both naturally and experimentally deformed olivine single crystals and olivine-rich rocks, as well as of experimentally deformed aggregates of natural olivine and single crystals of synthetic forsterite. Although some microstructural investigations have used TEM exclusively, many recent studies have been made almost entirely at the optical microscope level using the dislocation decoration technique first developed by Kohlstedt et al. (1976). One important feature of this technique is that large volumes of specimen can be examined. However, since Burgers vectors cannot be determined directly, its use is often supplemented by limited TEM observations. Consequently, microstructural studies such as that of Durham et al. (1977), that are based on the dislocation decoration technique, are not directly relevant to the main aim of this chapter, even though they have provided much important information about the deformation of olivine. Furthermore, the basic elements of the complex microstructures that develop and the processes involved are not fundamentally different from those already illustrated in preceding sections. Therefore, the microstructures observed by TEM in deformed olivine will be reviewed relatively briefly.

9.8.1 Deformation of pure synthetic forsterite.   The deformation of single crystals of pure synthetic forsterite has been investigated in creep by Durham, Froidevaux, and Jaoul (1979) and Darot and Gueguen (1981) and in constant strain-rate experiments by Kashima, Sunagawa, and Sumino (1983). In all cases, the as-grown crystals had low dislocation densities of the order of 10^3-10^4 cm^{-2}. The distribution of dislocations was fairly uniform, there being little or no subgrain development. Burgers vectors $\mathbf{b} = [100]$ and $[001]$ were identified by Kashima et al. (1983) using x-ray topography.

Durham et al. (1979) found that even at temperatures as high as $T = 1{,}750°C$, the as-grown crystals did not deform at stresses $\sigma < 10$ MPa, but that strain was readily induced at $\sigma > 10$ MPa. Furthermore, specimens that had been prestrained a few percent subsequently deformed readily at $\sigma < 10$ MPa. It appears, therefore, that dislocations are generated from the grown-in dislocations at $\sigma > 10$ MPa (presumably by a Frank–Read type mechanism) and that these fresh dislocations are readily mobile at $\sigma < 10$ MPa. The large yield drops observed by Kashima et al. (1983) in experiments at constant strain-rates of 10^{-4}-10^{-5} s^{-1} at $T \leq 1{,}500°C$ are consistent with the creep behavior. TEM observations showed that specimens stressed to the upper yield point developed a reasonably uniform density of dislocations and small loops. However, during the stress drop from the upper to the lower yield points, the dislocations tend to become arranged into very low-angle subgrain boundaries. Subgrain development becomes more pronounced at higher strains (≈ 30 percent). Little or no work-hardening is evident in the stress-strain curves.

The dislocation microstructures that develop during creep with stresses in the range $15 < \sigma < 110$ MPa at temperatures from $1{,}400°$ to $1{,}600°C$ were studied in detail by Darot and Gueguen (1981) on the optical microscope scale using the dislocation decoration technique as modified by Jaoul et al. (1979). However, TEM was used to determine Burgers vectors by application of the $\mathbf{g} \cdot \mathbf{b} = 0$ and $\mathbf{g} \cdot \mathbf{b} \times \mathbf{u} = 0$ invisibility criteria (Section 5.6).

The observed mircrostructures and rheological behavior are consistent with the suggestion by Durham, Goetze, and Blake (1977) that the dislocations glide essentially unhindered by other dislocations and that deformation is limited by the number of active dislocation sources.

9.8.2 Deformation induced by indentation.   The deformation induced in single crystals of San Carlos (Arizona, USA) olivine around the indenter in a Vickers microhardness test has been studied by Gaboriaud et al. (1981) and Gaboriaud (1986) over the temperature range $20° < T <$ 1,100°C. The crystals were oriented so that the slip systems $(100)\langle001\rangle$ and $(001)\langle100\rangle$ were symmetrically activated. With the loads used (50–500 g) the critical resolved shear stress ranged from about 200 MPa to 2 GPa and was applied for 20 seconds. The resulting microstructures in and around the indentation were examined by optical microscopy after decorating the dislocations and by TEM.

Both fracturing and plasticity were observed over the whole temperature range. However, at room temperature, the dislocation density was rather low and dislocations were not observed around the fractures. This implies that either dislocations are not nucleated at the tip of the crack or that the stress concentration is high enough to nucleate but not to propagate dislocations. Above 600°C, the fractures are surrounded by a plastic zone: dislocation loops form at crack terminations and glide into the surrounding crystal. TEM contrast experiments indicate that all the dislocations have $\mathbf{b} = [001]$, even though this is not the shortest lattice vector. Two slip systems operate: $(100)[001]$ and $\{110\}[001]$. The dislocation loops are elongated parallel to $[001]$, so it is the short edge segments that are gliding freely, leaving long screw segments. The dislocation geometry is similar to that observed at the margins of the microcrack-ladders in deformed natural quartz, but there the screw segments of the expanding loops were gliding, leaving long edge segments behind; see Figure 9.22(a).

9.8.3 Experimental deformation of single crystals of natural olivine. The first detailed TEM investigation of experimentally deformed single crystals of natural olivine was made by Phakey, Dollinger, and Christie (1972). Gem-quality single crystals in which the density of grown-in dislocations was very low ($\rho \approx 10$ cm^{-2}) were deformed to strains of 6–12 percent at strain-rates of 10^{-4} and 10^{-5} s^{-1} at temperatures of 600°, 800°, 1,000°, and 1,250°C under a confining pressure of 1 GPa. The dislocation densities in the deformed specimens were in the range 10^8–10^{10} cm^{-2}, generally $\rho \approx 5 \times 10^9$ cm^{-2}. Obviously, rapid nucleation and multiplication of dislocations took place during deformation, but the high strains eliminated any evidence of the dislocation sources. However, because fractures were observed in most specimens, it seems likely that dislocations were nucleated in association with fractures, as has been observed around indentations in microhardness tests (Section 9.8.2) and in natural quartz (Section 9.5). Dislocations with Burgers vectors $\mathbf{b} = [001]$, $[010]$, and $[100]$ were identified using the $\mathbf{g} \cdot \mathbf{b} = 0$ and $\mathbf{g} \cdot \mathbf{b} \times \mathbf{u} = 0$ invisibility criteria. In most of the specimens, the dislocations were long, straight screws. It is possible that these are the segments of loops that have expanded by the preferential glide of the edge segments. Where segments of edge dislocations were observed, it was often possible to determine the slip plane and hence the slip system. Tangles of dislocations were observed in specimens deformed at 800° and 1,000°C, as well as cross-slip of the screw segments. At 1,000°C, climb was suggested by the presence of curved dislocations and the formation of small loops and other debris. Enhanced climb at 1,250°C was indicated by the observation of helical dislocations (see Hull and Bacon 1984, p. 63) and subgrain boundaries. In summary, the TEM observations show that dislocation glide is general at 600°C; cross-slip occurs at 800°C and is marked at 1,000°C; climb begins somewhat below 1,000°C and leads to extensive recovery and recrystallization at 1,250°C.

More recently Mackwell, Kohlstedt, and Paterson (1985) studied the deformation of single crystals of San Carlos (Arizona) olivine deformed under hydrous conditions at 1,300°C, 300 MPa confining pressure, and 10^{-5} s^{-1} strain-rate and found they were a factor of 1.5–2 weaker than those deformed in an anhydrous environment. TEM observations showed that specimens deformed under dry conditions, in an orientation such that the slip systems (001)[100] and (100)[001] would be activated, were characterized by a microstructure of generally curved dislocations and dislocation loops, but no organization into walls. The dislocation density was 10^8–10^9 cm^{-2} compared with an initial value of $<10^6$ cm^{-2}. Most of the dislocations and the loops lie approximately in the (010) plane; because they are in contrast for $g = 004$, they probably have $b = [001]$; dislocations with $b = [010]$ and [100] would be out-of-contrast for this reflection. However, the slip system (010)[001] is not expected to be active. It is not clear, therefore, if these dislocations are actually involved in the deformation. The general geometry of the dislocation microstructure is not inconsistent with some climb mobility; in fact, on the basis of the observations of Phakey et al. (1972), climb is certainly expected at 1,300°C.

Mackwell et al. (1985) found that when specimens that had been deformed under anhydrous conditions were subsequently further deformed under wet conditions, there was a significant change in microstructure. TEM observations revealed enhanced formation of dislocation walls, despite the reduced stress levels. This observation was interpreted as due to enhanced dislocation climb under wet conditions. However, the two walls illustrated by Mackwell et al. (1985) could be interpreted as healed or partly healed fractures. One wall consists of a very irregular network of dislocations with many bubbles, particularly at dislocation intersections.

These bubbles (up to 0.3 μm in diameter) were observed only in specimens deformed under wet conditions, so they almost certainly contain water. It is difficult to conceive how this water could have penetrated the crystal other than along an open fracture that later healed. The second wall is characterized by typical fault fringes (Section 5.3) and a fine structure that could be due to moiré fringes (Section 6.6) or an array of closely spaced dislocations (Section 8.9). The extremely irregular shape of the wall suggests that it may originally have been a crack. An increase in the density of cracks in a specimen deformed under wet conditions is consistent with the recent observations of Fitz Gerald et al. (1990) of the microstructures in natural quartz deformed under wet and dry conditions (Section 9.5). However, in the olivine specimens under discussion, there does not appear to have been any significant mobility of dislocations away from the walls. The suggestion of Mackwell et al. (1985) – that the walls observed in these specimens are evidence for enhanced dislocation climb under wet conditions and that this is the cause of the relative weakness of such specimens – must be treated with caution in the absence of more detailed microstructural information (see also Section 9.8.4).

9.8.4 Experimental deformation of dunite and olivine aggregates. Deformation experiments have been carried out by Chopra and Paterson (1981, 1984) on dunites from Anita Bay (New Zealand) and from Aheim (Norway) at strain-rates of 10^{-3}–10^{-6} s^{-1} and temperatures from 1,000° to 1,400°C at 300 MPa confining pressure. Both rocks contain traces of layer silicate minerals that dehydrate at the test temperatures, so water is normally present during deformation. To investigate the deformation under dry conditions, the specimens were preheated at 1,200°C for periods of 60 hours or longer in a controlled oxygen-fugacity furnace in order to dehydrate the hydrous silicates and drive off adsorbed water. Subsequently, Fitz Gerald and Chopra (1982) and Doukhan et al. (1984) examined the olivine grains in the deformed specimens (as well as in the starting material before and after heat-treatment) by TEM and related the observed microstructures to those observed in the optical microscope.

The matrix olivine grains, as distinct from the porphyroclast grains (Boland 1977; Boland et al. 1971; see also Section 8.9), in the Anita Bay dunite have an average grain-size of 100 μm and show few signs by light microscopy of prior deformation. However, in TEM, the dislocation density is observed to vary from about 5×10^7 cm^{-2} to 3×10^9 cm^{-2}. The dislocations are rarely organized, apart from a few rough (100) walls, and they vary from straight (usually parallel to [001]) to sharply curved and tangled, with some small loops in places. Most dislocations have Burgers vectors $b = [100]$ and $[001]$, the latter usually in pure edge orientation. The olivine grains in the Aheim dunite have an average grain size of 900 μm and appear in the light microscope to be essentially undeformed. The dislocation density is probably about 10^6 cm^{-2}.

In the preheated specimens, TEM reveals many well-organized dislocation walls parallel to (100), with fewer parallel to (010) and (001), in which the dislocations are spaced at about 50 nm. Within the cells defined by these walls, the free dislocation density is about 10^8 cm^{-2}. Small bubbles (<300 nm in diameter) are observed in some of the walls, suggesting that the walls may be healed cracks.

Deformation produces microstructures that have totally overprinted those originally present in both the wet and dry specimens. The TEM observations show that deformation has occurred by the glide of dislocations of $b = [100]$ and $b = [001]$ in all specimens, wet or dry, and that the microstructures are far more complex than the kink bands observed by light microscopy would indicate. Presumably, the dislocations involved in the deformation have been derived from the preexisting dislocations. In the dry specimens, there is a fairly homogeneous distribution of free screw dislocations of $b = [100]$ and $[001]$, the densities of both types being around 10^9 cm^{-2}. In wet specimens deformed at 1,000°C, the microstructure is not homogeneous. Many grains exhibit a microstructure similar to that just described, but other grains show well-organized cell structures with very low densities of free dislocations between the cell walls. This latter microstructure is typical of the wet specimens deformed at higher temperatures and suggests that the water may have enhanced dislocation climb, though this has not been established by the present observations. Enhanced dislocation climb could partially account for the lower strength of the wet specimens, as observed by Chopra and Paterson (1981, 1984). However, their studies of the influence of water and grain-size on strength suggest that water may have more influence on grain-boundary behavior than on deformation processes within the grains. Unfortunately, the TEM observations provide no direct information on the former.

The influence of grain-size and water on the rheology of hot-pressed aggregates of natural olivine has been investigated by Karato, Paterson, and Fitz Gerald (1986). The rheological results suggest that two deformation mechanisms may be operating, depending on the stress level and grain size. At relatively high stress and large grain size, the strain-rate $\dot{\epsilon}$ is proportional to about the cube power of the stress σ and is nearly independent of grain size. At low stress and small grain size, $\dot{\epsilon}$ depends almost linearly on σ and decreases markedly with increase in grain size. In both regimes, water was found to enhance the creep rate.

In the grain-size insensitive regime (high stress, large grain size), the microstructures observed by light microscopy and by TEM indicate that the deformation is intragranular involving dislocation motion in both wet and dry specimens. The dominant structure is (100)-organization of dislocations of $\mathbf{b} = [100]$, though in the free dislocation population, numerous dislocations of $\mathbf{b} = [001]$ are also present in densities roughly equal to those of the dislocations with $\mathbf{b} = [100]$. Well-recovered structures, low-angle (100) and (001) tilt walls and (010) twist walls, develop in both wet and dry specimens, indicating dislocation climb motion. Also present are nonplanar walls in which there are variable densities of dislocations of mixed character with $\mathbf{b} = [100]$ and $[001]$; see Figure 8.33(b). Free dislocations, many in near-screw orientation, interact weakly and have gently curving shapes. TEM contrast experiments indicate that there is a low density of dislocations with $\mathbf{b} = [010]$, both free and in walls.

The microstructures observed by TEM in specimens deformed in the grain-size sensitive regime (low stress, small grain size) are different from those just described in some quite distinct ways. Dislocation microstructures have developed during deformation, but the dislocation density is extremely heterogeneous and, on average, a little lower than observed in specimens deformed in the grain-size insensitive regime. The majority of dislocations have $\mathbf{b} = [100]$ or $[001]$. There is evidence of recovery, subgrain formation, and mobility of dislocations. However, in both wet and dry specimens, unlike those from the grain-size insensitive regime, there are considerable volumes of olivine in which the dislocation density is relatively very low (5×10^6 cm^{-2}). If these volumes of low dislocation density had been present during deformation, as suggested by Karato et al. (1986), then some other deformation mechanism, in addition to dislocation mobility, must be operating in the low stress regime. Karato et al. (1986) suggest that this other mechanism is diffusion creep. The low stress exponent $n = 1$–2 ($\dot{\epsilon} \sim \sigma^n$) is consistent with this suggestion (Nicolas and Poirier 1976, pp. 143–7).

The water-weakening effect observed in the stress-strain curves of specimens deformed in either regime is not reflected in the observed microstructures. This may be because the small difference in strength of wet and dry specimens (only a factor of about 2) is a consequence of only a small change in the relative activities of different deformation mechanisms, such as glide versus climb or dislocation mobility versus diffusion-controlled creep. However, these observations are not consistent with the assertion of Mackwell et al. (1985) that recovery due to climb is significant only in specimens deformed under wet conditions (see Section 9.8.3). It is clear that the water-weakening in olivine is not yet fully understood.

9.9 Deformation of pyroxenes

The pyroxenes are the most abundant minerals, after olivine, in peridotites, which are the dominant constituents of the upper mantle. It is not surprising, therefore, that there has been considerable interest in the mechanical properties of the pyroxenes, and a review has recently been given by Doukhan et al. (1986). The orthorhombic pyroxenes deform by slip and by a shear transformation that produces monoclinic lamellae (one or a few unit cells thick) parallel to (100). Coe and Kirby (1975) and McLaren and Etheridge (1976) have shown that the shear transformation is achieved by the glide of partial dislocations of $b = 0.83[001]$ in (100), which leave partial dislocations of $b = 0.17[001]$ terminating the shear lamellae. The dominant slip system is (100)[001]. Recent TEM observations by van Duysen, Doukhan, and Doukhan (1985) suggest that the dislocations associated with this slip system may be dissociated into four partials and that the slip system (100)[010] may also be activated. These observations are discussed in Section 9.9.1.

The deformation microstructures of monoclinic pyroxenes are considered in Section 9.9.2. Optical microscope observations (Griggs, Turner, and Heard 1960; Raleigh 1965) indicate that the dominant slip system in monoclinic pyroxenes is also (100)[100]. However, van Duysen and Doukhan (1984) found by TEM that in naturally deformed α-spodumene the activated slip systems are $\{110\}[001]$ and $\{1\bar{1}0\}\frac{1}{2}\langle110\rangle$. In specimens of α-spodumene deformed by scratching, they also observed interesting microstructures of dislocations and faults that may be related to the twins observed in deformed diopside by Kirby and Christie (1977).

9.9.1 Dislocations in deformed orthopyroxene. Van Duysen et al. (1985) studied the dislocations generated in a gem-quality crystal of enstatite ($Mg_{1.8}Fe_{0.2}Si_2O_6$) by scratching with a Vickers diamond microhardness indenter. The initial dislocation density was less than 10^5 cm^{-2}. The scratched surface was (301), so the easy glide plane (100) was at 45° to the surface. TEM revealed two sets of dislocations lying in (100). One set is in contrast for $g = 020$, 060, and 440 and out-of-contrast for $g = 202$ and $20\bar{2}$, suggesting $b = [010]$. The second set of dislocations is in contrast for $g = 202$ and $20\bar{2}$ and out-of-contrast for $g = 020$, 060, and 440, suggesting $b = [001]$. These characteristics are illustrated in Figure 9.29(a, b). Note that the dislocations with $b = [001]$ shown in Figure 9.29(b) occur in pairs,

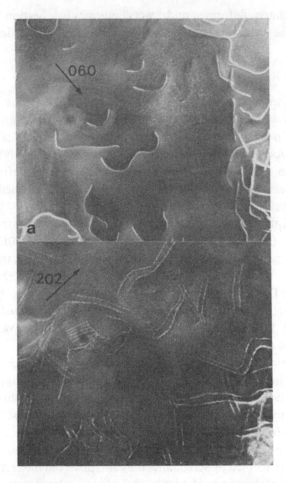

Figure 9.29. WBDF images of dislocations induced by scratching in a (301) foil of orthoenstatite. (a) Only dislocations of $b = [010]$ are in contrast with $g = 060$. (b) Same area imaged with $g = 202$. The dislocations of $b = [010]$ are out-of-contrast, and a second set of dislocations (which occur in pairs) are in contrast, suggesting $b = [001]$. (From van Duysen et al. 1985.)

the separation being about 540 nm. Higher magnification WBDF images, Figure 9.30(a), show that each dislocation of such a pair gives rise to a double image which, under these imaging conditions, almost certainly indicates two dislocations separated by about 10 nm. Van Duysen et al. (1985) interpreted these images as the dissociation of dislocations of $b = [001]$ into four partials bounding three stacking faults with fault vectors R_1, R_2, and R_3, as shown schematically in Figure 9.31. However, this interpretation is not justified on the basis of these images alone. It is at least

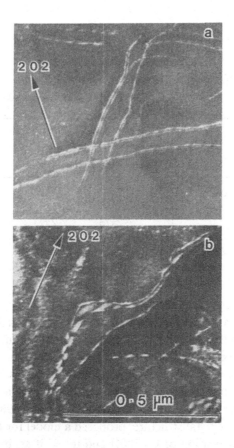

Figure 9.30. (a) Higher magnification WBDF images of the same type of dislocations as shown in Figure 9.29(b). Note that each dislocation of the pair consists of
two closely spaced dislocations. In (b), the separation between one pair of closely
spaced dislocations is large enough for fringes to be seen between them. This
suggests that the dislocations are partials. Compare with Figure 9.31.
(From van Duysen et al. 1985.)

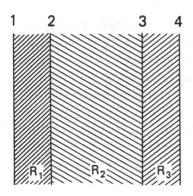

Figure 9.31. Schematic diagram showing
the dissociation of a dislocation into four
partial dislocations (1-4) bounding three
stacking faults with fault vectors R_1, R_2,
and R_3.

Table 9.1. *Calculated values of* $\alpha = 2\pi g \cdot R$ *for the stacking faults associated with the dissociation of a dislocation of* $b = [001]$ *in orthopyroxene*

g	$2\pi g \cdot R$		
	R_1	R_2	R_3
060	0	2π	2π
020	0	$2\pi/3$	$-2\pi/3$
440	0	$-2\pi/3$	$2\pi/3$
202	$4\pi\gamma \approx \pi$	2π	$2\pi(1-2\gamma) \approx \pi$
$20\bar{2}$	$\approx \pi$	2π	$\approx \pi$

necessary to image the faults separating the partial dislocations and to show that they have the characteristics of stacking faults; that is, they give rise to α-fringes (see Section 5.3.1). Furthermore, it is advisable to determine the fault vectors and show that they are compatible with the crystal structure. In the present case, this ideal procedure was not possible because only a limited number of suitable reflections were accessible in foils of the orientation used. Therefore, van Duysen et al. (1985), from an examination of the crystal structure, proposed a model for the dissociation in (100) of a dislocation of $b = [001]$ and tested it against the observed images.

It is clear that all fault vectors must be of the form $R = [0vw]$, and the proposed model (whose details need not concern us here) indicated that

$$R_1 = [0, \beta, \gamma]$$
$$R_2 = [0, v, \tfrac{1}{2}]$$
$$R_3 = [0, -\beta-v, \tfrac{1}{2}-\gamma]$$

with $v = \tfrac{1}{6}$, $\beta = 0$, and $\tfrac{1}{4} < \gamma < \tfrac{1}{3}$. Note that the operation $R_1 + R_2 + R_3$ brings the structure back into register, as required. Table 9.1 lists the calculated values of $\alpha = 2\pi g \cdot R$ for the diffraction vectors g that were used. The calculated values of α indicate that all faults will be out-of-contrast for $g = 060$, as observed in Figure 9.29(a), and that the central fault (R_2) will be out-of-contrast for $g = 202$, as observed in Figures 9.29(b) and 9.30(a). The outer faults (R_1 and R_3) should be in contrast for this reflection since $\alpha = \pi$. However, they are not observed in Figure 9.30(a). The reason may simply be that the partials in this part of the specimen are too close together. In some places, the separation is large enough for the fringes to be seen, as in Figure 9.30(b). The calculated values of α for $\mathbf{g} = 020$ and 044 indicate that the wide central stacking faults ($\mathbf{R}_2$) should be in strong contrast for these reflections. However, they were not observed. It is possible that fringes are not visible because these reflections are so weak that the extinction distance $1/\sigma$, even in WBDF images with large s, is greater than the thickness of the specimen (see Section 4.7). However, even if this were so, the stacking fault ribbon should still be observed as a change of contrast, just as APBs are visible with very weak superlattice reflections (Section 8.8). The failure of the model to predict the visibility of the relatively wide stacking faults suggests that other models should be considered. It is possible that the four partial dislocations arise from the dissociation of two parallel dislocations of $\mathbf{b} = [001]$ of opposite sign (dipoles) and that there is no stacking fault ($\mathbf{R}_2$) between the second and third partials.

9.9.2 Dislocations and faults in deformed monoclinic α-spodumene and diopside.

Van Duysen and Doukhan (1984) examined the dislocations in both naturally and experimentally deformed single crystals of α-spodumene. The naturally deformed specimens were characterized by a heterogeneous distribution of dislocations ($\rho \approx 10^8$ cm^{-2}). The dislocations are not dissociated and most have $\mathbf{b} = [001]$ and lie in the crystallographically equivalent planes (110) and (1$\bar{1}$0). Other dislocations, probably with $\mathbf{b} = \frac{1}{2}[110]$ and $\frac{1}{2}[\bar{1}10]$, were also observed and commonly interact to form irregular networks according to the reaction

$$\tfrac{1}{2}[110] + \tfrac{1}{2}[\bar{1}10] = [010]$$

Crystals deformed at a constant strain-rate ($\dot{\epsilon} = 10^{-6}$ s^{-1}) with a confining pressure of 300 MPa and 400°C in an orientation expected to activate the (100)[010] slip system, developed numerous microtwins in (100) and some dislocations that were not fully characterized. However, interesting dislocations and associated faults were observed in specimens scratched on a (110) surface. Figure 9.32 is typical of the dislocation microstructures observed in these specimens and shows segments of dislocation loops bounding planar defects on (100).

Van Duysen and Doukhan (1984) interpreted these features as stacking faults and partial dislocations due to the dissociation of dislocations of $\mathbf{b} = [010]$ in (100). This interpretation was based on the visibility of the dislocations and faults with different $\mathbf{g}$, as illustrated in Figure 9.33. They found that the dislocations were in contrast for $\mathbf{g} = 020$, 4$\bar{4}$0, 002, and 11$\bar{1}$,

Figure 9.32. WBDF image of the dislocations and faults on (100) observed in a (110) scratched foil of α-spodumene. (From van Duysen and Doukhan 1984.)

from which they concluded that $\mathbf{b} = x[0vw]$. The faults were in contrast for $\mathbf{g} = 002$ and $11\bar{1}$, but out-of-contrast for $\mathbf{g} = 020$ and $4\bar{4}0$, which is consistent with $\mathbf{R} = [0\frac{1}{2}w]$. Since, on this model, this fault vector must also be the Burgers vector of one of the partial dislocations, they concluded that dislocations with $\mathbf{b} = [010]$ were dissociated in (100) according to the reaction

$$[010] = [0\tfrac{1}{2}w] + [0\tfrac{1}{2}\bar{w}]$$

From the fringe contrast, they estimated w to be in the range $0.1 < w < 0.25$.

However, the observed fringe patterns could be due to microtwins (Sections 5.3.7 and 8.4) rather than stacking faults. Figure 9.34 shows how a fault vector $\mathbf{R} = \frac{1}{4}[001]$ can be introduced by a microtwin in (100) one unit cell wide, assuming that there is no relative displacement along [010] across the microtwin. The values of $\alpha = 2\pi\mathbf{g}\cdot\mathbf{R}$ calculated for the reflections used by van Duysen and Doukhan (1984) are listed in Table 9.2,

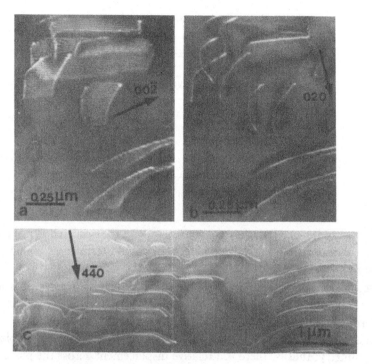

Figure 9.33. WBDF images of the dislocations and faults in α-spodumene observed with several different diffraction vectors. (From van Duysen and Doukhan 1984.)

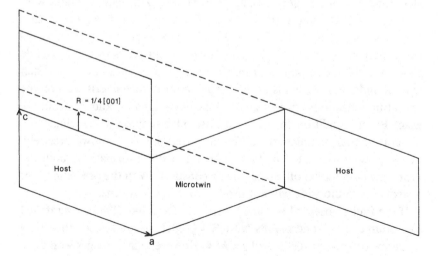

Figure 9.34. Diagram showing a (100) microtwin one unit cell wide in α-spodumene projected onto (010).

Table 9.2. *Contrast of* (100) *microtwins*
in α-*spodumene*

g	$\alpha = 2\pi \mathbf{g} \cdot \frac{1}{4}[001]$	Observed contrast*
020	0	O
002	π	X
$4\bar{4}0$	0	O
$11\bar{1}$	$-\pi/2$	X

* O = out-of-contrast; X = in contrast.

where it can be seen that they are consistent with the observed fringe contrast. A dislocation will form where a microtwin terminates within the crystal.

Because α-spodumene and diopside have similar structures, it is appropriate to examine the microtwin model of the faults in α-spodumene in terms of the model of (100) twins in diopside proposed by Kirby and Christie (1977). In their model, the twin boundary plane is a *b*-glide plane of symmetry; that is, there is a relative displacement across the plane of $\frac{1}{2}[010]$. Thus, the relative displacement along [010] across a microtwin one unit cell wide will be zero in both diopside and α-spodumene. However, in diopside there is no relative displacement along [001] across a microtwin. Hence, microtwins in diopside cannot give rise to α-fringes, and none were observed by Kirby and Christie (1977). The complex fringe patterns that they observed at microtwin lamellae are, in general, overlapping (α-δ)-fringe patterns (Section 5.3.6). The α-component arises from the *b*-glide plane and the δ-component from the change in orientation across each twin boundary. The details of the fringe patterns depend critically on the operating reflections ($\mathbf{g}_1$ and $\mathbf{g}_2$) and the deviations (s_1 and s_2) from the exact Bragg condition in twin and host. Overlapping (α-δ)-fringes are difficult to analyze unless the diffracting conditions are known precisely. In the present example, the fringes provide little information about the proposed twin model other than being consistent with the prediction that microtwins in diopside will not produce α-fringe patterns.

If the faults observed by van Duysen and Doukhan (1984) in scratched α-spodumene are microtwins with $\mathbf{R} = \frac{1}{4}[001]$, as suggested, then they would be out-of-contrast for all $\mathbf{g} = hk4$. However, since images with these reflections are not given, the microtwin model cannot be adequately tested without further observations. Although the nature of the faults observed in scratched α-spodumene is clearly in doubt, the microtwin model is certainly simpler than the stacking fault model and is consistent with the observation by van Duysen and Doukhan (1984) that microtwins are the predominant defect in the specimens deformed at a constant strain-rate.

9.10 Dislocations in experimentally deformed perovskites

It is now generally accepted that the Earth's lower mantle consists mostly of magnesio-wüstite, $(Mg,Fe)O$, and $(Mg,Fe)SiO_3$ with the perovskite structure. However, in spite of the widespread interest in mantle convection, very little is known about the mechanical properties and lattice defects of crystals with the perovskite structure. Because $(Mg,Fe)SiO_3$ perovskite is not stable under conditions easily achievable in the laboratory, experimental studies have been made using high-quality single crystals of the structural analogues currently available. Poirier et al. (1983) investigated the viscosity and conductivity of $KZnF_3$, and more recently Doukhan and Doukhan (1986) studied the dislocations produced by plastic deformation of $BaTiO_3$ and $CaTiO_3$. Two deformation processes were used: (i) scratching at room temperature and (ii) high-temperature creep of $BaTiO_3$; only the latter is considered here.

Specimens of $BaTiO_3$ deformed about 1 percent by creep at atmospheric pressure and 1400°C exhibited two quite distinct microstructures. The first, shown in Figure 9.35(a), corresponds to dislocation glide on the slip systems $\{110\}\langle\bar{1}10\rangle$.* These slip systems were also activated by scratching at room temperature. The second type of microstructure mostly consists of dislocation loops exhibiting a series of scallops all along their lines, as shown in Figure 9.35(b). Stereographic analysis showed that both the loops and the scallops lie in (010). Doukhan and Doukhan (1986) analyzed the scalloped loops in detail and proposed that the scallops are due to climb dissociation of segments of the loops.

Figure 9.36 is a series of WBDF images of the same part of a scalloped dislocation loop for several different diffraction vectors **g**, together with a sketch of the loop in which the segments A–F are defined. Application of the normal invisibility criteria indicates that the observed contrast of the various segments of the scalloped dislocation loop is consistent with the following Burgers vectors:

---

* The Bravais lattice of $BaTiO_3$ at room temperature is primitive tetragonal; but because the distortion from cubic is small, the structure can be considered to be primitive cubic with $a(pc) = a = b \approx c \approx 0.4$ nm.

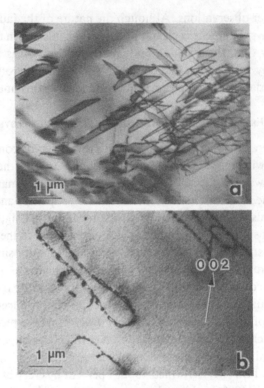

Figure 9.35. (a) BF image of dislocations associated with the {110}⟨1̄10⟩ slip systems in BaTiO₃ deformed 1 percent in creep at atmospheric pressure and 1400°C. (b) BF image of dislocation loops exhibiting scallops all along their length in a (110) foil of the same specimen. (From Doukhan and Doukhan 1986.)

$$A, D: \quad \mathbf{b} = [010]$$
$$B, E: \quad \mathbf{b}_1 = \tfrac{1}{2}[01\bar{1}]$$
$$C, F: \quad \mathbf{b}_2 = \tfrac{1}{2}[011]$$

Figure 9.36. *Facing page.* Analysis of a scalloped dislocation in a (110) foil of BaTiO₃ deformed 1 percent in creep at atmospheric pressure and 1400°C. (a, b, c) WBDF images under different diffracting conditions such that a different dislocation segment (the perfect dislocation or one of the partials) is out-of-contrast for each weak beam. (d) The dislocation is tilted so that the dissociation plane (010) is viewed edge-on. (e) Sketch of a scalloped dislocation loop in (010). Segments like A, D are undissociated with $\mathbf{b} = [010]$ normal to the loop. Segments like B, E are partials with $\mathbf{b}_1 = \tfrac{1}{2}[01\bar{1}]$ and have not moved during the dissociation process. Segments like C, F are partials with $\mathbf{b}_2 = \tfrac{1}{2}[011]$ that have climbed in (010), forming the small half-loops (scallops). (From Doukhan and Doukhan 1986.)

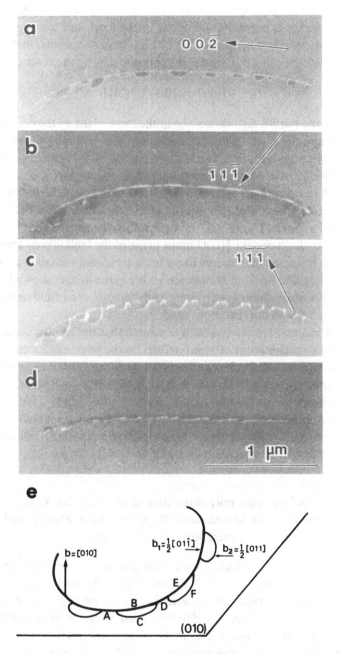

Figure 9.36. See caption on facing page.

Thus, the main loop (A, D) is a prismatic edge-dislocation loop (the Burgers vector is normal to the plane of the loop) and can expand in its plane only by climb. Segments of the loop have dissociated into pairs of partial dislocations (B, C and E, F), presumably by the reaction

$$[010] = \tfrac{1}{2}[011] + \tfrac{1}{2}[01\bar{1}]$$

Because the partial dislocations lie in the same plane as the original prismatic loop, the dissociation is presumably due to climb, as shown schematically in Figure 9.37. The leading partial of $\mathbf{b} = \tfrac{1}{2}[011]$ generates a fault ($\mathbf{R} = \tfrac{1}{2}[011]$) which is terminated by the trailing partial of $\mathbf{b} = \tfrac{1}{2}[01\bar{1}]$. Thus, $\mathbf{R}$ has components both normal and in the plane of the fault. The normal component $\tfrac{1}{2}[010]$ is directly due to the climb dissociation, which (as will be seen in Figure 9.37) brings two like-layers adjacent to one another. The displacement $\tfrac{1}{2}[001]$ in the plane of the fault is presumably necessary to lower the high energy likely to be associated with such a configuration. Note that no fault fringes are visible in the images shown in Figure 9.36. This observation is consistent with the proposed fault vector since $2\pi\mathbf{g}\cdot\mathbf{R} = 0$ for all the reflections used.

Doukhan and Doukhan (1986) suggested that the climb dissociation is due to the precipitation of point defects on the prismatic loops when the specimens are cooled. The equilibrium concentration and the mobility of the point defects are both expected to be very high at 1,400°C, thus favoring deformation by dislocation climb.

These observations and conclusions have recently been confirmed and extended by Beauchesne and Poirier (1989) in a detailed rheological and microstructural investigation of $BaTiO_3$.

9.11 Deformation microstructures in naturally deformed rocks and their use in estimating the mechanisms, history, and conditions of deformation

The previous sections have been mostly concerned with the dislocations and microstructures observed in single crystals deformed to various strains under known experimental conditions. In some minerals, notably quartz and olivine, the *macroscopic* deformational behavior, as revealed by the creep and stress-strain curves, can be understood in terms of the microstructural evolution during deformation; and, furthermore, certain quantifiable characteristics of the microstructure correlate with the imposed

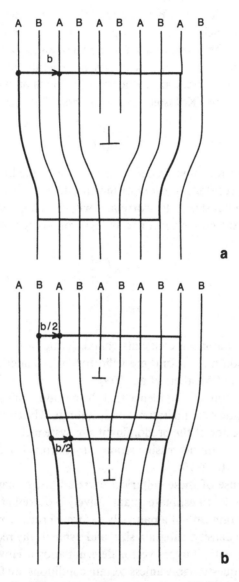

Figure 9.37. Schematic diagrams showing (a) a perfect edge dislocation in a structure consisting of alternating layers A and B, and (b) the climb dissociation of such a dislocation into two partial dislocations. As shown, the dissociation is due to the preferential precipitation of vacancies at A layers, or of interstitials at B layers. Burgers circuits are shown for the perfect dislocation (a) and the two partials (b). In $BaTiO_3$, the A and B layers could correspond to BaO and TiO_2 layers parallel to {100}.

experimental parameters. For example, the steady-state dislocation density ρ (usually reached after a few percent strain) is often observed to be proportional to the square of the differential stress σ. Two theoretical models – one based on the stress necessary to generate dislocations and the other on the stress necessary to cause dislocations on parallel slip planes to pass one another (Kohlstedt and Weathers 1980) – lead to the equation

$$\rho = \left(\frac{\sigma}{\alpha\mu b}\right)^2 \tag{9.2}$$

where μ is the shear modulus, b the magnitude of the Burgers vector, and α a materials constant of order of magnitude 1.

The differential stress also correlates with the subgrain size d and with the size D of the grains generated by dynamic recrystallization. The empirical relations

$$\sigma = K\left(\frac{\mu b}{d}\right) \tag{9.3}$$

$$\sigma = \frac{A}{D^m} \tag{9.4}$$

where K, A, and m are constants characteristic of the material and the recrystallization mechanism, are valid for many materials including olivine and quartz (Weathers et al. 1979).

There is experimental evidence that the predominant slip system in olivine is a function of temperature and strain-rate. The temperature at which the Burgers vector of the predominant slip system changes from [001] to [100] decreases with decreasing strain-rate (Skrotzki et al. 1990; Avé Lallement and Carter 1970).

Thus, because of these empirical correlations, it may be possible, at least in principle, to estimate quantitatively the stress, temperature, and perhaps the strain-rate of a naturally deformed rock from measurements of dislocation density, subgrain size, and dynamically recrystallized grain size, together with Burgers vector determinations. However, these estimates will be questionable unless certain conditions are fulfilled. Some of the more important of these conditions will now be discussed before considering specific examples of the application of microstructural observations to tectonic problems.

A transmission electron micrograph taken at a magnification of $\times 10,000$ presents information from a specimen volume of about $10~\mu m \times 10~\mu m \times 0.1~\mu m = 10^{-11}~cm^3$. It is obvious, therefore, that the microstructure observed in a region of this size will not be characteristic of the bulk specimen unless the microstructure is homogeneous on the order of 10 μm. This degree of homogeneity is rarely achieved even in experimentally deformed single crystals a few millimeters in size, let alone in a large geological structure. Therefore, TEM is best used to obtain detailed microstructural information about deformation-induced features that are characteristic of the deformed grains in petrological thin sections that are themselves representative of the much larger components of the geological structure under investigation.

Many TEM observations have been made of the dislocation microstructures in the grains of deformed rocks. Observations of the dislocations in quartz (Liddell, Phakey, and Wenk 1976; McLaren and Hobbs 1972; Weathers et al. 1979), feldspars (Gandais and Willaime 1984; Montardi and Mainprice 1987), olivine (Buiskool Toxopeus and Boland 1976; Green 1976), and orthopyroxenes (Naze et al. 1987), for example, have contributed significantly to our understanding of the deformation mechanisms in these minerals. In many cases, the microstructures are similar to those observed in experimentally deformed single crystals. When this similarity is observed, it is reasonable to assume that essentially the same dislocation mechanisms operate.

Stress estimates based on measurements of ρ, d, and D in naturally deformed crystals also assume (i) steady-state deformation, at the cessation of which the evolved microstructure is "frozen in," (ii) a simple deformation history, and (iii) experimental data that can be extrapolated to geological conditions.

If the flow law (strain-rate $\dot{\epsilon}$ as a function of stress σ and temperature T) is known from rheological measurements, then the stress estimated from the microstructure can be used to constrain the $(\dot{\epsilon}, T)$-regime. The strain-rate (or equivalent viscosity) can be estimated if the temperature is constrained by the mineral assemblages that are present. Conversely, the temperature can be estimated if it is assumed that $\dot{\epsilon}$ is in the range 10^{-12}–10^{-15} s^{-1}. The $(\dot{\epsilon}, T)$-regime may also be constrained by the Burgers vectors if the predominant slip system has been experimentally determined as a function of $\dot{\epsilon}$ and T and if these observations can be extrapolated to geological conditions.

However, in practice there are still a number of unresolved problems in the use of microstructural piezometers (Knipe 1989; Schmid 1982; White 1979) because the assumptions mentioned may not be valid. Nicolas and Poirier (1976, p. 128) consider that it is always questionable to estimate the applied stress from the measured dislocation density because the theoretical foundation of Eq. (9.2) rests on so many assumptions that extrapolation of experimental results to geological strain-rates may not be justified. The main limitation is probably the complex history of natural deformations, which are unlikely to proceed under steady-state conditions. Furthermore, microstructures can be reset by subsequent deformation. Consequently, dislocation densities "record" the stress during the last increment of plastic deformation, subgrains record the stress during the last 5–20 percent of strain, and recrystallized grains record the stress during the last 4–60 percent of strain, depending on the type of recrystallization mechanism that operated. All microstructures can be reset by static annealing, and small (1–20 μm) subgrains can be readily reset by later small-strain deformations.

In spite of these problems and the skepticism often expressed by field-oriented geologists (Schmid 1982), microstructural observations are being used increasingly to estimate the conditions of deformation in nature. Although the estimates are clearly very rough, advocates of the method point out that these estimates are often no worse than those obtained using more conventional geological and geophysical methods, and that it is useful to obtain even rough estimates by independent methods. Two specific examples are discussed in the following subsections.

9.11.1 Estimates of deformation conditions during crustal deformation from studies of quartz microstructures.   Blenkinsop and Drury (1988) studied the microstructures in quartzites associated with the development of a fault that forms part of a thrust belt in the Cantabrian Mountains of northwest Spain. The thickness of the overlying sediments suggests that the deformation occurred at shallow depths in the upper crust at pressures of 60–100 MPa. The deformation temperature was estimated to be 150°–250°C from the illite crystallinity.

Field and optical microscope studies showed that the fault plane is defined by a 2–3 cm wide cataclastic zone that is bounded laterally by a 1–3 m envelope of plastic and cataclastic deformation. Outside this envelope, the quartz microstructures displayed no evidence of significant deformation. Quartz in the fault envelope contains well-developed deformation bands, deformation lamellae, and intragranular healed fractures, which are visible at all scales of observation.

Figure 9.38 is a BF image of a healed fracture that is about 4 μm wide. The fracture walls are straight and separated by a band of new quartz, which has grown in the open crack. The bands of electron-beam damage centers (Section 8.13.1) parallel to the fracture walls probably mark the positions of trace impurity concentrations formed at incremental growth

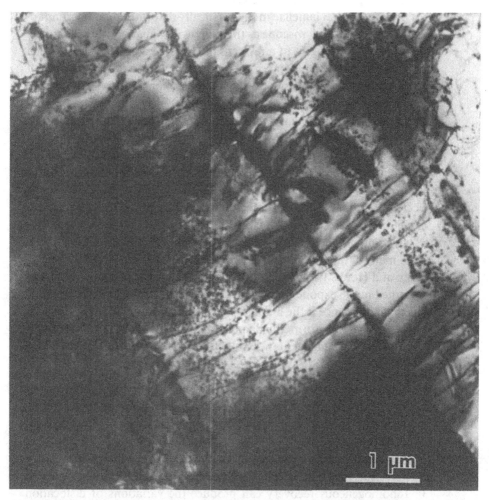

Figure 9.38. BF image of a healed fracture in the quartz of the deformed envelope of a narrow cataclastic fault zone, which forms part of a thrust belt in the Cantabrian Mountains of northwest Spain. (Courtesy of Martyn Drury.)

surfaces of the new quartz. The straight dislocations perpendicular to the wall are probably screw dislocations associated with the growth of the new quartz in the open crack. It is clear that quartz that is precipitated from solution has a significantly different microstructure from plastically deformed metamorphic quartz. Thus, TEM provides a powerful means of identifying the extent and scale of quartz precipitation during deformation by "pressure solution" creep.

The deformation lamellae in the plastically deformed quartz grains appear in the optical microscope as thin lines with apparent widths of a few micrometers and lengths of a few tens of micrometers. TEM showed that these optical microscope images are produced by bands of subgrains about 1.5 μm wide. Most of the subgrain boundaries (consisting of well-ordered networks of two or three sets of dislocations) have a sub-basal orientation, and the lamellae planes tend to be approximately parallel to $\{10\bar{1}2\}$. Within the subgrains, the dislocation density ranges from 4×10^8 to 2×10^9 cm^{-2}. The subgrains with the highest dislocation densities contain few fluid inclusions. In contrast, the subgrains with the lowest dislocation densities have a high content of fluid inclusions, which are usually located at the nodes of the three-dimensional network of dislocations. These dislocations are out-of-contrast for $\mathbf{g} = 0003$, suggesting that they all have Burgers vectors $\mathbf{b} = \frac{1}{3}\langle 11\bar{2}0 \rangle$. The structure of deformation lamellae in many naturally deformed quartzites (Christie and Ardell 1974; McLaren and Hobbs 1972; White 1973) is similar to that produced in metals deformed in the exponential creep regime (Drury and Humphreys 1986). The main difference is the inhomogeneous distribution of fluid inclusions within individual deformation lamellae in quartz. All of the main features of deformation lamellae in naturally deformed quartz can be explained by the following model:

1 A banded substructure is produced by the recovery of dislocations of $\mathbf{b} = \frac{1}{3}\langle 11\bar{2}0 \rangle$ in slip bands associated with several slip systems. During deformation, the water content of the quartz is present along dislocation cores.
2 After deformation, the free dislocations climb and form low-energy networks; water inclusions develop and grow at dislocation nodes.
3 Inhomogeneous recovery can produce the variations of dislocation density and fluid inclusion content of individual lamellae and may ultimately produce the extreme concentrations of fluid inclusions found to define some natural deformation lamellae (Christie and Ardell 1974).

The deformation lamellae can be used to provide an estimate of the stress during the formation of the deformed envelope of the fault. In metals and halides, deformation lamellae form only in the exponential creep regime, which occurs above a critical stress level (Tsenn and Carter 1987). Experimental data on quartz suggest that lamellae form only at high stress, but no particular association with exponential creep can be discerned in the available data. The normalized transition stress in different classes of materials increases with increasing crystal bond strength. In the absence of suitable experimental data on quartz, a lower estimate for the critical stress in quartz of 145 MPa can be obtained from the reliable data on ionic materials. The development of deformation lamellae, therefore, suggests that at some stage in the development of the fault, the stress exceeded 145 MPa at positions less than 3 m from the fault plane. The spacing of deformation lamellae can also be used as a piezometer (Koch and Christie 1981; McCormick 1977). This technique is questionable, however, because in many cases the objects that produce the deformation lamellae observed by optical microscopy have a smaller spacing than the resolution of the microscope (McLaren et al. 1970). Stress estimates obtained from the free-dislocation density are likely to be underestimates because the free dislocations have a recovered configuration. The free-dislocation density suggests that stresses exceeded 120 MPa in the deformed envelope adjacent to the fault.

The spatial distribution of plastic deformation adjacent to the fault plane can be explained by a model in which the propagating fault is preceded by a zone of stress concentration at its tip. Calculations by Blenkinsop and Drury (1988) indicate that, for remote applied stresses of less than 100.MPa, the zone of stress concentration, with stress greater than 145 MPa, will have a lateral width of 1–3 m once the initial fracture is longer than 12 m. The high stresses estimated from the quartz microstructures adjacent to the fault are consistent with a low average stress in the crust because very high transient values of stress may occur locally, particularly at the propagating termination of fractures, shear zones, and folds.

9.11.2 Deformation microstructures in naturally deformed peridotites and implications for mantle rheology. Mantle peridotites outcrop as huge massifs within mountain belts. Drury, Hoogerduyn Stratung, and Vissers (1990) and Drury (1990) studied the defect substructures in peridotites from the Voltri massif in northwest Italy. Field mapping and optical microscope studies indicate that two main deformation stages have affected these rocks. The temperatures and pressures during deformation were estimated from the phase assemblages and mineral chemistry. The first deformation stage occurred under estimated conditions of 900°–1,200°C and 1–3 GPa. The high-temperature deformation structures and microstructures are locally overprinted by narrow shear zones, which were active at estimated conditions of 600°–930°C and 0.5–1 GPa.

TEM studies of the high-temperature microstructures reveal inhomogeneous dislocation substructures within individual grains. The predominant subgrain boundaries are parallel to (100), and a moderate density of free dislocations ($\approx 10^8$ cm^{-2}) occurs within the subgrains. The Burgers vectors of the free dislocations were analyzed using the usual invisibility criteria. Images taken under strong beam conditions were often difficult to interpret. However, the contrast was rarely ambiguous in kinematical or WBDF images; so these were used to determine Burgers vectors. In most areas, 60 percent of the free dislocations had **b** = [100]; the remainder had **b** = [001]. Analysis of glide loops and tilt walls indicated the operation of the [100]{0*kl*} and [001](*hk*0) slip systems.

In olivine deformed in the lower temperature shear zones, a different substructure developed. The substructure consists of (100) tilt walls defined by dislocations with **b** = [100], spaced 3–10 μm apart. Free dislocations occur within the subgrains, with a density of about 10^9 cm^{-2}. Three types of dislocation can be distinguished: (1) undissociated dislocations with **b** = [100], (2) undissociated dislocations with **b** = [001], and (3) dislocations that show unusual contrast for olivine. Type 3 occur as bundles of 4–8 straight dislocations parallel to [100]. The total Burgers vector **b**$_t$ of these individual bundles was determined by counting the number of thickness fringes in WBDF images that terminate at the intersection of a bundle with the foil surface for three noncoplanar diffraction vectors **g** (see Section 5.6). It was found, for example, that for **g** = 06$\bar{2}$, $n = -2$; for **g** = $\bar{2}$40, $n = 0$; and for **g** = 2$\bar{2}$2, $n = 2$. Because $n = $ **g** · **b**, we have

$$0u + 6v - 2w = -2$$

$$-2u - 4v + 0 = 0$$

$$2u - 2v + 2w = 2$$

where u, v, w are the components of **b**$_t$. On solving these equations, we find **b**$_t$ = [001]. The observation of fringe patterns, defining planar defects on {021} and (001) between the dislocations of a bundle, suggests that the dislocations are partials. Although the fault vectors and the Burgers vectors have not been uniquely determined, there are indications that the fault vectors have a component normal to the fault plane, so that the partial dislocations may be due to climb dissociation. Drury (1990) has proposed that this dissociation may involve a layer of humite-type structure.

Both types of microstructure found in olivine are indicative of a significant component of dislocation climb during deformation. The dissociated dislocations present in the low-temperature microstructure have not been reproduced in any experiments nor have they been found in other naturally deformed olivines. The climb-dissociation may affect the type

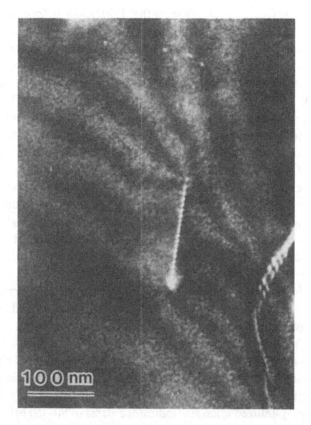

Figure 9.39. WBDF micrograph showing the termination of thickness fringes where a dislocation intersects the foil surfaces in olivine. In this micrograph, $g = 062$, and there are two terminating thickness fringes at each end of the dislocation. Hence, $g \cdot b = 2$, which is consistent with $b = [001]$.
(Courtesy of Martyn Drury.)

and kinetics of deformation mechanisms operating in lithospheric shear zones. Glide of climb-dissociated edge dislocations can occur only if the partial dislocations constrict to reform a unit dislocation or if the planar defect migrates by diffusion along with the partial dislocations. In both cases, the glide mobility of the dislocations will be controlled by diffusion; thus, glide will have similar kinetics to climb. This type of deformation mechanism is known as viscous glide (Weertman 1957), and materials that deform in this manner have distinct rheology and microstructural development compared with pure materials where glide is very much faster than climb (Drury and Humphreys 1986; Sherby and Burke 1968).

The types of free dislocations show that [100] slip was dominant during the high-temperature deformation, and [001] slip was dominant during the low-temperature deformation. Extrapolation of experimental data (Avé Lallement and Carter 1970) to natural deformation strain-rates suggests a transition temperature of 600°–800°C, which compares with a value of 700°–800°C inferred from the mineral assemblages. The stresses and strain-rates during the two stages of deformation can, in principle, be estimated from microstructural information, as already discussed.

Stresses of 4–11 MPa were estimated by Drury et al. (1990) from the recrystallized grain size in the high-temperature microstructure, assuming that the recrystallized grain size was produced by dynamic recrystallization. If static recrystallization had occurred, the grain size would provide a lower bound to the stress. Free-dislocation densities in the high-temperature microstructure indicate stresses of 42–55 MPa for the last increment of plastic deformation. A range of deformation histories could produce the observed microstructures. The most simple history would be a single deformation in which stress increased sharply in the last increment of strain. Because the temperature of deformation is constrained from petrological information, it is possible to estimate the strain-rates and equivalent viscosities for the natural deformation from experimental flow laws (Karato et al. 1986). Strain-rate estimates of 7.5×10^{-11} to 2.7×10^{-13} s^{-1} and equivalent viscosities of 10^{17}–10^{20} Pa·s can be obtained in this way for the high-temperature deformation. For the low-temperature deformation, the recrystallized grain size indicates a stress of 90–150 MPa, whereas the free-dislocation density indicates a stress of 340 MPa for the last increment of deformation. The estimated strain-rate is 1.4×10^{-10} to 2.3×10^{-12} s^{-1}, equivalent to a viscosity of 10^{17}–10^{19} Pa·s.

The strain-rate estimates suggest that the microstructures were formed by flow in the mantle related to large-scale tectonic processes. In the case of the Voltri peridotites, Drury et al. (1990) have related this deformation to extension and rifting of a continental lithospheric plate. The high-temperature deformation occurred under asthenospheric conditions, whereas the low-temperature deformation occurred within the lithosphere. The estimates of upper mantle viscosity obtained from the study of microstructures are lower than those for average viscosity obtained independently from other methods, such as analysis of glacial rebound of land surfaces (Nakada and Lambeck 1987), which are also based on a number of assumptions. All the samples studied are derived from the uppermost part of the mantle, so the viscosity estimates obtained from the microstructural study are consistent with the presence of a low viscosity zone between the lithosphere and the main part of the convecting upper mantle. The estimates of viscosity obtained for the lithosphere are surprisingly low; in fact, the lithosphere is estimated to have a viscosity similar to that of the underlying asthenosphere. This study suggests that when the lithosphere is deformed to large strains, deformation may become accommodated by a network of soft localized shear zones; and the lithosphere can then be as weak as the underlying convecting mantle.

References

Ahn, J. H., Burt, D. M., & Buseck, P. R. (1988). Alteration of andalusite to sheet silicates in a pegmatite. *Amer. Mineral.*, 73, 559-67.

Aines, R. D., & Rossman, G. R. (1984). Water in minerals: a peak in the infrared. *J. Geophys. Res.*, 89, 4059-71.

Akizuki, M. (1972). Electron-microscope investigation of microcline twinning. *Amer. Mineral.*, 57, 797-808.

Alexander, H., & Haasen, P. (1968). Dislocations and plastic flow in the diamond structure. *Solid State Physics*, 22, 27-158.

Amelinckx, S. (1964). *Direct Observation of Dislocations*. New York: Academic Press.

  (1970). The study of planar interfaces by means of electron microscopy. In *Modern Diffraction and Imaging Techniques in Materials Science*, edited by S. Amelinckx, R. Gevers, G. Remaut, & J. van Landuyt, pp. 257-94. Amsterdam: North-Holland.

Ardell, A. J., Christie, J. M., & McCormick, J. W. (1974). Dislocation images in quartz and the determination of Burgers vectors. *Phil. Mag.*, 29, 1399-411.

Ashby, M. F., & Brown, L. M. (1963a). Diffraction contrast from spherically symmetrical coherency strains. *Phil. Mag.*, 8, 1063-103.

  (1963b). On diffraction contrast from inclusions. *Phil. Mag.*, 8, 1649-76.

Avé Lallement, H. G., & Carter, N. L. (1970). Syntectonic recrystallization of olivine and modes of flow in the upper mantle. *Geological Society of America, Bulletin*, 81, 2203-20.

Bailey, J. C., Champness, P. E., Dunham, A. C., Esson, J., Fyfe, W. S., MacKenzie, W. S., Stumpfl, E. F., & Zussman, J. (1970). Mineralogy and petrology of Apollo 11 lunar samples. *Proc. Apollo 11 Lunar Sci. Conf. Geochim Cosmochim Acta, Suppl.* 1, Vol. 1, 169-94.

Bambauer, H. U., Krause, C., & Kroll, H. (1989). TEM-investigation of the sanidine/microcline transition across metamorphic zones: the K-feldspar varieties. *Eur. J. Mineral.*, 1, 47-58.

Barber, D. J. (1970). Thin foils of non-metals made for electron microscopy by sputter-etching. *J. Materials Sci.*, 5, 1-8.

  (1977). Defect microstructures in deformed and recovered dolomite. *Tectonophysics*, 39, 193-213.

Barber, D. J., Freeman, L. A., & Smith, D. J. (1983). Analysis of high-voltage, high-resolution images of lattice defects in experimentally-deformed dolomite. *Phys. Chem. Minerals*, 9, 102-8.

365

Barber, D. J., Heard, H. C., & Wenk, H.-R. (1981). Deformation of dolomite single crystals from 20–800°C. *Phys. Chem. Minerals, 7,* 271–86.

Barber, D. J., & Wenk, H.-R. (1979). Deformation twinning in calcite, dolomite, and other rhombohedral carbonates. *Phys. Chem. Minerals, 5,* 141–65.

Batterman, B. W. (1969). Detection of foreign atom sites by their x-ray fluorescence scattering. *Phys. Rev. Letters, 22,* 703–5.

Batterman, B. W., & Cole, H. (1964). Dynamical diffraction of x-rays by perfect crystals. *Rev. Mod. Phys., 36,* 681–717.

Beauchesne, S., & Poirier, J. P. (1989). Creep of barium titanate perovskite: a contribution to a systematic approach to the viscosity of the lower mantle. *Physics of the Earth and Planetary Interiors, 55,* 187–99.

Blacic, J. D., & Christie, J. M. (1984). Plasticity and hydrolytic weakening of quartz single crystals. *J. Geophys. Res., 89,* 4223–39.

Blenkinsop, T. G., & Drury, M. R. (1988). Stress estimates and fault history from quartz microstructures. *Journal of Structural Geology, 10,* 673–84.

Boland, J. N. (1977). Deformation mechanisms in alpine-type ultramafic rocks from New Zealand. *Tectonophysics, 39,* 215–30.

Boland, J. N., McLaren, A. C., & Hobbs, B. E. (1971). Dislocations associated with optical features in naturally-deformed olivine. *Contrib. Mineral. Petrol., 30,* 53–63.

Born, M., & Wolf, E. (1965). *Principles of Optics,* 3rd edition. Oxford: Pergamon Press.

Bristowe, P. D., & Sass, S. L. (1980). The atomic structure of a large angle (001) twist boundary in gold determined by a joint computer modelling and x-ray diffraction study. *Acta Metall., 28,* 575–88.

Brown, G. E. (1982). Olivine and silicate spinels. In *Reviews in Mineralogy, Vol. 5, Orthosilicates,* 2d ed., edited by P. H. Ribbe, pp. 275–365. Washington, DC: Mineralogical Society of America.

Brown, W. L., & Parson, I. (1984). Exsolution and coarsening mechanisms and kinetics in an ordered cryptoperthite series. *Contrib. Mineral. Petrol., 86,* 3–18.

Budai, J., Gaudig, W., & Sass, S. L. (1979). Measurement of grain boundary thickness using x-ray diffraction techniques. *Phil. Mag., 39,* 533–49.

Buiskool Toxopeus, J. M. A., & Boland, J. N. (1976). Several types of natural deformation in olivine, an electron microscope study. *Tectonophysics, 32,* 209–33.

Bursill, L. A. (1983). Small and extended defect structures in gem-quality Type 1 diamonds. *Endeavour,* New Series, 7, 70–7.

Bursill, L. A., & McLaren, A. C. (1966). Transmission electron microscope study of natural radiation damage in zircon (ZrSiO₄). *phys. stat. sol., 13,* 331–43.

Buseck, P. R., Nord, G. L., & Veblen, D. R. (1980). Subsolidus phenomena in pyroxenes. In *Reviews in Mineralogy, Vol. 7, Pyroxenes,* edited by C. T. Prewitt, pp. 117–204. Washington, DC: Mineralogical Society of America.

Cahn, J. W., (1961). On spinodal decomposition. *Acta Metall., 9,* 795–801.
(1968). Spinodal decomposition. *Trans. AIME, 242,* 166–80.

Carpenter, M. A. (1978). Nucleation of augite at antiphase boundaries in pigeonite. *Phys. Chem. Minerals, 2,* 237–51.

(1979). Contrasting properties and behaviour of antiphase domains in pyroxenes. *Phys. Chem. Minerals,* 5, 119-31.

Carter, C. B., & Kohlstedt, D. L. (1981). Electron irradiation damage in natural quartz grains. *Phys. Chem. Minerals,* 7, 110-16.

Carter, C. B., & Sass, S. L. (1981). Electron diffraction and microscopy techniques for studying grain-boundary structure. *J. Amer. Ceramic Soc.,* 64, 335-45.

Carter, N. L., Christie, J. M., & Griggs, D. T. (1964). Experimental deformation and recrystallization of quartz. *J. Geol.,* 72, 687-733.

Chadderton, L. T. (1964). Fission fragment damage to crystal lattices: track contrast. *Proc. Roy. Soc. (London),* A280, 110-22.

Chakoumakos, B. C., Murakami, T., Lumpkin, G. R., & Ewing, R. C. (1987). Alpha-decay-induced fracturing in zircon: The transition from the crystalline to the metamict state. *Science,* 236, 1556-9.

Champness, P. E. (1973). Speculation on an order-disorder transformation in omphacite. *Amer. Mineral.,* 58, 540-2.

(1977). Transmission electron microscopy in earth science. *Ann. Rev. Earth & Planet. Sci.,* 5, 203-26.

Champness, P. E., & Copley, P. A. (1976). The transformation of pigeonite to orthopyroxene. In *Electron Microscopy in Mineralogy,* edited by H.-R. Wenk, pp. 228-33. Berlin: Springer-Verlag.

Champness, P. E., & Lorimer, G. W. (1973). Precipitation (exsolution) in an orthopyroxene. *J. Materials Sci.,* 8, 467-74.

(1976). Exsolution in silicates. In *Electron Microscopy in Mineralogy,* edited by H.-R. Wenk, pp. 174-204. Berlin: Springer-Verlag.

Charai, A., & Boulesteix, C. (1983). Visualization and study of the shape and size of very small coherent precipitates by a phase contrast method. *phys. stat. sol.,* 80(a), 333-41.

Cherns, D., Howie, A., & Jacobs, M. H. (1973). Characteristic x-ray production in thin crystals. *Z. Naturforsch.,* 28a, 565-71.

Cherns, D., Jenkins, M. L., & White, S. (1980). The structure of dislocations in quartz under electron irradiation. *Proc. EMAGS 79,* 121-22. Bristol: Institute of Physics.

Chopra, P. N., & Paterson, M. S. (1981). Experimental deformation of dunite. *Tectonophysics,* 78, 453-73.

(1984). The role of water in the deformation of dunite. *J. Geophys. Res.,* 89, 7861-76.

Christie, J. M., & Ardell, A. J. (1974). Substructures of deformation lamellae in quartz. *Geology,* 2, 405-8.

Christie, J. M., Griggs, D. T., & Carter, N. L. (1964). Experimental evidence of basal slip in quartz. *J. Geol.,* 72, 734-56.

Christie, J. M., Heard, H. C., & La Mori, P. N. (1964). Experimental deformation of quartz single crystals at 27 to 30 kilobars confining pressure and 24°C. *Amer. J. Sci.,* 262, 26-55.

Christie, J. M., Lally, J. S., Heuer, A. H., Fisher, R. M., Griggs, D. T., & Radcliffe, S. V. (1971). Comparative electron petrography of Apollo 11, Apollo 12, and terrestrial rocks. *Proc. 2nd Lunar Sci. Conf. Geochim Cosmochim. Acta, Suppl.* 2, Vol. 1, 69-89.

Christy, A. G., & Putnis, A. (1988). Planar and line defects in the sapphirine poly-types. *Phys. Chem. Minerals,* 15, 548–58.

Clarke, D. R. (1979a). High-resolution techniques and application to nonoxide ceramics. *J. Amer. Ceramic Soc.,* 62, 236–46.

(1979b). On the detection of thin intergranular films by electron microscopy. *Ultramicroscopy,* 4, 33–44.

(1980). Observations of microcracks and thin intergranular films in ceramics by transmission electron microscopy. *J. Amer. Ceramic Soc.,* 63, 104–6.

(1987). On the equilibrium thickness of intergranular glass phases in ceramic materials. *J. Amer. Ceramic Soc.,* 70, 15–22.

Cliff, G., Champness, P. E., Nissen, H.-U., & Lorimer, G. W. (1976). Analytical electron microscopy of exsolution lamellae in plagioclase feldspars. In *Electron Microscopy in Mineralogy,* edited by H.-R. Wenk, pp. 258–65. Berlin: Springer-Verlag.

Cliff, G., & Lorimer, G. W. (1975). The quantitative analysis of thin specimens. *J. Microscopy,* 103, 203–7.

Cockayne, D. J. H. (1973). The principles and practice of the weak-beam method of electron microscopy. *J. Microscopy,* 98. 116–34.

Cockayne, D. J. H., Parsons, J. R., & Hoelke, C. W. (1971). A study of the relationship between lattice fringes and lattice planes in electron microscope images of crystals containing defects. *Phil. Mag.,* 24, 139–53.

Coe, R. S., & Kirby, S. H. (1975). The orthoenstatite to clinoenstatite transformation by shearing and reversion by annealing: mechanism and potential applications. *Contrib. Mineral. Petrol.,* 52, 29–55.

Comer, J. J. (1972). Electron microscope study of dauphine microtwins formed in synthetic quartz. *J. Cryst. Growth,* 15, 179–87.

Cordier, P., & Doukhan, J. C. (1989). Water solubility in quartz and its influence on ductility. *Eur. J. Mineral.,* 1, 221–37.

Cowley, J. M., & Moodie, A. F. (1957). The scattering of electrons by atoms and crystals. I. A new theoretical approach. *Acta. Crystallogr.,* 10, 609–19.

Czank, M., Landuyt, J. van, Schulz, H., Laves, F., & Amelinckx, S. (1973). Electron microscopic study of the structural changes as a function of temperature in anorthite. *Z. Krist.,* 138, 403–18.

Czank, M., Schulz, H., & Laves, F. (1972). Investigation of domains in anorthite by electron microscopy. *Naturwiss.,* 59, 77–8.

Darot, M., & Gueguen, Y. (1981). High-temperature creep in forsterite single crystals. *J. Geophys. Res.,* 86, 6219–34.

Das, G., & Mitchell, T. E. (1974). Electron irradiation damage in quartz. *Radiation Effects,* 23, 49–52.

Deer, W. A., Howie, R. A., & Zussman, J. (1963). *Rock-Forming Minerals.* London: Longmans.

Doukhan, J. C., Doukhan, N., Naze, L., & Duysen, J. C. van, (1986). Défauts de réseau et plasticité cristalline dans les pyroxènes: une revue. *Bull. Mineral.,* 109, 377–94.

Doukhan, J. C., & Trepied, L. (1985). Plastic deformation of quartz single crystals. *Bull. Mineral.,* 108, 97–123.

Doukhan, N., Doukhan, J. C., Fitz Gerald, J. D., Chopra, P. N., & Paterson, M. S. (1984). A TEM microstructural study of experimentally deformed Anita

Bay dunite. In *Deformation of Ceramics II,* edited by R. E. Tressler & R. C. Bradt, pp. 307-19. New York: Plenum.

Doukhan, N., & Doukhan, J. C. (1986). Dislocations in perovskites $BaTiO_3$ and $CaTiO_3$. *Phys. Chem. Minerals,* 13, 403-10.

Drury, M. R. (1990). Hydration induced climb dissociation in olivine. To be submitted to *Phys. Chem. Minerals.*

Drury, M. R., Hoogerduyn Stratung, E. H., & Vissers, R. L. M. (1990). Shear zone structures and microstructures in mantle peridotites from the Voltri Massif, Ligurian Alps, N.W. Italy. *Geologie en Mynbouw.,* 69, 3-17.

Drury, M. R., & Humphreys, F. J. (1986). The development of microstructure in Al-5% Mg during high temperature deformation. *Acta Metall.,* 34, 2259-71.

Duffy, D. M. (1986). Review article: Grain boundaries in ionic crystals. *J. Phys. C.,* 19, 4393-412.

Duncumb, P. (1962). Enhanced x-ray emission from extinction contours in a single-crystal gold film. *Phil. Mag.,* 7, 2101-5.

Durham, W. B., Froidevaux, C., & Jaoul, O. (1979). Transient and steady-state creep of pure forsterite at low stress. *Physics of the Earth and Planetary Interiors,* 19, 263-74.

Durham, W. B., Goetze, C., & Blake, B. (1977). Plastic flow of oriented single crystals of olivine. 2: Observations and interpretation of the dislocation structures. *J. Geophys. Res.,* 82, 5755-70.

Duysen, J. C. van, & Doukhan, J. C. (1984). Room temperature microplasticity of α-spodumene $LiAlSi_2O_6$. *Phys. Chem. Minerals,* 10, 125-32.

Duysen, J. C. van, Doukhan, N., & Doukhan, J. C. (1985). Transmission electron microscope study of dislocations in orthopyroxene $(Mg, Fe)_2Si_2O_6$. *Phys. Chem. Minerals,* 12, 39-44.

Edington, J. W. (1975). *Practical Electron Microscopy in Materials Science, Monograph Three: Interpretation of Transmission Electron Micrographs.* Eindhoven: Philips Technical Library.

Eggleton, R. A., & Boland, J. N. (1982). Weathering of enstatite to talc through a sequence of transitional phases. *Clays & Clay Min.,* 30, 11-20.

Faure, G. (1977). *Principles of Isotope Geology.* New York: Wiley.

Fitz Gerald, J. D. (1980). Crystallography and defect structures of some naturally-occurring feldspars. Ph.D. thesis, Monash University. Clayton, Victoria, Australia.

Fitz Gerald, J. D., Boland, J. N., McLaren, A. C., Ord, A., & Hobbs, B. E. (1990). Microstructures in water-weakened single cyrstals of quartz. Accepted for publication in *J. Geophys. Res.*

Fitz Gerald, J. D., & Chopra, P. N. (1982). Deformation mechanisms in dunite – the results of high temperature testing. In *Strength of Metals and Alloys,* edited by R. C. Gifkins, pp. 735-40. Oxford: Pergamon Press.

Fitz Gerald, J. D., & McLaren, A. C. (1982). The microstructures of microcline from some granitic rocks and pegmatites. *Contrib. Mineral. Petrol.,* 80, 219-29.

Fleet, S. G., & Ribbe, P. H. (1963). An electron-microscope investigation of a moonstone. *Phil. Mag.,* 8, 1179-87.

   (1965). An electron microscopic study of peristerite plagioclases. *Min. Mag.,* 35, 165-76.

Fowles, Grant R. (1968). *Introduction to Modern Optics*. New York: Holt, Rinehart and Winston.

Frondel, C. (1962). *Dana's System of Mineralogy, Vol. III, Silica Minerals*. New York: Wiley.

Frondel, C., & Collette, R. L. (1957). Hydrothermal synthesis of zircon, thorite and huttonite. *Amer. Mineral.*, 42, 759-65.

Fujino, K., Furo, K., & Momoi, H. (1988). Preferred orientation of antiphase boundaries in pigeonite as a cooling ratemeter. *Phys. Chem. Minerals*, 15, 329-35.

Gaboriaud, R. J. (1986). Dislocations in olivine single crystals indented between 25 and 1100°C. *Bull. Mineral.*, 109, 185-91.

Gaboriaud, R. J., Darot, M., Gueguen, Y., and Woirgard, J. (1981). Dislocations in olivine indented at low temperatures. *Phys. Chem. Minerals*, 7, 100-4.

Gandais, M., Gervais, A., Sacerdoti, M., & Strunk, H. (1987). Défauts plans de faible amplitude dans la sanidine (feldspath potassique). Etude par METHT. *J. Microsc. Spectrosc. Electron.*, 12, 201-8.

Gandais, M., & Strunk, H. (1983). Microstructure of plastically deformed sanidine (K-feldspar). In *Proc. 7th Int. Conf. on High Voltage Electron Microscopy*, Berkeley, edited by R. M. Fisher, R. Gronsky, & K. H. Westmacott, pp. 353-8. Berkeley (Laboratory), CA, and ENTIS, Springfield, VA.

Gandais, M., & Willaime, C. (1984). Mechanical properties of feldspars. In *Feldspars and Felspathoids: Structures, Properties and Occurrences*, edited by W. L. Brown, pp. 207-46. NATO ASI Series. Dordrecht: D. Reidel.

Gerretsen, J., McLaren, A. C., & Paterson, M. S. (1987). Evolution of water inclusions in wet synthetic quartz as a function of temperature and pressure. *EOS*, 68, 1453.

Gerretsen, J., Paterson, M. S., & McLaren, A. C. (1989). The uptake and solubility of water at elevated pressure and temperature in quartz. *Phys. Chem. Minerals*, 16, 334-42.

Gevers, R., Art, A., & Amelinckx, S. (1963). Electron microscope images of single and intersecting stacking faults in thick foils; Part I: Single faults. *phys. stat. sol.*, 3, 1563-93.

(1964). Electron microscope images of single and intersecting stacking faults in thick foils; Part II: Intersecting faults. *phys. stat. sol.*, 7, 605-32.

Gevers, R. Landuyt, J. van, & Amelinckx, S. (1965). Intensity profiles for fringe patterns due to planar interfaces as observed by electron microscopy. *phys. stat. sol.*, 11, 689-709.

Giancoli, Douglas C. (1984). *General Physics*. London: Prentice-Hall International.

Gillespie, P., McLaren, A. C., & Boland, J. N. (1971). Operating characteristics of an ion-bombardment apparatus for thinning non-metals for transmission electron microscopy. *J. Materials Sci.*, 6, 87-9.

Gjonnes, J., & Olsen, A. (1974). Study of exsolution structures in sodium feldspars by a combination of electron microscopy and energy dispersive x-ray analysis. In *Electron Microscopy 1974, Eighth International Congress on Electron Microscopy, Vol. 1*, edited by J. V. Sanders & D. J. Goodchild, pp. 504-5. Canberra, Australia: Australian Academy of Science.

Gleiter, H. (1982). Review paper: On the structure of grain boundaries in metals. *Mat. Sci. and Eng.,* 52, 91–131.

Godbeer, W. C., & Wilkins, R. W. T. (1977). The water content of a synthetic quartz. *Amer. Mineral.,* 62, 831–2.

Goodman, P., & Johnson, A. W. S. (1977). Identification of enantiomorphically-related space groups by electron diffraction – a second method. *Acta Crystallogr.,* A33, 997–1001.

Green, H. W. (1976). Plasticity of olivines in periodotites. In *Electron Microscopy in Mineralogy,* edited by H.-R. Wenk, pp. 443–64. Berlin: Springer-Verlag.

Griggs, D. T. (1974). A model of hydrolytic weakening in quartz. *J. Geophys. Res.,* 79, 1653–61.

Griggs, D. T., & Blacic, J. D. (1964). The strength of quartz in the ductile regime. *Trans. Am. Geophys. Union,* 45, 102–30.

(1965). Quartz: anomalous weakness of synthetic crystals. *Science,* 147, 292–5.

Griggs, D. T., Turner, F. J., & Heard, H. C. (1960). Deformation of rocks at 500 to 800°C. In *Rock Deformation,* Geological Society of America Memoir 79, edited by D. T. Griggs & J. Handin, pp. 39–104. Boulder, CO: Geological Society of America.

Ham, R. K. (1961). The determination of dislocation densities in thin films. *Phil. Mag.,* 6, 1183–4.

Ham, R. K., & Sharpe, N. G. (1961). A systematic error in the determination of dislocation densities in thin films. *Phil. Mag.,* 6, 1193–4.

Hampar, M. S. (1971). Thermal and electron beam induced breakdown of topaz. In *Electron Microscopy and Structure of Materials,* edited by Gareth Thomas, pp. 1256–66. Berkeley and Los Angeles: Univ. of California Press.

Hashimoto, H., Howie, A., & Whelan, M. J. (1962). Anomalous electron absorption effects in metal foils: theory and comparison with experiment. *Proc. Roy. Soc. (London),* A269, 80–103.

Hay, R. S., & Evans, B. (1988). Intergranular distribution of pore fluid and the nature of high-angle grain boundaries in limestone and marble. *J. Geophys. Res.,* 93, 8959–74.

Head, A. K. (1967a). Computer generation of electron microscope pictures of dislocations. *Aust. J. Phys.,* 20, 557–66.

(1967b). Unstable dislocations in anisotropic crystals. *phys. stat. sol.,* 19, 185–92.

Head, A. K., Humble, P., Clarebough, L. M., Morton, A. J., & Forwood, C. T. (1973). *Computed Electron Micrographs and Defect Identification.* Amsterdam: North-Holland.

Heggie, M. I., & Zheng, Y. (1987). Planar defects and dissociation of dislocations in a K-feldspar. *Phil. Mag.,* A56, 681–8.

Heidenreich, Robert D. (1964). *Fundamentals of Transmission Electron Microscopy.* New York: Wiley.

Heuer, A. H., Lally, J. S., Christie, J. M., & Radcliffe, S. V. (1972). Phase transformations and exsolution in lunar and terrestrial calcic plagioclases. *Phil. Mag.,* 26, 465–82.

Hirsch, P. B., Howie, A., Nicholson, R. B., Pashley, D. W., & Whelan, M. J. (1965). *Electron Microscopy of Thin Crystals.* London: Butterworths.

Hirsch, P. B., Howie, A., & Whelan, M. J. (1960). A kinematical theory of diffraction contrast of electron transmission microscope images of dislocations and other defects. *Phil. Trans. Roy. Soc. (London)*, A252, 499–529.

Hobbs, B. E. (1968). Recrystallization of single crystals of quartz. *Tectonophysics*, 6, 353–401.

(1985). The hydrolytic weakening effect in quartz. In *Point Defects in Minerals*, Geophysical Monograph 31, edited by R. N. Schock, pp. 151–70. Washington, DC: American Geophysical Union.

Hobbs, B. E., McLaren, A. C., & Paterson, M. S. (1972). Plasticity of single crystals of synthetic quartz. In *Flow and Fracture of Rocks*, Geophysical Monograph 16, edited by H. C. Heard, I. Y. Borg, N. L. Carter, & C. B. Raleigh, pp. 29–53. Washington, DC: American Geophysical Union.

Hobbs, B. E., Means, W. D., & Williams, P. F. (1976). *An Outline of Structural Geology*. New York: Wiley.

Holland, H. D., & Gottfried, D. (1955). The effect of nuclear radiation on the structure of zircon. *Acta Crystallogr.*, 8, 291–300.

Howie, A., & Whelan, M. J. (1962). Diffraction contrast of electron microscope images of crystal lattice defects. III. Results and experimental confirmation of the dynamical theory of dislocation image contrast. *Proc. Roy. Soc., (London)*, A267, 206–30.

Hull, D., & Bacon, D. J. (1984). *Introduction to Dislocations*. Oxford: Pergamon Press.

Humphreys, C. J., Howie, A., & Booker, G. R. (1967). Some electron diffraction contrast effects at planar defects in crystals. *Phil. Mag.*, 15, 507–22.

Hutchison, J. L., Irusteta, M. C., & Whittaker, E. J. W. (1975). High-resolution electron microscopy and diffraction studies of fibrous amphiboles. *Acta Crystallogr.*, A31, 794–801.

Hutchison, J. L., & McLaren, A. C. (1976). Two-dimensional lattice images of stacking disorder in wollastonite. *Contrib. Mineral. Petrol.*, 55, 303–9.

(1977). Stacking disorder in wollastonite and its relationship to twinning and the structure of parawollastonite. *Contrib. Mineral. Petrol.*, 61, 11–13.

Hyde, B. G., & Bursill, L. A. (1970). Point, line and planar defects in some non-stoichiometric compounds. In *The Chemistry of Extended Defects in Non-metallic Solids*, edited by L. Eyring & M. O'Keefe, pp. 347–78. Amsterdam: North Holland.

Ishida, Y., Ishida, H., Kohra, K., & Ichinose, H. (1980). Determination of the Burgers vector of a dislocation by weak-beam imaging in a HVEM. *Phil. Mag.*, A42, 453–62.

Jaoul, O., Gueguen, Y., Michaut, M., & Ricoult, D. (1979). A technique for decorating dislocations in forsterite. *Phys. Chem. Minerals*, 5, 15–19.

Jefferson, D. A., & Thomas, J. M. (1975). Electron-microscope analysis of disorder in wollastonite. *Mat. Res. Bull.*, 10, 761–8.

Johnson, W. G. (1962). Yield points and delay times in single crystals. *J. Appl. Phys.*, 33, 2716–30.

Karato, S.-I., Paterson, M. S., & Fitz Gerald, J. D. (1986). Rheology of synthetic olivine aggregates: influence of grain size and water. *J. Geophys. Res.*, 91, 8151–76.

Kashima, K., Sunagawa, I., & Sumino, K. (1983). Plastic deformation of olivine single crystals. *Science Reports of the Tohoku University,* series III, 15, 281-407.

Kekulawala, K. R. S. S., Paterson, M. S., & Boland, J. N. (1981). An experimental study of the role of water in quartz deformation. In *Mechanical Behaviour of Crustal Rocks, The Handin Volume,* Geophysical Monograph 24, edited by N. L. Carter, M. Friedman, J. M. Logan, & D. W. Stearns, pp. 49-60. Washington, DC: American Geophysical Union.

Kirby, S. H., & Christie, J. M. (1977). Mechanical twinning in diopside Ca(Mg, Fe)Si$_2$O$_6$: structural mechanism and associated crystal defects. *Phys. Chem. Minerals,* 1, 137-63.

Kirby, S. H., & McCormick, J. W. (1979). Creep of hydrolytically weakened synthetic quartz crystals oriented to promote {2$\overline{1}$10}⟨0001⟩ slip: a brief summary of work to date. *Bull. Mineral.,* 102, 124-37.

Kittel, C. (1968). *Introduction to Solid State Physics,* 3d ed. New York: Wiley.

Knipe, R. J. (1980). Distribution of impurities in deformed quartz and its implications for deformation studies. *Tectonophysics,* 64, T11-18.

(1989). Deformation mechanisms - recognition from natural tectonites. *J. Structural Geol.,* 11, 127-46.

Koch, P. S., & Christie, J. M. (1981). Spacing of deformation lamellae as a paleopiezometer. *EOS Transactions of the American Geophysical Union,* 62, 1030.

Kohlstedt, D. L., Goetze, C., Durham, W. B., & Vander Sande, J. B. (1976). A new technique for decorating dislocations in olivine. *Science,* 191, 1045-6.

Kohlstedt, D. L., & Weathers, M. S. (1980). Deformation induced microstructures, paleopiezometers and differential stress in deeply eroded fault zones. *J. Geophys. Res.,* 85, 6269-85.

Konings, R. J. M., Boland, J. N., Vriend, S. P., & Jansen, J. B. H. (1988). Chemistry of biotites and muscovites in Abas granite, northern Portugal. *Amer. Mineral.,* 73, 754-65.

Kouh Simpson, Y., Carter, C. B., Morrissey, K. J., Angelini, P., & Bentley, J. (1986). The identification of thin amorphous films at grain boundaries in Al$_2$O$_3$. *J. Materials Sci.,* 21, 2689-96.

Kovacs, M. P., & Gandais, M. (1980). Transmission electron microscope study of experimentally deformed K-feldspar single crystals. *Phys. Chem. Minerals,* 6, 61-76.

Kronenberg, A. K., Kirby, S. H., & Rossman, G. R. (1986). Solubility and diffusional uptake of hydrogen in quartz at high water pressures: implications for hydrolytic weakening. *J. Geophys. Res.,* 91, 12723-44.

Landuyt, J. van, Gevers, R., & Amelinckx, S. (1964). Fringe patterns at antiphase boundaries with $\alpha = \pi$ observed in the electron microscope. *phys. stat. sol.,* 7, 519-46.

(1965). Dynamical theory of the images of microtwins as observed in the electron microscope. I: overlapping twins. *phys. stat. sol.,* 9, 135-55.

Lee, F. (1976). The submicroscopic structure of wenkite. In *Electron Microscopy in Mineralogy,* edited by H.-R. Wenk, pp. 361-70. Berlin: Springer-Verlag.

374   *References*

Liddell, N. A., Phakey, P. P., & Wenk, H.-R. (1976). The microstructures of some naturally deformed quartzites. In *Electron Microscopy in Mineralogy,* edited by H.-R. Wenk, pp. 419–27. Berlin: Springer-Verlag.

Linker, M. F., Kirby, S. H., Ord, A., & Christie, J. M. (1984). Effects of compression direction on the plasticity and rheology of hydrolytically weakened synthetic quartz crystals at atmospheric pressure. *J. Geophys. Res., 89,* 4241–55.

Livi, K. J. T., & Veblen, D. R. (1987). "Eastonite" from Easton, Pennsylvania: a mixture of phlogopite and a new form of serpentine. *Amer. Mineral., 72,* 113–25.

Lloyd, G. E. (1985). Review of instrumentation, techniques and applications of SEM in mineralogy. In *Short Course in Applications of Electron Microscopy in the Earth Sciences,* edited by J. C. White, pp. 151–88. Toronto: Mineralogical Association of Canada.

Loberg, B., & Norden, H. (1976). High resolution microscopy of grain boundary structure. In *Grain Boundary Structure and Properties,* edited by G. A. Chadwick & D. A. Smith, pp. 1–43. London: Academic Press.

Loretto, M. H. (1984). *Electron Beam Analysis of Materials.* London: Chapman and Hall.

McConnell, J. D. C. (1965). Electron optical study of effects associated with partial inversion in a silicate phase. *Phil. Mag., 11,* 1289–301.

(1969a). Electron optical study of incipient exsolution and inversion phenomena in the system $NaAlSi_3O_8$–$KAlSi_3O_8$. *Phil. Mag., 19,* 221–9.

(1969b). Photochemical degredation of a silicate in the beam of the electron microscope. *Phil. Mag., 20,* 1195–202.

McCormick, J. W. (1977). Transmission electron microscopy of experimentally deformed synthetic quartz. PhD thesis, University of California, Los Angeles.

McCormick, T. C., Smyth, J. R., & Lofgren, G. E. (1987). Site occupancies of minor elements in synthetic olivines as determined by channelling-enhanced x-ray emission. *Phys. Chem. Minerals, 14,* 368–72.

MacGillavry, C. H., & Rieck, G. D. (1983). *International Tables for X-ray Crystallography. Vol. III: Physical and Chemical Tables,* International Union of Crystallography. Dordrecht: D. Reidel.

MacKenzie, J. K. (1949). PhD thesis, Bristol University.

McLaren, A. C. (1973). The domain structure of a transitional anorthite; a study by direct lattice-resolution electron microscopy. *Contrib. Mineral. Petrol., 41,* 47–52.

(1974). Transmission electron microscopy of feldspars. In *The Feldspars,* edited by W. S. MacKenzie and J. Zussman, pp. 378–423. Manchester: Manchester University Press.

(1978). Defects and microstructures in feldspars. In *Chemical Physics of Solids and Their Surfaces, Vol. 7,* edited by M. W. Roberts & J. M. Thomas, pp. 1–30. London: The Chemical Society.

(1984). Transmission electron microscope investigations of the microstructures of microclines. In *Feldspars and Feldspathoids, Structures, Properties and Occurrences,* edited by W. L. Brown, pp. 373–409. Dordrecht: D. Reidel.

(1986). Some speculations on the nature of high-angle grain boundaries in quartz rocks. In *Mineral and Rock Deformation: Laboratory Studies, The Paterson*

Volume, Geophysical Monograph 36, edited by B. E. Hobbs & H. C. Heard, pp. 233–47. Washington, DC: American Geophysical Union.

McLaren, A. C., Cook, R. F., Hyde, S. T., & Tobin, R. C. (1983). The mechanisms of the formation and growth of water bubbles and associated dislocation loops in synthetic quartz. *Phys. Chem. Minerals,* 9, 79–94.

McLaren, A. C., & Etheridge, M. A. (1976). A transmission electron microscope study of naturally deformed orthopyroxene. I. Slip mechanisms. *Contrib. Mineral. Petrol.,* 57, 163–77.

(1980). A transmission electron microscope study of naturally deformed orthopyroxene. II: mechanisms of kinking. *Bull. Mineral.,* 103, 558–63.

McLaren, A. C., & Fitz Gerald, J. D. (1987). CBED and ALCHEMI investigation of local symmetry and Al, Si ordering in K-feldspars. *Phys. Chem. Minerals,* 14, 281–92.

McLaren, A. C., Fitz Gerald, J. D., & Gerretsen, J. (1989). Dislocation nucleation and multiplication in synthetic quartz: relevance to water weakening. *Phys. Chem. Minerals,* 16, 465–82.

McLaren, A. C., & Hobbs, B. E. (1972). Transmission electron microscope investigation of some naturally deformed quartzites. In *Flow and Fracture of Rocks,* Geophysical Monograph 16, edited by H. C. Heard, I. Y. Borg, N. L. Carter, & C. B. Raleigh, pp. 55–66. Washington, DC: American Geophysical Union.

McLaren, A. C., & MacKenzie, W. S. (1976). The spatial coherence of x-ray and electron beams and its influence on the diffraction patterns from materials with long-period superlattices. *phys. stat. sol.,* (a)33, 491–5.

McLaren, A. C., & Marshall, D. B. (1974). Transmission electron microscope study of the domain structure associated with the b-, c-, d-, e- and f-reflections in plagioclase feldspars. *Contrib. Mineral. Petrol.,* 44, 237–49.

McLaren, A. C., Osborne, C. F., & Saunders, L. A. (1971). X-ray topographic study of dislocations in synthetic quartz. *phys. stat. sol.,* (a)4, 235–47.

McLaren, A. C., & Phakey, P. P. (1965a). A transmission electron microscope study of amethyst and citrine. *Aust. J. Phys.,* 18, 135–41.

(1965b). Dislocations in quartz observed by transmission electron microscopy. *J. Appl. Phys.,* 36, 3244–6.

(1966). Electron microscope study of Brazil twin boundaries in amethyst quartz. *phys. stat. sol.,* 13, 413–22.

(1969). Diffraction contrast from Dauphiné twin boundaries in quartz. *phys. stat. sol.,* 31, 723–37.

McLaren, A. C., & Pitkethly, D. R. (1982). The twinning microstructure and growth of amethyst quartz. *Phys. Chem. Minerals,* 8, 128–35.

McLaren, A. C., & Retchford, J. A. (1969). Transmission electron microscope study of the dislocations in plastically deformed synthetic quartz. *phys. stat. sol.,* 33, 657–68.

McLaren, A. C., Retchford, J. A., Griggs, D. T., & Christie, J. M. (1967). Transmission electron microscope study of Brazil twins and dislocations experimentally produced in natural quartz. *phys. stat. sol.,* 19, 631–44.

McLaren, A. C., Turner, R. G., Boland, J. N., & Hobbs, B. E. (1970). Dislocation structure of the deformation lamellae in synthetic quartz; a study by electron and optical microscopy. *Contrib. Mineral. Petrol.,* 29, 104–15.

Mackwell, S. J., Kohlstedt, D. L., & Paterson, M. S. (1985). The role of water in the deformation of olivine single crystals. *J. Geophys. Res.,* 90, 11319–33.

Mackwell, S. J., & Paterson, M. S. (1985). Water-related diffusion and deformation effects in quartz at pressures of 1500 and 300 MPa. In *Point Defects in Minerals,* Geophysical Monograph 31, edited by R. N. Schock, pp. 141–50. Washington, DC: American Geophysical Union.

Marshall, D. B., & McLaren, A. C. (1974). The structure of albite and pericline twin boundaries in anorthite. *Electron Microscopy 1974, Eighth International Congress on Electron Microscopy, Vol.* 1, edited by J. V. Sanders & D. J. Goodchild, pp. 490–1. Canberra, Australia: Australian Academy of Science.

    (1977a). Deformation mechanisms in experimentally deformed plagioclase feldspars. *Phys. Chem. Minerals,* 1, 351–70.

    (1977b). The direct observation and analysis of dislocations in experimentally deformed plagioclase feldspars. *J. Materials Sci.,* 12, 893–903.

Mellini, M., Ferraris, G., & Compagnoni, R. (1985). Carlosturanite: HRTEM evidence of a polysomatic series including serpentine. *Amer. Mineral.,* 70, 773–81.

Mellini, M., Merlino, S., & Pasero, M. (1986). X-ray and HRTEM structure analysis of orientite. *Amer. Mineral.,* 71, 176–87.

Montardi, Y., & Mainprice, D. (1987). A transmission electron microscopic study of the natural plastic deformation of calcic plagioclases (An68-70). *Bull. Mineral.,* 110, 1–14.

Morrison-Smith, J. D., Paterson, M. S., & Hobbs, B. E. (1976). An electron microscope study of plastic deformation in single crystals of synthetic quartz. *Tectonophysics,* 33, 43–79.

Müller, W. F. (1976). On stacking disorder and polytypism in pectolite and serandite. *Z. Krist.,* 144, 401–8.

Müller, W. F., Vojdan-Shemshadi, Y., & Pentinghaus, H. (1987). Transmission electron microscope study of antiphase domains in $CaAl_2Ge_2O_8$-feldspar. *Phys. Chem. Minerals,* 14, 235–7.

Müller, W. F., & Wenk, H.-R. (1975). Transmission electron microscopic study of wollastonite ($CaSiO_3$). *Acta Crystallogr.,* A31, suppl. S294.

Müller, W. F., Wenk, H.-R., Bell, W. L., & Thomas, G. (1973). Analysis of the displacement vectors of antiphase domain boundaries in anorthites ($CaAl_2Si_2O_8$). *Contrib. Mineral. Petrol.,* 40, 63–74.

Müller, W. F., Wenk, H.-R., & Thomas, G. (1972). Structural variations in anorthites. *Contrib. Mineral. Petrol.,* 34, 304–14.

Nakada, M., & Lambeck, K. (1987). Glacial rebound and relative sea level variations: a new appraisal. *Geophysics Journal of the Royal Astronomical Society,* 90, 171–224.

Naze, L., Doukhan, N., Doukhan, J. C., & Latrous, K. (1987). A TEM study of lattice defects in naturally and experimentally deformed orthopyroxenes. *Bull. Mineral.,* 110, 497–512.

Nelson, R. S., Mazey, D. J., & Barnes, R. S. (1965). The thermal equilibrium shape and size of holes in solids. *Phil. Mag.,* 11, 91–111.

Neuber, H. (1958). *Kerbspannungslehre,* 2d ed. Berlin, Heidelberg, New York: Springer-Verlag.

Nicolas, A., & Poirier, J. P. (1976). *Crystalline Plasticity and Solid State Flow in Metamorphic Rocks.* London: Wiley.

Nissen, H.-U. (1967). Direct electron-microscope proof of domain texture in orthoclase. *Contrib. Mineral. Petrol.,* 16, 354-60.

Nissen, H.-U., Champness, P. E., Cliff, G., & Lorimer, G. (1973). Chemical evidence for exsolution in a labradorite. *Nature Physical Sci.,* 245, 135-7.

Nord, G. L., Heuer, A. H., & Lally, J. S. (1974). Transmission electron microscopy of substructures in Stillwater bytownites. In *The Feldspars,* edited by W. S. MacKenzie and J. Zussman, pp. 522-35. Manchester: Manchester University Press.

(1976). Pigeonite exsolution from augite. In *Electron Microscopy in Mineralogy,* edited by H.-R. Wenk, pp. 220-7. Berlin: Springer-Verlag.

Nord, G. L., Lally, J. S., Heuer, A. H., Christie, J. M., Radcliffe, S. V., Griggs, D. T., & Fisher, R. M. (1973). Petrologic study of igneous and metaigneous rocks from Apollo 15 and 16 using high-voltage transmission electron microscopy. *Proc. 4th Lunar Sci. Conf. Geochim. Cosmochim. Acta. Suppl. 4,* Vol. 1, 953-70.

Olsen, T. S., & Kohlstedt, D. L. (1984). Analysis of dislocations in some naturally deformed plagioclase feldspars. *Phys. Chem. Minerals,* 11, 153-60.

Ord, A., & Hobbs, B. E. (1986). Experimental control of the water-weakening effect in quartz. In *Mineral and Rock Deformation: Laboratory Studies, The Paterson Volume,* Geophysical Monograph 36, edited by B. E. Hobbs & H. C. Heard, pp. 51-72. Washington, DC: American Geophysical Union.

Orowan, E. (1934). Plasticity of crystals. *Z. Phys.,* 89, 605-59.

Otten, M. T. (1989). A practical guide to ALCHEMI. *Philips Electron Optics Bulletin,* 126, 21-8.

Owen, D. C., & McConnell, J. D. C. (1971). Spinodal behaviour in an alkali feldspar. *Nature Physical Sci.,* 230, 118-9.

(1974). Spinodal unmixing in an alkali feldspar. In *The Feldspars,* edited by W. S. MacKenzie & J. Zussman, pp. 424-39. Manchester: Manchester University Press.

Pascucci, M. R., Hutchison, J. L., & Hobbs, L. W. (1983). The metamict transformation in alpha-quartz. *Radiation Effects,* 74, 219-26.

Paterson, M. S. (1982). The determination of hydroxyl by infrared absorption in quartz, silicate glasses, and similar materials. *Bull. Mineral.,* 105, 20-9.

(1989). The interaction of water with quartz and its influence in dislocation flow - an overview. In *Rheology of Solids and of the Earth,* edited by S. Karato and M. Toriumi, pp. 107-42. London: Oxford University Press.

Paterson, M. S., & Kekulawala, K. R. S. S. (1979). The role of water in quartz deformation. *Bull. Mineral.,* 102, 92-8.

Phakey, P. P. (1967). Defects in quartz studied by transmission electron microscopy and x-ray diffraction topography. PhD. thesis, Monash University, Clayton, Victoria, Australia.

Phakey, P. P., Dollinger, G., & Christie, J. M. (1972). Transmission electron microscopy of experimentally deformed olivine crystals. In *Flow and Fracture of Rocks,* Geophysical Monograph 16, edited by H. C. Heard, I. Y. Borg, N. L. Carter, & C. B. Raleigh, pp. 117-38. Washington, DC: American Geophysical Union.

Phakey, P. P., & Ghose, S. (1972). Scapolite: observation of anti-phase domain structure. *Nature Physical Sci.,* 38, 78–80.

    (1973). Direct observation of anti-phase domain structure in omphacite. *Contrib. Mineral. Petrol.,* 39, 239–45.

Poirier, J. P. (1985). *Creep of Crystals.* Cambridge: Cambridge University Press.

Poirier, J. P., Peyronneau, J., Gesland, J. Y., & Brebec, G. (1983). Viscosity and conductivity of the lower mantle; an experimental study on a MgSiO$_3$ perovskite analogue KZnF$_3$. *Physics of the Earth and Planetary Interiors,* 32, 273–87.

Polanyi, M. (1934). Lattice distortion which originates plastic flow. *Z. Phys.,* 89, 660.

Price, P. B., & Walker, R. M. (1963). A simple method of measuring low uranium concentrations in natural crystals. *Appl. Phys. Letters,* 2, 23–5.

Putnis, A., & McConnell, J. D. C. (1980). *Principles of Mineral Behaviour.* Oxford: Blackwell Scientific Publications.

Raleigh, C. B. (1965). Glide mechanisms of experimentally deformed minerals. *Science,* 150, 739–41.

Read, W. T. (1953). *Dislocations in Crystals.* New York: McGraw-Hill.

Reeder, R. J., & Nakajima, Y. (1982). The nature of ordering and ordering defects in dolomite. *Phys. Chem. Minerals,* 8, 29–35.

Remaut, G., Gevers, R., Lagasse, A., & Amelinckx, S. (1965). Dynamical theory of the images of microtwins as observed in the electron microscope. II: overlapping domain wall boundaries. *phys. stat. sol.,* 10, 121–39.

    (1966). Dynamical theory of the images of microtwins as observed in the electron microscope. III: observations and results of numerical calculations. *phys. stat. sol.,* 13, 125–40.

Resnick, R., & Halliday, D. (1966). *Physics.* New York: Wiley.

Ricoult, D. L., & Kohlstedt, D. L. (1983). Structural width of low-angle grain boundaries in olivine. *Phys. Chem. Minerals,* 9, 133–8.

Ross, M., & Huebner, J. S. (1979). Temperature-composition relationships between naturally occurring augite, pigeonite and orthopyroxene at one bar pressure. *Amer. Mineral.,* 64, 1133–55.

Rossouw, C. J., & Maslen, V. W. (1987). Localization and ALCHEMI for zone axis orientations. *Ultramicroscopy,* 21, 277–90.

Rossouw, C. J., Turner, P. S., & White, T. J. (1988). Axial electron-channelling analysis of perovskite. I Theory and experiment for CaTiO$_3$; II site identification of Sr, Zr and U impurities, *Phil. Mag.,* 57B, 209–41.

Rovetta, M. R., Holloway, J. R., & Blacic, J. D. (1986). Solubility of hydroxyl in natural quartz annealed in water at 900°C and 1.5 GPa. *Geophys. Res. Letters,* 13, 145–8.

Schmid, S. M. (1982). Microfabric studies as indicators of deformation mechanisms and flow laws operative in mountain building. In *Mountain Building Processes,* edited by K. J. Hsü, pp. 95–110. London: Academic Press.

Self, P. & O'Keefe, M. A. (1988). Calculation of diffraction patterns and images for fast electrons. In *High Resolution Transmission Electron Microscopy and Associated Techniques,* edited by P. R. Buseck, J. M. Cowley, & L. Eyring, pp. 244–307. New York: Oxford University Press.

Sherby, O. D., & Burke, P. M. (1968). Mechanical behaviour of crystalline solids at elevated temperatures. *Progress in Materials Science*, 13, 325–90.

Skrotzki, W., Wedel, A., Weber, K., & Müller, W. F. (1990). Microstructure and texture in lherzolites of the Balmuccia massif and their significance regarding the thermomechanical history. *Tectonophysics*, 179, 227–51.

Smith, E. (1979). Dislocations and cracks. In *Dislocations in Solids, Vol. 4, Dislocations in Metallurgy*, edited by F. R. N. Nabarro, pp. 363–448. Amsterdam: North-Holland.

Smith, J. V. (1974). *The Feldspars, Vol. 1*, 1st ed. Berlin: Springer-Verlag.

Smith, J. V., & Brown, W. L. (1988). *Feldspar Minerals, Vol. 1*, 2d rev. and extended ed. Berlin: Springer-Verlag.

Smith, K. L., McLaren, A. C., & O'Donnell, R. G. (1987). Optical and electron microscope investigation of temperature-dependent microstructures in anorthoclase, *Can. J. Earth Sci.*, 24, 528–43.

Smith, P. P. K. (1986). Direct imaging of tunnel cations in zinkenite by high-resolution electron microscopy. *Amer. Mineral.*, 71, 194–201.

Southworth, H. N. (1975). *Introduction to Modern Microscopy*. London and Winchester: Wykeham.

Speer, J. A. (1982). Zircon. In *Reviews in Mineralogy, Vol. 5, Orthosilicates*, 2d ed., edited by P. H. Ribbe. Washington, DC: Mineralogical Society of America.

Spence, J. C. H. (1981) *Experimental High-resolution Electron Microscopy*. Oxford: Clarendon Press.

Spence, J. C. H., & Tafto, J. (1982). Atomic site and species determination using the channelling effect in electron diffraction. In *Scanning Electron Microscopy 1982, II*, pp. 523–31. Chicago: SEM Inc., AMF O'Hare.

(1983). ALCHEMI: a new technique for locating atoms in crystals. *J. Microscopy*, 130, 147–54.

Steeds, J. W. (1979). Convergent beam electron diffraction: In *Introduction to Analytical Electron Microscopy*, edited by J. J. Hren, J. I. Goldstein, & D. C. Joy. New York: Plenum.

Sun, C. P., & Balluffi, R. W. (1982). Secondary grain boundary dislocations in [001] twist boundaries in MgO. *Phil. Mag.*, A46, 49–62.

Tafto, J. (1982). The cation-atom distribution in a $(Cr, Fe, Al, Mg)_3O_4$ spinel as revealed from the channelling effect in electron-induced x-ray emission. *J. Appl. Cryst.*, 15, 378–81.

Tafto, J., & Buseck, P. R. (1983). Quantitative study of Al-Si ordering in an orthoclase feldspar using an analytical transmission electron microscope. *Amer. Mineral.*, 68, 944–50.

Tafto, J., & Spence, J. C. H. (1982). Crystal site location of iron and trace elements in a magnesium-iron olivine by a new crystallographic technique. *Science*, 218, 49–51.

Taylor, G. I. (1934). The mechanism of plastic deformation of crystals. *Proc. Roy. Soc., (London)*, 145, 362–404.

Tendeloo, G. van, Lunduyt, J. van, & Amelinckx, S. (1976). The α-β phase transition in quartz and $AlPO_4$ as studied by electron microscopy and diffraction. *phys. stat. sol.*, (a)33, 723–35.

Tendeloo, G. van, Wenk, H.-R., & Gronsky, R. (1985). Modulated structures in calcian dolomite: a study by electron microscopy. *Phys. Chem. Minerals,* 12, 333–41.

Tholen, A. R. (1970). On the ambiguity between moiré fringes and the electron diffraction contrast from closely spaced dislocations. *phys. stat. sol.,* (a)2, 537–50.

Thompson, J. B. (1978). Biopyriboles and polysomatic series. *Amer. Mineral.,* 63, 239–49.

Tibbals, J. E., & Olsen, A. (1977). An electron microscope study of some twinning and exsolution textures in microcline amazonites. *Phys. Chem. Minerals,* 1, 313–24.

Trepied, L., & Doukhan, J. C. (1978). Dissociated 'a' dislocations in quartz. *J. Materials Sci.,* 13, 492–8.

Trojer, F. J. (1968). The crystal structure of parawollastonite. *Z. Krist.,* 127, 291–308.

Tsenn, M. C., & Carter, N. L. (1987). Upper limits of power law creep of rocks. *Tectonophysics,* 136, 1–26.

Tullis, J., & Yund, R. A. (1989). Hydrolytic weakening of quartz aggregates: the role of "water" and pressure on recovery. *Geophys. Res. Letters,* 16, 1343–6.

Tunstall, W. J., Hirsch, P. B., & Steeds, J. (1964). Effects of surface stress relaxation on the electron microscope images of dislocations normal to thin metal foils. *Phil. Mag.,* 9, 99–119.

Turner, F. J., & Weiss, L. E. (1963). *Structural Analysis of Metamorphic Tectonites.* New York: McGraw-Hill.

Vance, E. R., & Boland, J. N. (1975). Fission fragment damage in zircon. *Radiation Effects,* 26, 135–9.

Vaughan, P. J., & Kohlstedt, D. L. (1982). Distribution of glass phase in hotpressed, olivine-basalt aggregates: an electron microscopy study. *Contrib. Mineral. Petrol.,* 81, 253–61.

Veblen, D. R. (1985a). High-resolution transmission electron microscopy. In *Short Course in Applications of Electron Microscopy in the Earth Sciences,* edited by J. C. White, pp. 63–90. Toronto: Mineralogical Association of Canada.
   (1985b). Direct TEM imaging of complex structures and defects in silicates. *Ann. Rev. Earth & Planet. Sci.,* 13, 119–46.

Vitek, J. M., & Rühle, M. (1986). Diffraction effects from internal interfaces – I. general considerations and grain boundary effects. *Acta. Metall.,* 34, 2085–94.

Weathers, M. S., Bird, J. M., Cooper, R. F., & Kohlstedt, D. L. (1979). Differential stress determined from deformation-induced microstructures of the Moine thrust zone, *J. Geophys. Res.,* 84, 7495–509.

Weertman, J. (1957). Steady state creep of crystals. *J. Appl. Phys.,* 28, 1185–9.

Weissmann, S., & Nakajima, K. (1963). Defect structure and density decrease in neutron-irradiated quartz. *J. Appl. Phys.,* 34, 611–8.

Wenk, H.-R. (1976). *Electron Microscopy in Mineralogy.* Berlin: Springer-Verlag.
   (1979). Some roots of experimental rock deformation. *Bull. Mineral.,* 102, 195–202.
   (1985). *Preferred Orientation in Deformed Metals and Rocks: An Introduction to Modern Texture Analysis.* Orlando, Florida: Academic Press.

Wenk, H.-R., Barber, D. J., & Reeder, R. J. (1983). Microstructures in carbonates. In *Carbonates: Mineralogy and Chemistry, Reviews in Mineralogy Vol. 11,* edited by R. J. Reeder, pp. 301–67. Mineralogical Society of America.

Wenk, H.-R., Müller, W. F., Liddell, N. A., & Phakey, P. P. (1976). Polytypism in wollastonite. In *Electron Microscopy in Mineralogy,* edited by H.-R. Wenk, pp. 324–31. Berlin: Springer-Verlag.

Wenk, H.-R., & Nakajima, Y. (1980). Structure, formation, and decomposition of APBs in calcic plagioclase. *Phys. Chem. Minerals,* 6, 169–86.

White, S. H. (1973). Deformation lamellae in naturally deformed quartz. *Nature Physical Sciences,* 245, 26–8.

(1979). Grain and sub-grain size variations across a mylonite zone. *Contrib. Mineral. Petrol.,* 70, 193–202.

Willaime, C. Brown, W. L., & Gandais, M. (1976). Physical aspects of exsolution in natural alkali feldspars. In *Electron Microscopy in Mineralogy,* edited by H.-R. Wenk, pp. 248–57. Berlin: Springer-Verlag.

Willaime, C., & Gandais, M. (1972). Study of exsolution in alkali feldspars. Calculation of elastic stresses inducing periodic twins. *phys. stat. sol.,* (a)9, 529–39.

Williams, D. B. (1984). *Practical Analytical Electron Microscopy in Materials Science.* Mahwah, NJ: Philips Electronic Instruments, Inc., Electron Optics Publishing Group.

Williams, T. B., & Hyde, B. G. (1988). Electron microscopy of cylindrite and franckeite. *Phys. Chem. Minerals,* 15, 521–44.

Yada, K., Tanji, T., & Sunagawa, I. (1981). Application of lattice imagery to radiation damage investigation in natural zircon. *Phys. Chem. Minerals,* 7, 47–52.

Young, F. W., & Savage, J. R. (1964). Growth of copper crystals of low dislocation density. *J. Appl. Phys.,* 35, 1917–24.

Yund, R. A., McLaren, A. C., & Hobbs, B. E. (1974). Coarsening kinetics of the exsolution microstructure in alkali feldspar. *Contrib. Mineral. Petrol.,* 48, 45–55.

Zhing, Y., & Gandais, M. (1987a). Fine structure of (010)[001] dislocations in K-feldspars. *Phil. Mag.,* A55, 329–38.

(1987b). Modèles de structure des dislocations (010)[001] dans les feldspaths alcalins. *Bull. Mineral.,* 110, 15–24.

Index

Abbe, Ernst, 4, 11, 38, 43
aberrations, *see* lens aberrations
absorption length, 117–18
adularia, *see* feldspars
albite, *see* feldspars
albite twin, *see* feldspars
ALCHEMI, 6, 193–6, 233, 267–8
alkali feldspar, *see* feldspars
aluminum, 287
amethyst, *see* quartz
amphibole, 221, 265
andalusite, 267
angular frequency, 12
anomalous absorption, *see* anomalous
    transmission
anomalous transmission, 117
anorthite, *see* feldspars
anorthoclase, *see* feldspars
antiphase boundary (APB), *see* planar
    defects
apatite, 274–5
astigmatism, *see* lens aberrations
atomic scattering factor, 63–9
augite, *see* pyroxenes

bend contours, 85, 111–12, 119–22
bright-field (BF) image, 31, 50, 85–7,
    128–9, 136
biotite, 267
Bormann effect, 117
Bragg angle, deviation from exact, 69–74,
    77–82, 104–5, 109
Bragg law, 56–7
Brazil twin, *see* quartz
brazilianite, 274
bremsstrahlung, 185, 188
bubbles, 133, 164–7, 171, 273, 282, 300–11,
    337–9
Burgers vector, determination of, 161–3,
    301–5, 349–52, 360

calcite
    deformation of, 328–30
    grain boundaries in, 246–7
camera constant, 75
carlosturanite, 267
cataclastic deformation, 356
chain-width defects, 265
chesterite, 265
chromatic aberration, *see* lens aberrations
chromium, 92, 110–11
Cliff–Lorimer factor, 191
climb, *see* dislocations
climb dissociation, *see* dislocations
clinopyroxene, *see* pyroxenes
coherence
    area, 35–6, 144, 250–2
    length, 33
    time, 33
column approximation, 71, 89, 125
coma, *see* lens aberrations
complex amplitude, 12
convergent beam electron diffraction
    (CBED), 40, 82–4
cross-hatched twinning, *see* feldspars
cylindrite, 266

dark-field (DF) image, 31, 51, 84–7, 129,
    136, 146
Darwin–Howie–Whelan equations, 122–6
Dauphiné twin, *see* quartz
defect of focus, 44–9, 172, 174–6, 178–9
deformation lamellae, 22, 284, 296, 311,
    356–9
diamond, 167, 294
diffuse-scattering image, 146, 246
diposide, *see* pyroxenes
dislocation loops, *see* dislocations
dislocations
    climb dissociation of, 349–52, 360
    climb of, 287, 291–6

Printed in the United States
By Bookmasters